GW00985792

Student Support Materials for

AQA

AS CHEMISTRY

Module 3: Introduction to Organic Chemistry

John Bentham
Colin Chambers
Graham Curtis

This booklet has been designed to support the AQA Chemistry AS specification. It contains some material which has been added in order to clarify the specification.
The examination will be limited to material set out in the specification document.

Published by HarperCollins*Publishers* Limited
77–85 Fulham Palace Road
Hammersmith
London
W6 8JB

www.**Collins**Education.com
Online support for schools and colleges

10 9 8 7

ISBN 0 00 327703 8

British Library Cataloguing in Publication Data
A catalogue record for this publication is available from the British Library

Writing team: John Bentham, Colin Chambers, Graham Curtis, Geoffrey Hallas, David Nicholls, Andrew Maczek
Front cover designed by Chi Leung
Editorial, design and production by Gecko Limited, Cambridge
Printed and bound by Scotprint, Haddington

The publisher wishes to thank the Assessment and Qualifications Alliance for permission to reproduce the examination questions.

Other useful texts

Full colour textbooks
Collins Advanced Modular Sciences: Chemistry AS
Collins Advanced Science: Chemistry

Student Support Booklets
AQA Chemistry 1: Atomic Structure, Bonding and Periodicity
AQA Chemistry 2: Foundation Physical and Inorganic Chemistry

To the student

What books do I need to study this course?

You will probably use a range of resources during your course. Some will be produced by the centre where you are studying, some by a commercial publisher and others may be borrowed from libraries or study centres. Different resources have different uses – but remember, owning a book is not enough – it must be *used.*

What does this booklet cover?

This *Student Support Booklet* covers the content you need to know and understand to pass the module test for AQA Chemistry Module 3: Introduction to Organic Chemistry. It is very concise and you will need to study it carefully to make sure you can remember all of the material.

How can I remember all this material?

Reading the booklet is an essential first step – but reading by itself is not a good way to get stuff into your memory. If you have bought the booklet and can write on it, you could try the following techniques to help you to memorise the material:

- underline or highlight the most important words in every paragraph
- underline or highlight scientific jargon – write a note of the meaning in the margin if you are unsure
- remember the number of items in a list – then you can tell if you have forgotten one when you try to remember it later
- tick sections when you are sure you know them – and then concentrate on the sections you do not yet know.

How can I check my progress?

The module test at the end is a useful check on your progress – you may want to wait until you have nearly completed the module and use it as a mock exam or try questions one by one as you progress. The answers show you how much you need to do to get the marks.

What if I get stuck?

A colour textbook such as *Collins Advanced Modular Sciences: Chemistry AS* provides more explanation than this booklet. It may help you to make progress if you get stuck.

Any other good advice?

- You will not learn well if you are tired or stressed. Set aside time for work (and play!) and try to stick to it.
- Don't leave everything until the last minute – whatever your friends may tell you it doesn't work.
- You are most effective if you work hard for shorter periods of time and then take a (short!) break. 30 minutes of work followed by a five or ten minute break is a useful pattern. Then get back to work.
- Some people work better in the morning, some in the evening. Find out which works better for you and do that whenever possible.
- Do not suffer in silence – ask friends and your teacher for help.
- Stay calm, enjoy it and ... good luck!

The main text gives a very concise explanation of the ideas in your course. You must study all of it – none is spare or not needed.

There are rigorous definitions of the main terms used in your examination – memorise these exactly.

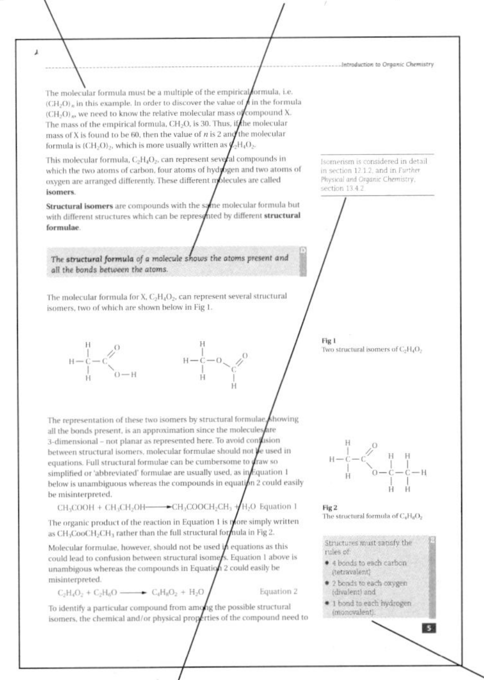

Introduction to Organic Chemistry

The molecular formula must be a multiple of the empirical formula, i.e. $(CH_2O)_n$ in this example. In order to discover the value of n in the formula $(CH_2O)_n$, we need to know the relative molecular mass of compound X. The mass of the empirical formula, CH_2O, is 30. Thus, if the molecular mass of X is found to be 60, then the value of n is 2 and the molecular formula is $(CH_2O)_2$, which is more usually written as $C_2H_4O_2$.

This molecular formula, $C_2H_4O_2$, can represent several compounds in which the two atoms of carbon, four atoms of hydrogen and two atoms of oxygen are arranged differently. These different molecules are called **isomers**.

Isomerism is considered in detail in section 12.1.2, and in *Further Physical and Organic Chemistry*, section 13.4.2.

Structural isomers are compounds with the same molecular formula but with different structures which can be represented by different **structural formulae**.

***The structural formula** of a molecule shows the atoms present and all the bonds between the atoms.*

The molecular formula for X, $C_2H_4O_2$, can represent several structural isomers, two of which are shown below in Fig 1.

Fig 1 Two structural isomers of $C_2H_4O_2$

The representation of these two isomers by structural formulae, showing all the bonds present, is an approximation since the molecules are 3-dimensional – not planar as represented here. To avoid confusion between structural isomers, molecular formulae should not be used in equations. Full structural formulae can be cumbersome to draw so simplified or 'abbreviated' formulae are usually used, as in Equation 1 below is unambiguous whereas the compounds in equation 2 could easily be misinterpreted.

$$CH_3COOH + CH_3CH_2OH \longrightarrow CH_3COOCH_2CH_3 + H_2O \quad \text{Equation 1}$$

The organic product of the reaction in Equation 1 is more simply written as $CH_3CooCH_2CH_3$ rather than the full structural formula in Fig 2.

Fig 2 The structural formula of $C_4H_8O_2$

Molecular formulae, however, should not be used in equations as this could lead to confusion between structural isomers. Equation 1 above is unambigous whereas the compounds in Equation 2 could easily be misinterpreted.

$$C_2H_4O_2 + C_2H_6O \longrightarrow C_4H_8O_2 + H_2O \quad \text{Equation 2}$$

To identify a particular compound from among the possible structural isomers, the chemical and/or physical properties of the compound need to

Structures must satisfy the rules of:

- 4 bonds to each carbon (tetravalent)
- 2 bonds to each oxygen (divalent) and
- 1 bond to each hydrogen (monovalent).

5

Further explanation references give a little extra detail, or direct you to other texts if you need more help or need to read around a topic.

The examiner's notes are always useful – make sure you read them because they will help with your module test.

12.1 Nomenclature and isomerism

Organic chemistry is the study of the many millions of covalent compounds of the element carbon. These compounds constitute an enormous variety of materials, ranging from molecules in living systems to synthetic materials made from petroleum, such as drugs, medicines and plastics. In this module we consider mainly the chemistry of molecules derived from the hydrocarbons (compounds containing only C and H atoms) present in petroleum.

12.1.1 *Nomenclature*

Whenever a new compound is studied, it is first analysed to determine the percentage composition by mass of each of the elements present. From the data obtained, the **empirical formula** of the substance can be derived (see also *Atomic Structure, Bonding and Periodicity*, section 10.2.4).

D *The* ***empirical formula*** *gives the simplest ratio of atoms of each element in a compound.*

For example, if a compound X is found to contain 40.0% carbon, 6.7% hydrogen and 53.3% oxygen by mass, then the empirical formula can be calculated as in Table 1.

Table 1
Calculation of the empirical formula

	Carbon	Hydrogen	Oxygen
% by mass	40.0	6.67	53.3
Divide by A_r	$\frac{40.0}{12} = 3.33$	$\frac{6.67}{1} = 6.67$	$\frac{53.3}{16} = 3.33$
Divide by the smallest number	$\frac{3.33}{3.33} = 1$	$\frac{6.67}{3.33} = 2$	$\frac{3.33}{3.33} = 1$
Simplest ratio	1	2	1
∴ Empirical formula = CH_2O			

E The percentage of oxygen is not usually determined experimentally. The percentages of carbon and hydrogen are found by analysing the combustion products, carbon dioxide and water. The difference between 100% and the sum of the carbon and hydrogen percentages is taken to be the percentage of oxygen.

Although the empirical formula for X, CH_2O, gives the simplest ratio of atoms, this ratio can be found in many different molecules. From the empirical formula we can determine the **molecular formula**.

D *The* ***molecular formula*** *gives the actual number of atoms of each element present in a molecule.*

The molecular formula must be a multiple of the empirical formula, i.e. $(CH_2O)_n$ in this example. In order to discover the value of n in the formula $(CH_2O)_n$, we need to know the relative molecular mass of compound X. The mass of the empirical formula, CH_2O, is 30. Thus, if the molecular mass of X is found to be 60, then the value of n is 2 and the molecular formula is $(CH_2O)_2$, which is more usually written as $C_2H_4O_2$.

This molecular formula, $C_2H_4O_2$, can represent several compounds in which the two atoms of carbon, four atoms of hydrogen and two atoms of oxygen are arranged differently. These different molecules are called **isomers**.

Isomerism is considered in detail in section 12.1.2, and in *Further Physical and Organic Chemistry*, section 13.4.2.

Structural isomers are compounds with the same molecular formula but with different structures which can be represented by different **structural formulae**.

D

*The **structural formula** of a molecule shows the atoms present and all the bonds between the atoms.*

The molecular formula for X, $C_2H_4O_2$, can represent several structural isomers, two of which are shown below in Fig 1.

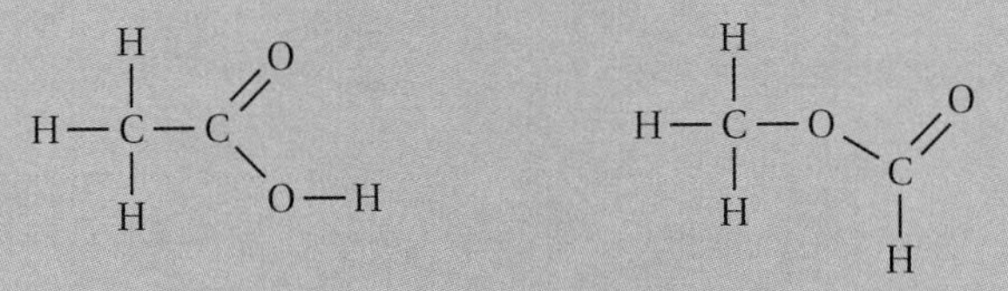

Fig 1
Two structural isomers of $C_2H_4O_2$

The representation of these two isomers by structural formulae, showing all the bonds present, is an approximation since the molecules are 3-dimensional – not planar as represented here. Full structural formulae can be cumbersome to draw so simplified or 'abbreviated' formulae are usually used, as in Equation 1.

$$CH_3COOH + CH_3CH_2OH \longrightarrow CH_3COOCH_2CH_3 + H_2O \quad \text{Equation 1}$$

The organic product of the reaction in Equation 1 is more simply written as $CH_3COOCH_2CH_3$ rather than the full structural formula in Fig 2.

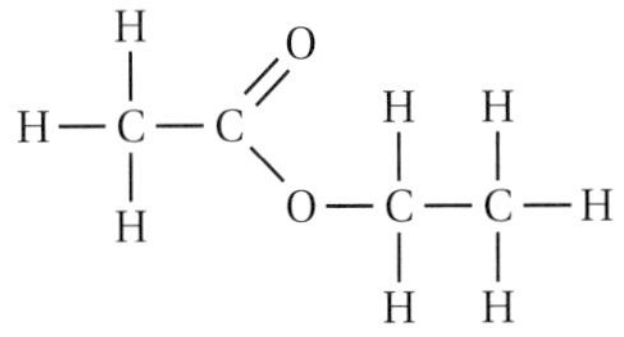

Fig 2
The structural formula of $C_4H_8O_2$

Molecular formulae, however, should not be used in equations as this could lead to confusion between structural isomers. Equation 1 above is unambigous whereas the compounds in Equation 2 could easily be misinterpreted.

$$C_2H_4O_2 + C_2H_6O \longrightarrow C_4H_8O_2 + H_2O \quad \text{Equation 2}$$

E

Structures must satisfy the rules of:

- 4 bonds to each carbon (tetravalent)
- 2 bonds to each oxygen (divalent) and
- 1 bond to each hydrogen (monovalent).

To identify a particular compound from among the possible structural isomers, the chemical and/or physical properties of the compound need to be studied. Nowadays the traditional technique of elemental analysis has largely been superseded by instrumental methods such as infra-red spectroscopy, nuclear magnetic resonance spectroscopy and mass spectrometry. These more modern methods are considered in *Further Physical and Organic Chemistry*, section 13.11.

The organic compounds in this module are almost all based on the series of hydrocarbons called alkanes (see section 12.2) in which one of the hydrogen atoms may be replaced by an atom or group of atoms called a **functional group**.

D

*A **functional group** is an atom or group of atoms which, when present in different molecules, causes them to have similar chemical properties*

The functional group is the reactive part of a molecule; the properties of the molecule are largely determined by the nature of the functional group. Functional groups which appear in this module are given in Table 3.

A family of molecules which all contain the same functional group, but an increasing number of carbon atoms, is called an **homologous series**, and can be represented by a general formula. For example:

- C_nH_{2n+2} represents alkanes
- C_nH_{2n} represents alkenes (and cyclic alkanes)
- $C_nH_{2n+1}OH$ represents alcohols

E

In the molecule $C_nH_{2n+1}X$, C_nH_{2n+1} is called an alkyl group ('ane' in the name is replaced by 'yl'). Thus CH_3 is called methyl. Alkyl groups are represented by the letter R.

All members of the same homologous series have similar chemical properties since these properties are determined by the functional group present; their physical properties gradually change as the carbon chain gets longer. The boiling points of alkanes, for example, increase along the homologous series as the number of carbon atoms increases.

Each successive molecule in a homologous series contains an additional $-CH_2-$group.

Rules for naming organic compounds

Organic compounds are named according to the rules of the International Union of Pure and Applied Chemistry (IUPAC). These systematic or IUPAC names are based on the names of the parent alkanes. The first six alkanes are shown in Table 2.

Table 2
The first six alkanes

Number of carbon atoms	1	2	3	4	5	6
Name	methane	ethane	propane	butane	pentane	hexane
Formula	CH_4	C_2H_6	C_3H_8	C_4H_{10}	C_5H_{12}	C_6H_{14}

To assign the name of an alkane derivative, first look for the longest carbon chain in the skeleton; the number of carbons in this chain determines the stem of the name. Thus, if there are two carbon atoms in the longest chain, the stem name will be ethan-; if there are five carbon atoms, the stem name will be pentan-.

In many compounds the carbon skeleton is branched. The names of the side chains also depend on the number of carbon atoms in them, so that:

- a one-carbon branch is called methyl (CH_3-)
- a two-carbon branch is called ethyl (CH_3CH_2-)
- a three-carbon branch is called propyl ($CH_3CH_2CH_2-$)

The position of any branch on the chain must also be made clear. This is achieved by numbering the carbon atoms in the skeleton so as to keep the numbers used as low as possible when indicating the position of any branches. For example, the molecule in Fig 3 is called 2-methylpentane (numbering from the right) and not 4-methylpentane (numbering from the left).

Note that, in these examples, the molecules are really three-dimensional with each carbon in the alkyl groups surrounded tetrahedrally by four bonds with bond angles of 109.5° (see Fig 4).

$CH_3-CH(CH_3)-CH_3$

is called methylpropane as the longest chain has 3 carbon atoms (propane) with a one-carbon branch (methyl). No number is needed as no other methylpropane is possible.

$CH_3-CH_2-CH_2-CH(CH_3)-CH_3$

Fig 3
2-methylpentane

Table 3
Homologous series and functional groups

Homologous series	Name: prefix or suffix	Functional group	Example
alkanes	*suffix* -ane	$\geq C-H$	ethane C_2H_6
alkenes	*suffix* -ene	$>C=C<$	ethene C_2H_4
haloalkanes	*prefix* halo-	—Cl, —Br, —I	chloroethane CH_3CH_2Cl
alcohols	*suffix* -ol *prefix* hydroxy-	—OH	ethanol CH_3CH_2OH 2-hydroxy-propanoic acid $CH_3CH(OH)COOH$
ethers	*prefix* alkoxy-	—OR	methoxymethane CH_3OCH_3
aldehydes	*suffix* -al	$-C(=O)H$	ethanal CH_3CHO
ketones	*suffix* -one *prefix* oxo-	$>C=O$	propanone CH_3COCH_3 3-oxobutanoic acid CH_3COCH_2COOH
carboxylic acids	*suffix* -oic acid	$-C(=O)OH$	ethanoic acid CH_3COOH
amines	*prefix* amino- *suffix* -amine	$-NH_2$	aminoethanoic acid H_2NCH_2COOH methylamine CH_3NH_2
amides	*suffix* -amide	$-C(=O)NH_2$	ethanamide CH_3CONH_2
nitriles	*suffix* -nitrile	$-C\equiv N$	propanenitrile CH_3CH_2CN

Prefix means added before the rest of the name; suffix means the ending of the name.

The list in Table 3 covers the functional groups you will meet in this module.

When a suffix begins with a consonant, the 'e' in -ane is retained.

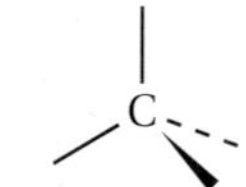

Fig 4
The tetrahedral arrangement of bonds around carbon

However, on paper it is usually simpler to represent a structure as if it had bond angles of 90° or 180°. For example, the three structures in Fig 5 all represent 3-methylpentane.

$CH_3—CH_2—CH—CH_2—CH_3$ (with CH_3 attached below the CH)

$CH_3—CH_2—CH—CH_3$ (with CH_2 attached below the CH, and CH_3 attached below that CH_2)

$CH_2—CH—CH_2$ (with CH_3 attached below each carbon)

Fig 5
Different representations of 3-methylpentane

Naming molecules containing functional groups

- The *type* of functional group present is indicated by either a prefix or a suffix on the alkane stem (see Table 3).
- The *position* of the functional group is usually indicated by a number; e.g. 2-chloropropane (see Fig 6) has the chlorine atom on the second (middle) carbon atom.
- For aldehydes, carboxylic acids and nitriles, the functional group must be at the end of the carbon chain; therefore it is automatically numbered carbon-1 so the number need not be included in the name, e.g. CH_3CH_2CHO is propanal.
- When two or more of a specific functional group are present, the *number* of substituents is shown by using the multipliers *di* for two, *tri* for three or *tetra* for four, e.g. dichloromethane, CH_2Cl_2 and tetrachloromethane, CCl_4.
- Numbers 1,2, etc. must also be used to show the position of each functional group. Commas are used between numbers and hyphens between numbers and letters, for example:

  CH_3CCl_3 is called 1,1,1-trichloroethane

  $CH_2ClCHCl_2$ is called 1,1,2-trichloroethane

  $HOCH_2CH_2OH$ is called ethane-1,2-diol
- If more than one type of functional group is present, the positions and names as prefixes are listed in alphabetical order, for example:

  $CH_3CHBrCH_2Cl$ is 2-bromo-1-chloropropane
- Multipliers are ignored when ordering substituents alphabetically; tribromo- will always come before dichloro-.
- Where two names which are usually suffixes are needed, the ending for acid takes precedence over aldehyde or ketone which, in turn, takes precedence over alcohol, for example:

  $CH_3CH(OH)COOH$ is 2-hydroxypropanoic acid

  CH_3COCH_2COOH is 3-oxobutanoic acid
- The suffix -ene for alkenes can be placed in front of other suffixes, for example:

  $H_2C{=}CH—CHCl—CHO$ is 2-chlorobut-3-enal

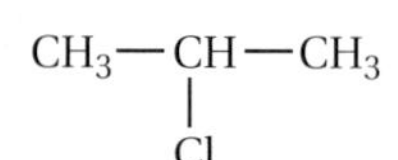

Fig 6
The structure of 2-chloropropane

E When naming compounds, look for:

- the longest carbon chain
- functional group(s)
- the number of substituents
- where they are.

When drawing a structure from a given name:

- use the name to identify the number of carbons atoms in the longest chain
- draw this carbon skeleton and number the carbon atoms
- add any functional groups in the correct positions
- add hydrogen atoms to make sure that every carbon atom has four bonds.

For example, the structure of 3-methylpentan-2-ol can be deduced as follows:

pentan- skeleton

```
C—C—C—C—C
```

pentan-2-ol skeleton

```
1   2   3   4   5
C—C—C—C—C
    |
    OH
```

3-methylpentan-2-ol skeleton

```
        CH3
        |
C—C—C—C—C
    |
    OH
```

3-methylpentan-2-ol complete

```
   H   H   CH3 H   H
   |   |   |   |   |
H—C—C—C—C—C—H
   |   |   |   |   |
   H   OH  H   H   H
```

which can be written more simply as: $CH_3CH(OH)CH(CH_3)CH_2CH_3$

Where systematic names become very complicated, trivial names are often used; e.g. glucose (see Fig 7 below) is one of the isomers of 2,3,4,5,6-pentahydroxyhexanal.

```
CHO
|
CHOH
|
CHOH
|
CHOH
|
CHOH
|
CH2OH
```

Fig 7
Structure of a glucose molecule

12.1.2 *Isomerism*

Isomerism occurs where molecules with the same molecular formula have their atoms arranged in different ways. Isomerism is divided into two main types which are also themselves subdivided:

- **structural isomerism**
- **stereoisomerism** (see also *Further Physical and Organic Chemistry*, section 13.4.2)

Structural isomerism

D

***Structural isomers** are compounds with the same molecular formula but with different structures.*

The different structures can arise in any of three different ways:

- chain isomerism
- position isomerism
- functional group isomerism.

The number 2 is optional in 2-methylpropane (see page 7).

Chain isomerism

This type occurs when there are two or more ways of arranging the carbon skeleton of a molecule. For example, C_4H_{10} can be butane or 2-methylpropane, as shown in Fig 8.

Fig 8
The isomers of C_4H_{10}

```
    H   H   H   H                 H       H       H
    |   |   |   |                 |       |       |
H — C — C — C — C — H         H — C ————— C ————— C — H
    |   |   |   |                 |       |       |
    H   H   H   H                 H   H — C — H   H
                                          |
                                          H
       butane                      2-methylpropane
```

The numbers 2 and 2,2 are optional (see page 7).

The three isomers of C_5H_{12} are pentane, 2-methylbutane and 2,2-dimethylpropane, as shown in Fig 9.

Fig 9
The isomers of C_5H_{12}

```
                                                                                  H
                                                                                  |
    H   H   H   H   H             H       H       H   H               H   H — C — H   H
    |   |   |   |   |             |       |       |   |               |       |       |
H — C — C — C — C — C — H     H — C ————— C ————— C — C — H       H — C ————— C ————— C — H
    |   |   |   |   |             |       |       |   |               |       |       |
    H   H   H   H   H             H   H — C — H   H   H               H   H — C — H   H
                                          |                                   |
                                          H                                   H
         pentane                     2-methylbutane                    2,2-dimethylpropane
```

These isomers have similar chemical properties but slightly different physical properties. Branched isomers have smaller volumes, weaker van der Waals' forces (see *Atomic Structure, Bonding and Periodicity,* section 10.3.3) and therefore lower boiling points.

The number of structural isomers of alkanes rises steeply as the number of carbon atoms increases, as shown in Table 4.

Table 4
The number of structural isomers of some alkanes

Number of carbon atoms	Number of isomers
1	1
2	1
3	1
4	2
5	3
6	5
7	9
8	18
9	35
10	75
11	159
12	355
13	802
14	1,858
15	4,347
20	366,319
25	36,797,588
30	4,111,846,763
40	62,491,178,805,831

Position isomerism

These isomers have the same carbon skeleton and the same functional group, but the functional group is joined at different places on the carbon skeleton. For example:

$CH_3CH_2CH_2Br$ 1-bromopropane	$CH_3CHBrCH_3$ 2-bromopropane
$CH_2{=}CHCH_2CH_3$ but-1-ene	$CH_3CH{=}CHCH_3$ but-2-ene

Again, such isomers have similar chemistry because they have the same functional group, but the different positions can cause some differences in properties.

Functional group isomerism

These isomers have different functional groups and so have different chemical and physical properties. There are three common examples:

- *aldehydes and ketones,* e.g. C_3H_6O can be

  CH_3CH_2CHO or CH_3COCH_3
  propanal propanone

- *carboxylic acids and esters,* e.g. $C_3H_6O_2$ can be

  CH_3CH_2COOH or CH_3COOCH_3 or $HCOOCH_2CH_3$
  propanoic acid methyl ethanoate ethyl methanoate

- *alcohols and ethers,* e.g. C_2H_6O can be

  CH_3CH_2OH or CH_3OCH_3
  ethanol methoxymethane

Stereoisomerism

D

***Stereoisomers** are molecules which have the same structural formula but their bonds are arranged differently in space.*

The two types of stereoisomerism are **geometrical** and **optical** (see *Further Physical and Organic Chemistry,* section 13.4.2).

Only geometrical isomerism is studied in this module.

Geometrical isomerism (also known as ***cis–trans*** isomerism)

Carbon–carbon double bonds in alkenes cannot rotate because of the π electron clouds present above and below the plane of the bond. When an alkene has 2 different groups at each end of the double bond, two different geometrical or *cis–trans* isomers result. These compounds have the same structural formula, but their bonds are arranged differently.

E

Such isomers are possible because rotation at a double bond requires a significant input of energy (too much energy to be supplied at room temperature); this situation is referred to as *restricted rotation.*

Cis means that two groups, often identical, are on the *same* side of the double bond and ***trans*** means they are on *opposite* sides; e.g. but-2-ene exists as two forms, identical except in the arrangement of the bonds in space (see Fig 10).

Fig 10
The two geometrical isomers of but-2-ene

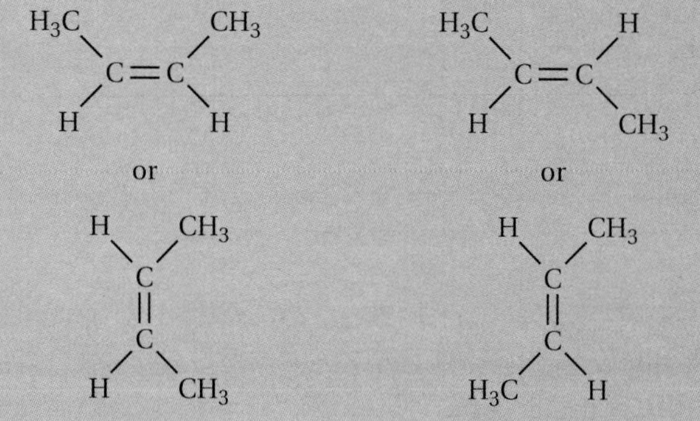

H_3C CH_3
C=C
H H

or

H CH_3
C
||
C
H CH_3

cis-but-2-ene

H_3C H
C=C
H CH_3

or

H CH_3
C
||
C
H_3C H

trans-but-2-ene

E Geometrical isomers have different physical and chemical properties. Cis-isomers usually have slightly higher boiling points because they have some polarity, whereas *trans*-isomers are less polar. For example, the boiling point of cis-but-2-ene is 4 °C but the boiling point of *trans*-but-2-ene is 1 °C. However, *trans*-isomers have higher melting points because they pack together more closely; e.g. the melting point of *trans*-but-2-ene is −106 °C, whereas that of cis-but-2-ene is −139 °C.

It is not possible to have geometrical isomerism when there are two identical groups joined to the same carbon atom in a double bond.

Methylpropene (see Fig 11) has two methyl groups on one carbon atom and two hydrogen atoms on the other. Therefore it is a structural isomer, but not a geometrical isomer, of but-2-ene.

Fig 11
The structure of methylpropene

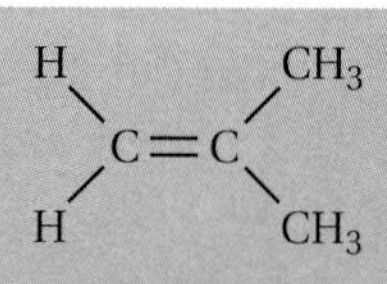

H CH_3
C=C
H CH_3

methylpropene

Other structural isomers of but-2-ene are but-1-ene and cyclobutane (a ring, or cyclic structure, containing four carbon atoms), shown in Fig 12.

Fig 12
Other structural isomers of but-2-ene

CH_3CH_2 H
C=C
H H

but-1-ene

$H_2C—CH_2$
| |
$H_2C—CH_2$

cyclobutane

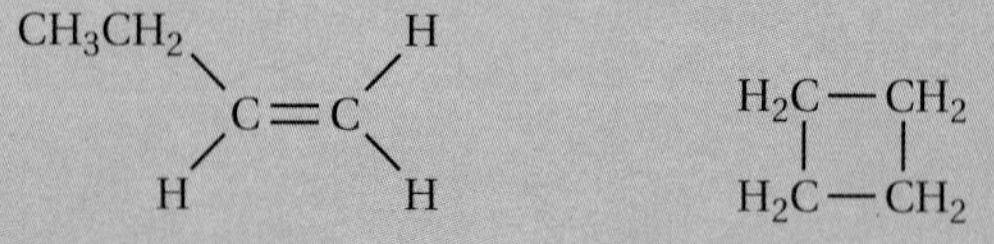

12.2 Petroleum and alkanes

12.2.1 *Petroleum: fractional distillation*

Petroleum

Petroleum or crude oil is a complex mixture of hydrocarbons, mainly alkanes (see below); it is derived from the remains of sea creatures and plants which sank to the bottom of the oceans millions of years ago. Subsequent deposits compressed this material and the high pressures and temperatures which developed – and also the absence of air – converted it into oil and gas.

Alkanes

Alkanes are the homologous series of saturated hydrocarbons with the general formula C_nH_{2n+2}. The first six alkanes are listed in Table 2 on p.6. The lower alkanes are gases at room temperature; their boiling points increase with the number of carbon atoms because the strength of van der Waals' forces between the molecules increases. This increase in boiling points allows crude oil to be separated by fractional distillation (see below).

Alkanes contain only carbon–carbon and carbon–hydrogen bonds; these bonds are relatively strong and are non-polar. Consequently, alkanes are unreactive towards acids, alkalis, electrophiles and nucleophiles. In common with all hydrocarbons, however, they burn in air or oxygen with highly exothermic reactions; hence, they are important for use as fuels (see section 12.2.3).

Electrophiles are electron-pair acceptors and seek electron-rich sites (see section 12.3.2).

Nucleophiles are electron-pair donors and seek electron-deficient sites (see section 12.4.1).

E

Fractional distillation

The complex mixture of hydrocarbons in crude oil, mainly alkanes, is separated into less complicated mixtures, or fractions, by fractional distillation (primary distillation – so called as it is the first stage in the separation process).

The crude oil is heated and the vapour/liquid mixture passed into a tower. The top of the tower is cooler than the bottom, i.e. there is a temperature gradient and this separates the petroleum mixture into fractions depending on the boiling points of the hydrocarbons present. Only the most volatile components, those with low boiling points, reach the top; others condense in trays at different levels up the tower and are drawn off.

Tower or column is the name given to the long vertical tube used in fractional distillation.

The residue from primary distillation still contains useful materials, such as lubricating oil and waxes; these boil above 350°C at atmospheric pressure. At such high temperatures, some of the components in the residue decompose. To avoid this, the residue is further distilled under reduced pressure (vacuum distillation). Using this method, the remaining hydrocarbons can be distilled at a lower temperature, i.e. one at which they do not decompose.

Fractional distillation is a *physical* process. Energy is needed only to separate molecules from each other, that is to overcome the van der Waals' forces.

E

Table 5
Fractions from crude oil

Name of fraction	Boiling range / °C (approx)	Uses	Length of carbon chain (approx)
LPG (liquefied petroleum gas)	up to 25	Calor Gas, Camping Gaz	1–4
petrol (gasoline)	40–100	petrol	4–12
naphtha	100–150	petrochemicals	7–14
kerosine (paraffin)	150–250	jet fuel, petrochemicals	11–15
gas oil (diesel)	220–350	central heating fuel, petrochemicals	15–19
mineral oil (lubricating oil)	over 350	lubricating oil, petrochemicals	20–30
fuel oil	over 400	fuel for ships and power stations	30–40
wax, grease	over 400	candles, grease for bearings, polish	40–50
bitumen	over 400	roofing, road surfacing	above 50

The major fractions and their uses are shown in Table 5.

The composition of crude oil varies from place to place. In general, however, the amount of each fraction produced by distillation does not match the demand (see Table 6).

Table 6
Supply and demand for oil fractions

	Approximate %	
Fraction	**Crude oil**	**Demand**
gases	2	4
petrol and naphtha	16	27
kerosine	13	8
gas oil	19	23
fuel oil and bitumen	50	38

E Cracking is a *chemical* process. Energy is needed to break C—C bonds.

A higher proportion of the high-value products (such as petrol) is used commercially than occurs naturally, while there is not enough demand for some of the heavier fractions. To solve this imbalance, larger alkane molecules are broken up into smaller molecules in a process called **cracking**.

12.2.2 *Petroleum: cracking*

Cracking

Hydrocarbon cracking involves breaking carbon–carbon and carbon–hydrogen bonds. Two main processes are used, **thermal** cracking and **catalytic** cracking. In general, large alkanes are cracked to form smaller alkanes, alkenes and sometimes also hydrogen:

$$\text{high } M_r \text{ alkanes} \longrightarrow \text{smaller } M_r \text{ alkanes} + \text{alkenes} (+ \text{hydrogen})$$

When cracked, molecules may break up in several different ways to form a mixture of products which can be separated by fractional distillation. For example, two possible fragmentations of the $C_{14}H_{30}$ molecule are:

$$C_{14}H_{30} \longrightarrow C_7H_{16} + C_3H_6 + 2C_2H_4$$

$$C_{14}H_{30} \longrightarrow C_{12}H_{24} + C_2H_4 + H_2$$

Steam is used in the process to increase the yield of alkenes. By diluting the naphtha feedstock with steam, the formation of carbon is reduced and the transfer of heat to the reactants is improved.

Thermal cracking

Thermal cracking results in the formation of a high proportion of alkenes. The energy required for bond breaking is provided by heat; the temperatures employed range from 400 °C to 900 °C at pressures of up to 7000 kPa. At the lower end of this temperature range, carbon chains break preferentially part way along the carbon chain of the molecule. With increasing temperature, the cracking shifts towards the end of the chain, leading to a greater percentage of low M_r alkenes. In order to avoid decomposition into the constituent elements, the length of exposure to high temperatures (residence time) has to be short, of the order of one second.

Homolytic fission occurs when a bond breaks equally and each atom takes one of the shared pair of electrons (see also section 12.2.4).

Bond enthalpy /kJ mol^{-1}

C—C 348

C—H 412

The C—C bonds are weaker and break more easily.

The thermal cracking process is initiated by homolytic fission of a C—C bond to form two alkyl radicals (compare this with the chlorination of methane in section 12.2.4). Each alkyl radical can abstract a hydrogen atom from an alkane molecule to produce a different alkyl radical and a shorter alkane. For example:

$$CH_3(CH_2)_6CH_3 \longrightarrow CH_3CH_2CH_2CH_2CH_2\bullet + \bullet CH_2CH_2CH_3$$

$$CH_3(CH_2)_6CH_3 + CH_3CH_2CH_2CH_2CH_2\bullet \longrightarrow CH_3(CH_2)_5\dot{C}HCH_3 + CH_3(CH_2)_3CH_3$$

Alternatively, the alkyl radical can undergo further bond cleavage to form an alkene and a shorter alkyl radical. For example:

$$CH_3CH_2CH_2—CH_2CH_2\bullet \longrightarrow CH_3CH_2CH_2\bullet + CH_2{=}CH_2$$

Other reactions which take place include dehydrogenation, isomerisation and cyclisation. Mixtures of products are separated by fractional distillation.

Dehydrogenation involves the loss of molecular hydrogen to produce alkenes from alkanes and aromatic hydrocarbons from cycloalkanes. *Isomerisation* occurs when unbranched alkanes are converted into branched isomers and when cycloalkanes undergo rearrangement, e.g. methylcyclopentane into cyclohexane. *Cyclisation* takes place, often with loss of hydrogen, when alkanes rearrange to form cycloalkanes and aromatic hydrocarbons, e.g. heptane via methylcyclohexane into methylbenzene.

A radical is a species which results from the homolytic fission of a covalent bond. Radicals contain an odd number of electrons with one unpaired electron. This condition is represented by writing a dot (as in Cl•) to indicate the unpaired electron in the radical.

Catalytic cracking

Catalytic cracking involves the use of zeolite catalysts (crystalline aluminosilicates), with a slight excess pressure and a temperature of about 450 °C. By this means, large alkanes are converted mainly into branched alkanes, cycloalkanes and aromatic hydrocarbons. For example:

$$C_{14}H_{30} \longrightarrow C_8H_{18} + C_6H_{12}$$

Aromatic compounds contain a benzene ring (see *Further Physical and Organic Chemistry*, section 13.6).

The proportion of alkenes is small so that catalytic cracking is primarily used for producing motor fuels. Branched-chain alkanes burn more smoothly than unbranched chains. In an engine, because of the pressures involved, the fuel–air mixture may ignite before the spark is produced, causing 'knocking'. This problem is prevented by using branched-chain alkanes.

When cracking is carried out in the presence of hydrogen (hydrocracking), the resulting mixture is free from impurities of sulphur (converted into hydrogen sulphide) and nitrogen (converted into ammonia) and also from alkenes (converted into alkanes).

A Lewis acid is an electron pair acceptor (see *Thermodynamics and Further Inorganic Chemistry*, section 14.5.1).

A carbocation is a carbon with a positive charge and only three covalent bonds (see section 12.3.2).

The mechanism of catalytic cracking, where the catalyst acts as a Lewis acid, involves the formation of carbocations.

Benzene and its methyl and dimethyl derivatives arise by a process of cyclisation and dehydrogenation. For example:

$$C_7H_{16} \longrightarrow C_6H_5CH_3 + 4H_2$$

12.2.3 *Petroleum: combustion*

Combustion of petroleum fractions

Fractions obtained from petroleum are used as fuels (see Table 5 on page 14). The hydrocarbons present in the fractions burn in air or oxygen in very exothermic reactions (see below). However, sulphur-containing impurities occur with hydrocarbons in petroleum fractions, and these burn when the hydrocarbons are burned; this produces oxides of sulphur, such as sulphur dioxide and sulphur trioxide. These are toxic and, being soluble in water, can cause acid rain.

Combustion of alkanes

E Complete combustion of hydrocarbons produces carbon dioxide and water.

In common with all hydrocarbons, alkanes burn in air or oxygen in very exothermic reactions. In the presence of a plentiful supply of oxygen, *complete combustion* of alkanes occurs to form carbon dioxide and water,. For example:

$$CH_4 + 2O_2 \longrightarrow CO_2 + 2H_2O \qquad \Delta H^{\ominus} = -890 \text{ kJ mol}^{-1}$$

$$C_4H_{10} + 6\tfrac{1}{2}O_2 \longrightarrow 4CO_2 + 5H_2O \qquad \Delta H^{\ominus} = -2880 \text{ kJ mol}^{-1}$$

As the number of carbon atoms increases, more oxygen is required per mole of hydrocarbon for complete combustion, and more energy is released.

E Incomplete combustion of hydrocarbons produces water and carbon or carbon monoxide.

When insufficient oxygen is available, *incomplete combustion* occurs. Water is formed together with carbon monoxide or carbon. For example, if a bunsen burner is used with the air-hole closed, the flame is not blue but yellow and luminous because of the carbon particles it contains. Any apparatus heated in a luminous flame becomes coated in black soot.

$$CH_4 + O_2 \longrightarrow C + 2H_2O$$

Incomplete combustion forming carbon monoxide is, however, much more of a hazard. Badly maintained gas central heating boilers may produce carbon monoxide because of an inadequate supply of air, and can cause accidental death by carbon monoxide poisoning.

$$CH_4 + 1\tfrac{1}{2}O_2 \longrightarrow CO + 2H_2O$$

Internal combustion engines

Carbon monoxide is also formed by the incomplete combustion of petrol vapour in a car engine.

$$C_8H_{18} + 8\frac{1}{2}O_2 \longrightarrow 8CO + 9H_2O$$

Cars with petrol engines produce carbon monoxide but diesel engines produce only carbon.

E

Motor car engines also produce other pollutants, notably oxides of nitrogen and unburned hydrocarbons. Oxides of nitrogen are formed when the air/petrol mixture is sparked and explodes. The temperature of burning petrol vapour can reach 2500 °C and this provides sufficient activation energy for nitrogen to react with oxygen to form nitrogen monoxide.

$$N_2 + O_2 \longrightarrow 2NO$$

On cooling, nitrogen monoxide reacts easily with more oxygen to form nitrogen dioxide. With water and more oxygen, nitric acid is formed, which can lead to acid rain.

$$2NO + O_2 \longrightarrow 2NO_2$$

$$4NO_2 + 2H_2O + O_2 \longrightarrow 4HNO_3$$

Nitrogen dioxide also reacts with oxygen or hydrocarbons in the presence of sunlight to form an irritating photochemical smog.

Catalytic converters

These devices help to remove carbon monoxide, nitrogen oxides and hydrocarbons from car exhausts (see Fig 13). Converters contain a honeycomb of ceramic material onto which metals such as platinum, palladium and rhodium are spread in a thin layer. These metals catalyse reactions between the pollutants and help to remove up to 90% of the harmful gases. For example:

$$2CO + 2NO \longrightarrow 2CO_2 + N_2$$

$$C_8H_{18} + 25NO \longrightarrow 8CO_2 + 12\frac{1}{2}N_2 + 9H_2O$$

Overall, the pollutant gases – CO and NO_x and hydrocarbons – are replaced by CO_2, N_2 and H_2O, which are harmless.

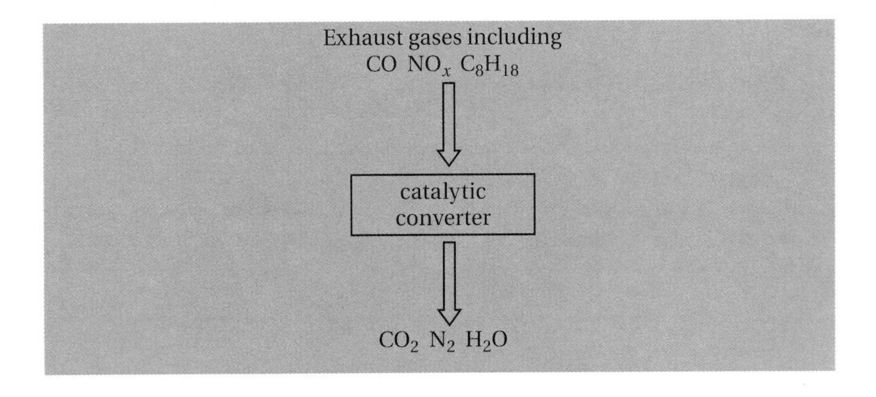
Exhaust gases including
CO NO_x C_8H_{18}
catalytic converter
CO_2 N_2 H_2O

Fig 13
The action of a catalytic converter

12.2.4 *Alkanes: chlorination*

Alkanes such as methane do not react with chlorine at room temperature or in the dark. In the presence of ultraviolet light, however, a mixture of methane and chlorine will react explosively even at room temperature, forming hydrogen chloride and a mixture of chlorinated methanes.

E Ultra-violet light consists of very high energy radiation, enough to break the Cl—Cl bond.

This process is a **free-radical substitution reaction** which occurs in several steps:

1. Initiation step

$$Cl_2 \longrightarrow 2Cl\bullet$$

The ultraviolet light provides the energy needed to start the reaction by splitting some chlorine molecules into atoms (radicals). This process occurs first because the Cl—Cl bond in chlorine is weaker than the C—H bond in methane.

2. Propagation steps

$$Cl\bullet + CH_4 \longrightarrow CH_3\bullet + HCl$$

$$CH_3\bullet + Cl_2 \longrightarrow CH_3Cl + Cl\bullet$$

In each step, a radical is used and a new radical is formed, so that the process continues and leads to a chain reaction. Each step is exothermic so that the chain reaction might produce an explosion. The overall reaction, which is the sum of the two propagation steps, can be represented by the equation:

$$CH_4 + Cl_2 \longrightarrow CH_3Cl + HCl$$

3. Termination steps

When two radicals combine, they form a stable molecule and the sequence of reactions stops; the unpaired electrons in the radicals pair up to form a covalent bond. Two possible termination steps are:

$$Cl\bullet + CH_3\bullet \longrightarrow CH_3Cl$$

$$CH_3\bullet + CH_3\bullet \longrightarrow CH_3CH_3$$

Such termination steps can lead to trace amounts of impurities, such as ethane, in the final product.

E Chlorination of propane produces two monochloropropanes, 1-chloropropane and 2-chloropropane, in an approximate ratio of 3 : 1 because of the ratio of 6 CH_3 hydrogens to 2 CH_2 hydrogens in propane.

Further substitution

The reaction of a chlorine radical with methane extracts a hydrogen radical to form HCl, as in the first propagation step above. Chloromethane still contains three hydrogen atoms, so further pairs of propagation steps are possible, leading to dichloromethane (CH_2Cl_2), trichloromethane ($CHCl_3$) and finally to tetrachloromethane (CCl_4). The propagation steps to form CH_2Cl_2 are shown below:

$$Cl\bullet + CH_3Cl \longrightarrow CH_2Cl\bullet + HCl$$

$$CH_2Cl\bullet + Cl_2 \longrightarrow CH_2Cl_2 + Cl\bullet$$

The likelihood of further substitution beyond the formation of CH_3Cl can be reduced if an excess of methane is used.

12.3 Alkenes and epoxyethane

12.3.1 *Alkenes: structure and bonding*

The alkenes are a homologous series of hydrocarbons with the general formula C_nH_{2n}. The first three members are:

- ethene C_2H_4
- propene C_3H_6
- butene C_4H_8

Alkenes contain two hydrogen atoms fewer than their parent alkanes and are said to be unsaturated. This term is used because alkenes contain a double bond between two carbon atoms (see Fig 14). This bond is an area of high electron density and is the cause of the high reactivity of alkenes.

E The general formula for alkenes, C_nH_{2n}, can also represent cyclic alkanes. For example, C_6H_{12} can represent several isomeric hexenes but also cyclohexane and alkyl-substituted smaller cycloalkanes.

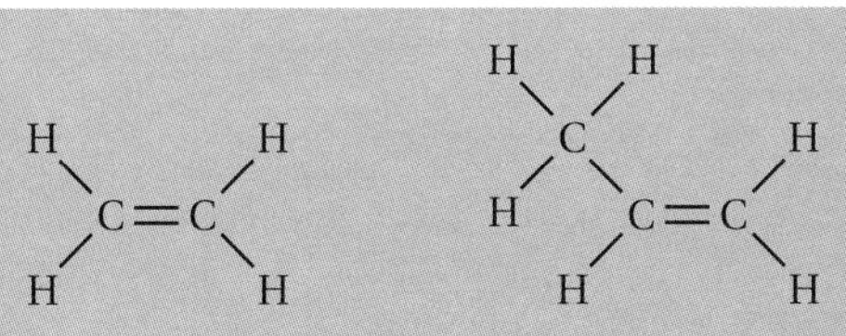

H H C=C H H
ethene

H H C H H C=C H H
propene

Fig 14
The structure of ethene and propene

E The double bond results from the overlap of a spare unbonded, singly-filled p-orbital present on each carbon atom in the bond (see Fig 15). This overlap produces a cloud of electrons above and below the molecule (a π bond), as shown in Fig 16. The two carbon atoms of the double bond and the four atoms attached to the double bond must lie in the same plane; i.e. the >C=C< arrangement is planar, so that ethene is a planar molecule.

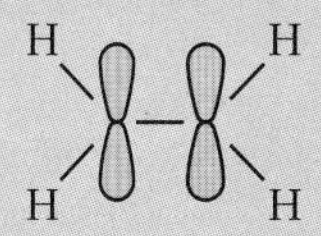

H H H H

Fig 15
The p-orbitals in ethene

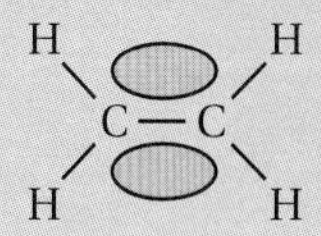

H H C—C H H

Fig 16
The π bond in ethene

12.3.2 *Alkene reactions*

Alkenes can become saturated by the addition of small molecules across the double bond. In these reactions the carbon–carbon double bond becomes a carbon–carbon single bond.

Hydrogenation (addition of hydrogen)

Alkenes react with hydrogen in the presence of a finely-divided nickel catalyst at about 150 °C to form alkanes. On the catalyst surface, the hydrogen molecule is split into two atoms which add to the same side of the double bond (see Fig 17).

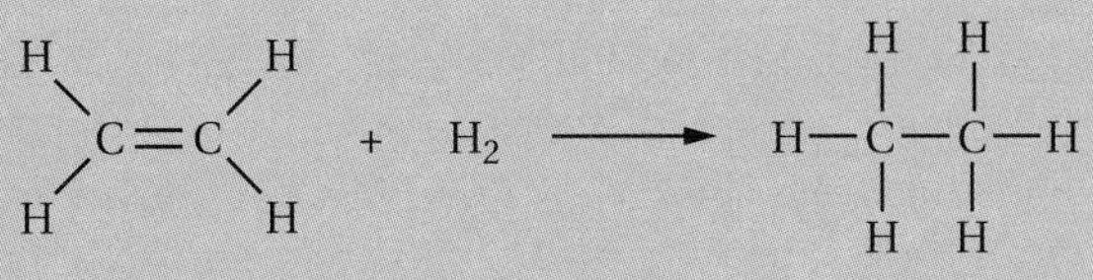

H H C=C H H + H_2 ⟶ H H H—C—C—H H H
ethane

Fig 17
The hydrogenation of ethene

Vegetable oils contain mainly unsaturated compounds and are hardened by hydrogenation during the manufacture of margarines. The polyunsaturated liquid oils are partly saturated by reacting them with hydrogen in the presence of a nickel catalyst, to form solid fats. The amount of hydrogenation can be controlled, so that hard or soft margarines can be produced as required.

Electrophilic additions

1. Bromine water (a test for alkenes)

Alkenes decolourise solutions of bromine in water or in an organic solvent. Removal of the red-brown colour of bromine shows the presence of unsaturation, typically a carbon–carbon double bond. Alkanes do not react with bromine under these conditions, so this reaction can be used to distinguish between alkanes and alkenes.

E The degree of unsaturation in a margarine or butter can be found by dissolving a sample in hexane and adding drops of bromine. The more drops of bromine that are decolourised, the more unsaturation is present. A surprisingly large amount of bromine is needed to determine the unsaturation in butter.

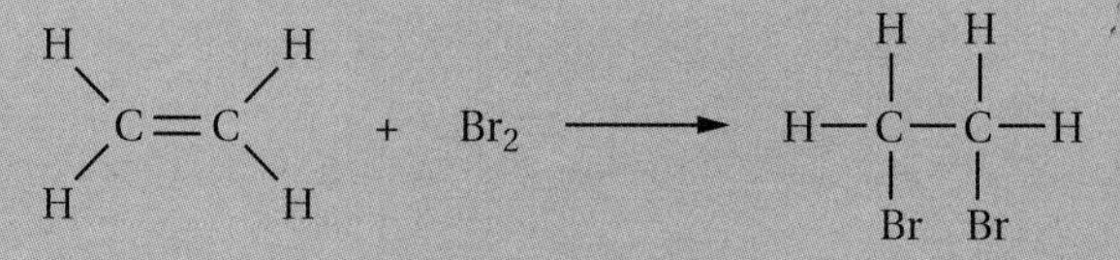

Ethene reacts with bromine to form a colourless, saturated product (Fig 18).

```
H         H                         H  H
 \       /                          |  |
  C = C       +   Br2  ------>   H—C—C—H
 /       \                          |  |
H         H                         Br Br
                                1,2-dibromoethane
```

Fig 18
The reaction of ethene with bromine

Electrophiles are positive ions or electron-deficient atoms and act as electron-pair acceptors; they seek electron-rich sites.

The mechanism of this reaction is **electrophilic addition**. The overall process involves addition across a carbon–carbon double bond as shown below in Fig 19.

```
H        H                  H  H                    H  H
 \      /                   |  |                    |  |
  C = C      ------>     H—C—C—H     ------>     H—C—C—H
 /      \                   +  |                    |  |
H   Br^δ+ H                    Br                   Br Br
     |                  :Br⁻
    Br^δ−
```

Fig 19
The mechanism of electrophilic addition of bromine to ethene

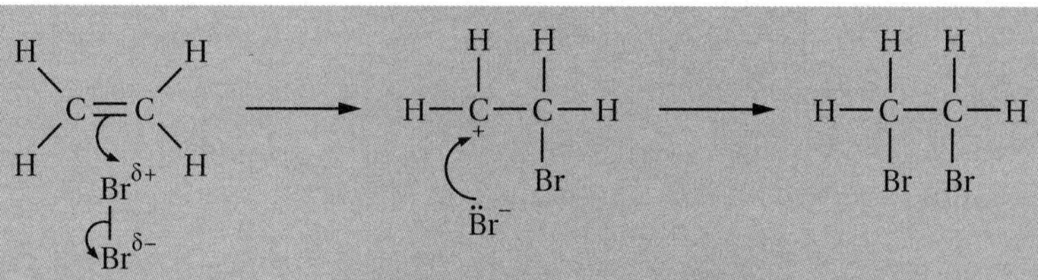

A curly arrow in the mechanism above represents the movement of a *pair* of electrons. Arrows start either at a lone pair or at the middle of a bond. Arrows end either between the atoms where the new covalent bond forms, or as a lone pair on an atom. The stages in the process are as follows:

- The electron-rich area of the double bond induces a dipole in the bromine molecule. The electron deficient (or δ+) bromine atom is the electrophile (*electron-seeking species*).
- Electrons from the double bond begin to form a new carbon–bromine bond with the δ+ bromine atom. The other carbon atom in the double bond becomes an electron-deficient carbocation (carbonium ion). The electrons in the bromine–bromine bond shift towards the more distant bromine atom and the Br—Br bond breaks, releasing a bromide ion.
- This bromide ion then acts as a nucleophile, i.e. it uses a lone pair of electrons to form a new bond with the carbocation.

E The electrons in the Br—Br bond are normally distributed symmetrically between the bromine atoms. An induced dipole results when the symmetry is distorted.

E In bromine water, the alternative product (below) is also formed by attack of the nucleophile water on the carbocation.

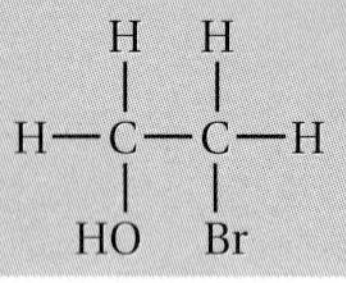

```
   H  H
   |  |
H—C—C—H
   |  |
  HO  Br
```

2. Hydrogen bromide

Alkenes react with hydrogen bromide in the gas phase or in concentrated aqueous solution to form bromoalkanes. For example, Fig 20 shows how ethene reacts to form bromoethane.

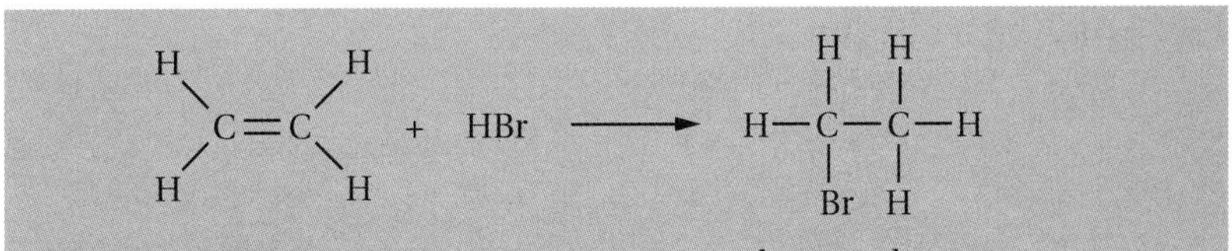

bromoethane

Fig 20
The reaction of ethene with hydrogen bromide

The mechanism is similar to that involved in the addition of bromine, except that H—Br possesses a permanent dipole; the hydrogen atom is electron deficient. When the reaction is performed in aqueous solution, H^+ (aq) ions act as electrophiles. In the gas phase, the mechanism can be written as in Fig 21.

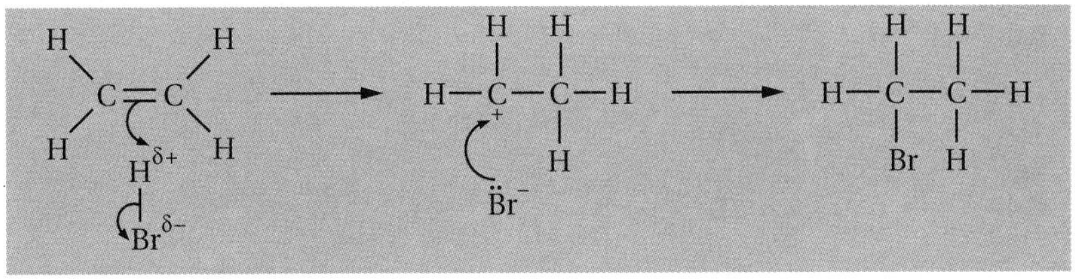

Fig 21
Electrophilic addition in the gas phase

3. Sulphuric acid

Alkenes are absorbed by cold, concentrated sulphuric acid to form alkyl hydrogensulphates, e.g. ethene forms ethyl hydrogensulphate (see Fig 22).

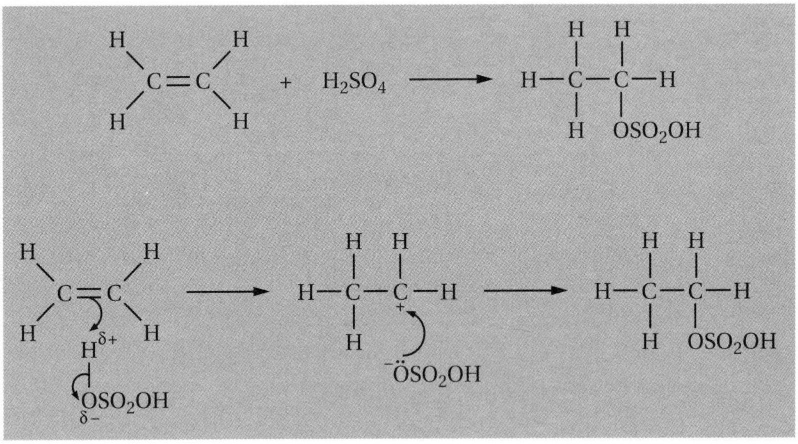

Fig 22
The reaction of ethene with concentrated sulphuric acid

Warming ethyl hydrogensulphate in dilute sulphuric acid causes hydrolysis and produces ethanol by the reaction shown in Fig 23.

Ethanol is produced industrially by the direct hydration of ethene (see p.23) or by fermentation (see section 12.5.1). E

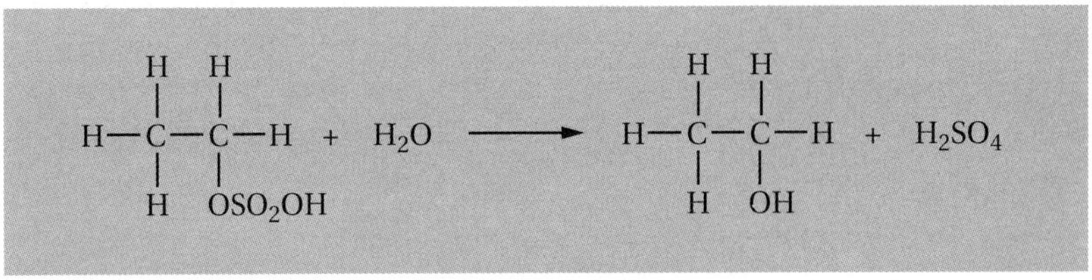

Fig 23
Hydrolysis of ethyl hydrogensulphate

These two reactions result in the overall addition of water to an alkene, forming an alcohol.

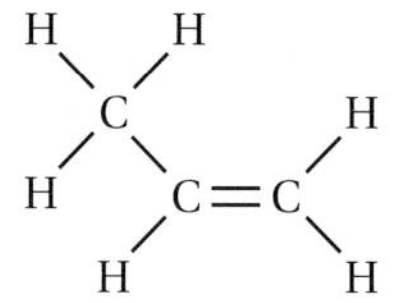

Fig 24
The structure of propene

> **E** A primary carbocation has one alkyl group attached to C^+.
>
> A secondary carbocation has two alkyl groups attached to C^+.
>
> A tertiary carbocation has three alkyl groups attached to C^+.

Electrophilic addition to unsymmetrical alkenes

If the alkene is unsymmetrical, such as propene (see Fig 24), and the molecule being added is also unsymmetrical, such as hydrogen bromide (H—Br) or sulphuric acid ($H—OSO_2OH$), two possible products can form. The major product is the one formed via the *more stable* carbocation (carbonium ion).

The order of stability of carbocations is tertiary (3°) > secondary (2°) > primary (1°) due to the inductive (electron releasing) effect of the attached alkyl groups. The more alkyl groups around the carbocation, the more stable it is and the more likely it is to be formed.

The carbocations of $C_4H_9^+$ can have four possible structures as shown below in Fig 25.

primary		secondary		tertiary
$CH_3CH_2CH_2\overset{+}{C}H_2$ $(CH_3)_2CH\overset{+}{C}H_2$	<	$CH_3CH_2\overset{+}{C}HCH_3$	<	$(CH_3)_3\overset{+}{C}$

increasing stability of carbocations →

Fig 25
Possible structures of $C_4H_9^+$

Propene reacts with hydrogen bromide to form mostly a secondary carbocation. This species then reacts with the bromide ion to produce 2-bromopropane (see Fig 26).

Fig 26
The mechanism of the reaction of propene with hydrogen bromide

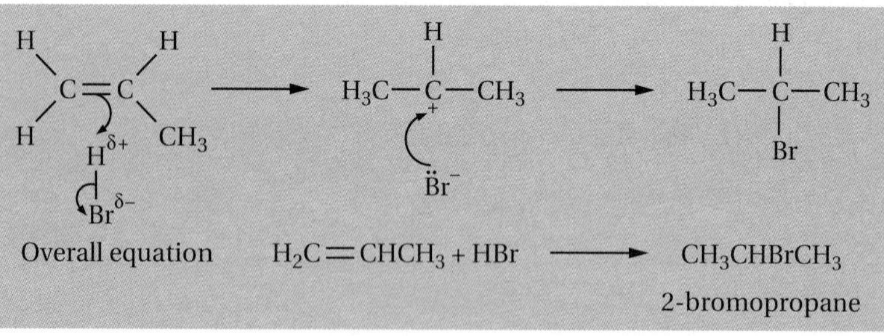

$$H_2C{=}CHCH_3 + H^{\delta+}{-}Br^{\delta-} \longrightarrow H_3C{-}\overset{+}{C}H{-}CH_3 + :Br^- \longrightarrow H_3C{-}CHBr{-}CH_3$$

Overall equation $H_2C{=}CHCH_3 + HBr \longrightarrow CH_3CHBrCH_3$
2-bromopropane

A little 1-bromopropane will also be formed as the minor product via the less stable primary carbocation, by the mechanism shown in Fig 27.

Fig 27
The formation of 1-bromopropane

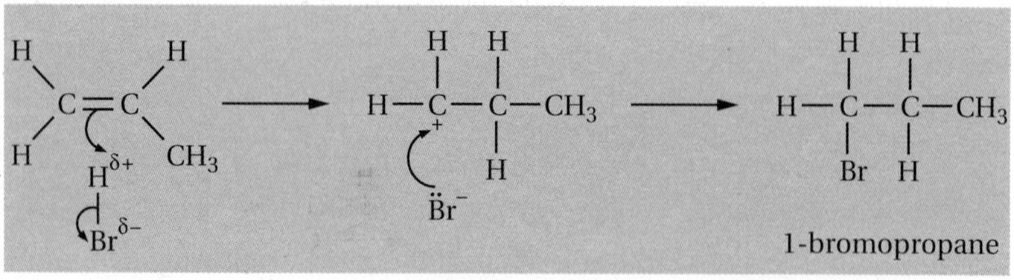

$$H_2C{=}CHCH_3 + H^{\delta+}{-}Br^{\delta-} \longrightarrow H{-}\overset{+}{C}H_2{-}CH_2{-}CH_3 + :Br^- \longrightarrow H{-}CHBr{-}CH_2{-}CH_3$$

1-bromopropane

The reaction of propene with concentrated sulphuric acid produces mainly 2-propyl hydrogensulphate, via the secondary carbocation. Hydrolysis of this hydrogensulphate forms propan-2-ol (see Fig 28).

$$H_2C=CH(CH_3) + H^{\delta+}\text{—}OSO_2OH^{\delta-} \longrightarrow H_3C\text{—}\overset{+}{C}H\text{—}CH_3 + {}^{-}\ddot{O}SO_2OH \longrightarrow H_3C\text{—}CH(OSO_2OH)\text{—}CH_3$$

$$CH_3CH(OSO_2OH)CH_3 + H_2O \longrightarrow CH_3CH(OH)CH_3 + H_2SO_4$$

propan-2-ol

Fig 28
The reaction of propene with concentrated sulphuric acid

Direct hydration of ethene

Ethene reacts with steam at a temperature of 300 °C and a pressure of 6.5×10^3 kPa, in the presence of phosphoric acid (H_3PO_4) as catalyst, to form ethanol by the reaction shown in Fig 29. This illustrates the industrial production of alcohols by the hydration of alkenes in the presence of an acid catalyst.

6.5×10^3 kPa = 6.5 MPa = 65 bar

See section 12.5.1 for the production of ethanol by fermentation.

$$H_2C=CH_2 + H_2O \longrightarrow H\text{—}CH_2\text{—}CH_2\text{—}OH$$

ethene, steam, ethanol

Fig 29
Production of ethanol by the hydration of ethene

The bond energies of C=C and C—C bonds are 612 and 348 kJ mol^{-1}, respectively. Polymerisation, which involves forming two C—C bonds from one C=C bond, is therefore an exothermic process.

Polymerisation (self addition)

Ethene molecules link together in the presence of a catalyst to form addition polymers which are saturated, e.g. poly(ethene). A section of the polymer (formed from eight ethene molecules) is shown below in Fig 30.

$$\text{—}CH_2\text{—}CH_2\text{—}CH_2\text{—}CH_2\text{—}CH_2\text{—}CH_2\text{—}CH_2\text{—}CH_2\text{—}CH_2\text{—}CH_2\text{—}CH_2\text{—}CH_2\text{—}CH_2\text{—}CH_2\text{—}CH_2\text{—}CH_2\text{—}$$

Fig 30
The structure of poly(ethene)

Such polymers are usually represented using a repeating unit, such as that for ethene shown in Fig 31.

The polymerisation of ethene can therefore be represented by the equation in Fig 32.

$$\text{—}CH_2\text{—}CH_2\text{—}$$

Fig 31
The repeating unit of poly(ethene)

$$n\, H_2C=CH_2 \longrightarrow \left(\!\text{—}CH_2\text{—}CH_2\text{—}\!\right)_n$$

poly(ethene)

Fig 32
The formation of poly(ethene)

n represents a large whole number which is the number of individual molecules (monomers) which join together to form the polymer.

For further information about addition polymers see *Further Physical and Organic Chemistry*, section 13.9.1.

Many polymers, including poly(propene), poly(chloroethene) – called *PVC* from poly(*vinyl chloride*), poly(phenylethene) (*polystyrene*) and poly(tetrafluoroethene) (*PTFE*) can be formed from monomers in which some or all of the hydrogen atoms in ethene have been replaced (Fig 33).

Fig 33
The formation of poly(propene) and poly(phenylethene)

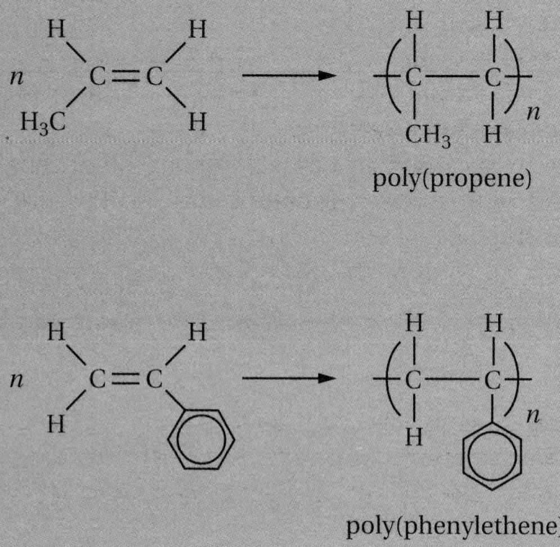

$$n\, H(H_3C)C{=}CH_2 \longrightarrow \left[-CH(CH_3)-CH_2- \right]_n$$

poly(propene)

$$n\, H_2C{=}CH(C_6H_5) \longrightarrow \left[-CH_2-CH(C_6H_5)- \right]_n$$

poly(phenylethene)

(hexagon symbol) represents a benzene ring

12.3.3 *Epoxyethane*

E Epoxyethane contains the C—O—C linkage. This is the ether functional group.

Epoxyethane, commonly known as ethylene oxide, is a highly reactive compound which is manufactured on a large scale. It is mainly used in the synthesis of important products such as ethane-1,2-diol and also in the manufacture of non-ionic surfactants (the active components in detergents).

Production of epoxyethane

Epoxyethane is produced commercially by the direct partial oxidation of ethene, using either oxygen or air, in the presence of a silver-based catalyst. The reaction, which is exothermic, is shown in Fig 34.

Fig 34
The direct partial oxidation of ethene

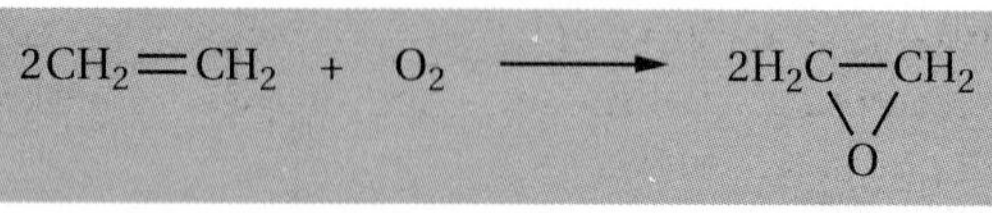

$$2CH_2{=}CH_2 + O_2 \longrightarrow 2H_2C\overset{O}{—}CH_2$$

E The catalyst is present as a finely-divided layer on a support such as alumina. The reaction is carried out at 250–300 °C and 1–2 MPa pressure. About 20% of the ethene is lost by oxidation to carbon dioxide and water. Small quantities of oxidation inhibitors – such as 1,2-dichloroethane – are added to minimise further oxidation of the epoxyethane, which is purified by fractional distillation.

Care has to be taken in the manufacture and handling of epoxyethane. The product is a colourless gas, with a boiling point of 10 °C, which is both flammable and explosive. Epoxyethane is also toxic and may cause respiratory system irritation and neurological effects. The gas is, however, an excellent sterilising agent against bacteria.

Reactions of epoxyethane

The strained three-membered ring in epoxyethane makes the molecule highly reactive towards nucleophiles. These undergo exothermic reactions with epoxyethane and cause the ring to open. The primary products of such reactions contain 2-hydroxyethyl groups. Further reaction of these primary products with epoxyethane is often possible.

—CH_2CH_2OH is named the 2-hydroxyethyl group.

Reaction with water

About half the epoxyethane produced industrially is converted into ethane-1,2-diol (ethylene glycol) by reacting it with water (see Fig 35). The exothermic reaction between epoxyethane and water is slow at room temperature in the absence of an acid catalyst. In industry, epoxyethane is treated with a ten-fold molar excess of water at 60 °C in the presence of sulphuric acid.

$$\underset{\text{(ring via O)}}{H_2C{-}CH_2} + H_2O \longrightarrow \underset{OH\ \ OH}{H_2C{-}CH_2}$$

Fig 35
The formation of ethane-1,2-diol

The resulting aqueous solution is concentrated by evaporating most of the water, followed by fractional distillation. Despite the large excess of water, the yield is only about 90%. The other product is largely the dihydroxyether $HOCH_2CH_2OCH_2CH_2OH$.

The major uses for ethane-1,2-diol are as a raw material for poly(ethylene terephthalate) (Terylene) and as an antifreeze in cars, for which its properties of a low freezing point (−12 °C), a high boiling point (198 °C) and complete miscibility with water are ideal.

If the proportion of water is lowered in the hydration of epoxyethane, more complex polymeric products are formed in a stepwise manner, as shown in Fig 36.

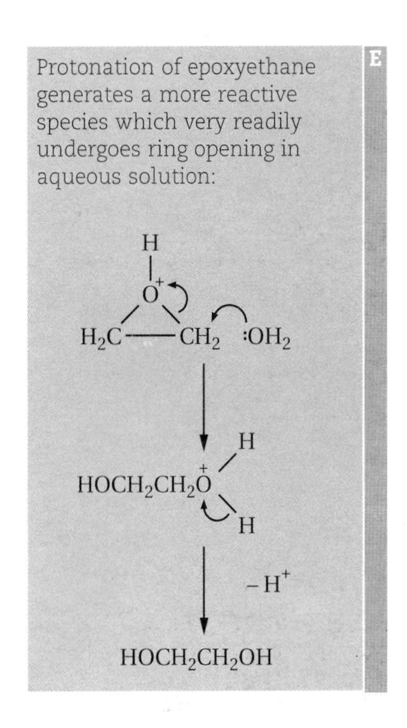
Protonation of epoxyethane generates a more reactive species which very readily undergoes ring opening in aqueous solution:
H
O+
H2C
CH2
:OH2
H
HOCH2CH2O
H
-H+
HOCH2CH2OH

Fig 36
The formation of polyethylene glycols

$$H_2O + n\,H_2C{-}CH_2\ (\text{ring via O}) \longrightarrow HO(CH_2CH_2O)_nH$$

Depending on the value of n, these so-called polyethylene glycols are used as solvents and lubricants, and in the manufacture of plasticisers, polyurethanes and polyester resins.

Reactions with alcohols

The reactions between epoxyethane and alcohols parallel those of epoxyethane with water. The primary products are monoalkyl ethers of ethane-1,2-diol (see Fig 37).

$$ROH + H_2C{-}CH_2\ (\text{ring via O}) \longrightarrow ROCH_2CH_2OH$$

Fig 37
The formation of monoalkyl ethers

The alcohols most frequently employed in the industrial reactions of epoxyethane are methanol, ethanol and butan-1-ol. Even with a large excess of the alcohol, secondary products are still formed, as shown in Fig 38.

Fig 38
The formation of secondary products

$$ROH + nH_2C{-}CH_2 \text{ (epoxide, bridged by O)} \longrightarrow RO(CH_2CH_2O)_nH$$

The monoethyl ether is manufactured at 180 °C and 1 MPa pressure. Such compounds are widely used as solvents in the paint industry and in printing inks. The more complex products are used in brake fluids and as plasticisers.

E These detergents are more biodegradable than traditional detergents.

Non-ionic detergents are produced by treating long-chain (C_{12} to C_{18}) alcohols with a large excess of epoxyethane. The structure of such a compound produced using the unbranched alcohol with 12 carbons is $CH_3(CH_2)_{10}CH_2O(CH_2CH_2O)_9H$.

E The hydrophobic hydrocarbon 'tail' dissolves in grease and the oxygen-rich hydrophilic 'head' forms hydrogen bonds with water molecules. This hydrogen bonding gives the detergent good solubility in water, carrying the grease with it. They are so effective as surfactants that clothes can be washed at a low temperature, e.g. 35 °C, saving energy and avoiding damage to dyes and fibres caused by hot washing.

12.4 Haloalkanes

12.4.1 *Nucleophilic substitution*

The haloalkanes are the homologous series of compounds with the general formula $C_nH_{2n+1}X$, where X is a halogen, i.e. F, Cl, Br or I. For example:

- CH_3CH_2Cl chloroethane
- $CH_3CHBrCH_3$ 2-bromopropane

Halogen atoms are electronegative (see Table 7) so carbon–halogen bonds are polar. The electrons in the C–X bond are attracted towards the halogen atom which gains a slight negative charge, δ−, leaving the carbon atom electron deficient or with a slight positive charge, δ+.

Table 7
Electronegativity values

Element	Electronegativity
C	2.5
F	4.0
Cl	3.0
Br	2.8
I	2.5

The δ+ carbon is then susceptible to attack by nucleophiles, i.e. ions or molecules with a lone pair of electrons. When nucleophilic attack occurs, the carbon–halogen bond breaks and a halide ion is released. The nucleophile replaces the halogen atom in a **nucleophilic substitution** reaction, as shown in Fig 39.

$$\ddot{Nu} \quad \overset{\delta+}{CH_3}-\overset{\delta-}{Br} \longrightarrow \overset{+}{Nu}-CH_3 + :Br^-$$

Fig 39
Nücleophilic substitution

N̈u represents any nucleophile, an electron pair donor.

The rate of such reactions is influenced by the strength of the carbon–halogen bond (see Table 8). Although the C—F bond is very polar, fluoroalkanes are very unreactive because the bond is so strong; chloroalkanes are also fairly slow to react. Carbon–bromine bonds, however, are more easily broken so that bromoalkanes react at a reasonable rate.

Bond	C—F	C—Cl	C—Br	C—I
Bond strength /kJ mol^{-1}	484	338	276	238

Table 8
Carbon–halogen bond strengths

Nucleophilic substitution reactions

1. Hydroxide ions

When haloalkanes are warmed with aqueous sodium hydroxide or potassium hydroxide, alcohols are formed (see Fig 40).

This process is sometimes called *hydrolysis*.

$$CH_3CH_2Br + OH^- \longrightarrow CH_3CH_2OH + Br^-$$
ethanol

$$H_3C-\overset{\delta+}{C}H_2-\overset{\delta-}{Br} \quad (H\overset{-}{O}:) \longrightarrow CH_3CH_2OH + :Br^-$$

Fig 40
The equation and mechanism for the formation of ethanol

Nucleophilic substitution with cyanide ions adds an extra carbon to the chain. Compounds of the homologous series RCN are called nitriles. See Table 3 on p. 7.

2. Cyanide ions

When haloalkanes are warmed with an aqueous/alcoholic solution of potassium cyanide, nitriles are formed. For example, see Fig 41.

$$CH_3CH_2Br + CN^- \longrightarrow CH_3CH_2CN + Br^-$$

propanenitrile

$$CH_3\overset{\delta+}{C}H_2-Br^{\delta-} \longrightarrow CH_3CH_2CN + :Br^-$$

$N\bar{C}:$

Fig 41
The equation and mechanism for the formation of propanenitrile

Nitriles are hydrolysed to carboxylic acids by heating under reflux either with aqueous alkali or with mineral acid; an amide is formed as an intermediate. For example:

$$CH_3CH_2CN + H_2O \longrightarrow CH_3CH_2CONH_2$$

propanenitrile → propanamide

$$CH_3CH_2CONH_2 + H_2O \longrightarrow CH_3CH_2COOH + NH_3$$

propanamide → propanoic acid

The overall reaction in acidic conditions is:

$$CH_3CH_2CN + 2H_2O + H^+ \longrightarrow CH_3CH_2COOH + NH_4^+$$

The overall reaction in alkaline conditions is:

$$CH_3CH_2CN + H_2O + OH^- \longrightarrow CH_3CH_2COO^- + NH_3$$

3. Ammonia

E Primary amines contain the $—NH_2$ functional group.

When haloalkanes are warmed with an excess of ammonia in a sealed container, primary amines are formed. For example, bromoethane forms ethylamine:

$$CH_3CH_2Br + NH_3 \longrightarrow CH_3CH_2NH_2 + HBr$$

ethylamine

Reduction of a nitrile using hydrogen in the presence of a nickel catalyst also forms a primary amine (see *Further Physical and Organic Chemistry*, section 13.7.3).

Since the acid HBr immediately reacts with the base NH_3, this equation is more correctly written as:

$$CH_3CH_2Br + 2NH_3 \longrightarrow CH_3CH_2NH_2 + NH_4Br$$

The mechanism has two steps, as shown in Fig 42.

$$CH_3\overset{\delta+}{C}H_2-Br^{\delta-} \longrightarrow CH_3CH_2-\overset{+}{N}H_3 + :Br^-$$

$H_3N:$

$$CH_3CH_2-\overset{+}{N}H_2-H \longrightarrow CH_3CH_2-\ddot{N}H_2 + NH_4^+$$

$H_3N:$

Fig 42
The mechanism for the formation of ethylamine

The excess of ammonia minimises the chance of further reaction of the primary amines to form secondary or tertiary amines, or quaternary ammonium salts (see *Further Physical and Organic Chemistry*, section 13.7.2).

12.4.2 *Elimination*

In the reaction between aqueous sodium hydroxide and a haloalkane, the hydroxide ion acts as a nucleophile and an alcohol is formed by nucleophilic substitution. However, the hydroxide ion can also function as a base, so an alternative **elimination** reaction also takes place in which hydrogen and bromine are eliminated from the haloalkane and an alkene is formed. For example:

$$CH_3CHBrCH_3 + OH^- \longrightarrow CH_3CH{=}CH_2 + H_2O + Br^-$$

2-bromopropane → propene

The mechanism is shown in Fig 43.

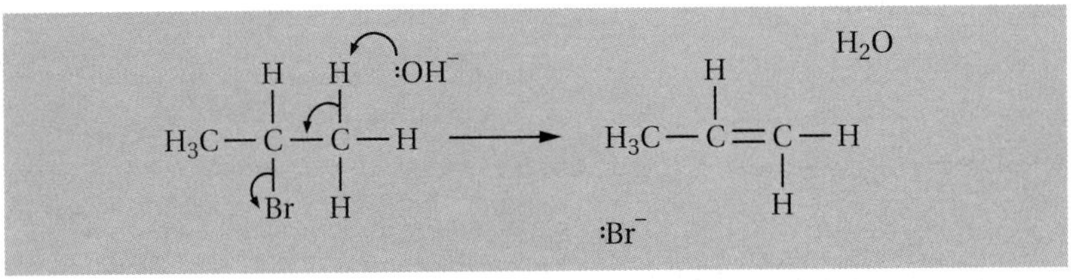

H H :OH^- → H_2O

$H_3C-C-C-H$ (Br, H) → $H_3C-C=C-H$ (H, H) + $:Br^-$

Fig 43
The formation of propene by an elimination reaction

The relative importance of substitution and elimination depends on several factors:

- **the structure of the haloalkane** – primary haloalkanes (RCH_2X) give predominantly *substitution* products whereas tertiary haloakanes (R_3CX) generally favour *elimination.* With secondary haloalkanes (R_2CHX), both *substitution* and *elimination* take place at the same time (concurrently)
- **the base strength of the nucleophile** – the likelihood of elimination increases as the base strength of the nucleophile increases
- **the reaction conditions** – higher reaction temperatures lead to a greater proportion of elimination.

In the reaction of 2-bromopropane with potassium hydroxide (see above), both substitution and elimination reactions occur together:

- elimination is favoured by hot ethanolic conditions
- substitution is favoured by warm aqueous conditions.

E

Depending on the structure of the haloalkane, more than one alkene may be formed. Thus for example, elimination from 2-bromobutane produces both but-1-ene and but-2-ene. The latter compound can, of course, exist as two geometrical isomers; cis-but-2-ene and *trans*-but-2-ene.

Aqueous NaOH contains the nucleophile OH^-. NaOH in ethanol also contains the stronger base $CH_3CH_2O^-$.

12.5 Alcohols

The most common alcohol is ethanol which is present in 'alcoholic' drinks.

Alcohols are the homologous series with the general formula $C_nH_{2n+1}OH$. They all contain the functional group, OH, which is called the hydroxyl group. The first two alcohols are:

- methanol CH_3OH
- ethanol CH_3CH_2OH

12.5.1 *Ethanol production*

Fermentation

Fermentation produces an aqueous solution of ethanol at a concentration of between 3% and 15%. Beers usually contain about 3–7% ethanol and wines about 9–13%. Fermentation rarely produces higher concentrations of ethanol because high alcohol concentrations kill the yeast. More concentrated solutions in spirits such as whisky, brandy or gin, which are about 40% ethanol, are made by fractional distillation of the fermented products.

Alcohol is produced by the process of fermentation, which uses living yeast cells to convert sugars such as glucose into ethanol and carbon dioxide:

$$C_6H_{12}O_6 \xrightarrow{\text{yeast}} 2C_2H_5OH + 2CO_2$$

At low temperatures the reaction is slow, as the enzymes (natural catalysts) in yeast are inactivated; at high temperatures, the yeast cannot survive. The process is therefore normally carried out at a compromise temperature of about 35 °C.

Direct hydration

Ethanol is also produced industrially by the direct hydration of ethene using steam and a phosphoric acid catalyst at 300 °C and 6.5×10^3 kPa pressure (see also section 12.3.2).

$$C_2H_4(g) + H_2O(g) \rightleftharpoons C_2H_5OH(g)$$

Direct hydration is currently preferred for the production of ethanol for industrial use in the UK. However, as this method uses ethene as a raw material, it may become less popular compared to fermentation when oil supplies begin to run out. Table 9 compares the two methods of production.

Table 9
Comparison of methods used to produce ethanol

Method	Rate of reaction	Quality of product	Raw material	Type of process
hydration	fast	pure	ethene from oil (a finite resource)	continuous (cheap on manpower) (expensive equipment)
fermentation	slow	impure	sugars (a renewable resource)	batch (expensive on manpower) (cheap equipment)

12.5.2 *Classifications and reactions*

Alcohols can be classified as primary, secondary or tertiary, depending on the carbon skeleton to which the hydroxyl group is attached, as shown in Fig 44, where R is any alkyl group.

Many reactions of the OH functional group are the same in all alcohols, independent of where the OH group is attached to the carbon skeleton.

primary	secondary	tertiary
$R-\overset{\displaystyle H}{\overset{\vert}{\underset{\underset{\displaystyle H}{\vert}}{C}}}-OH$	$R-\overset{\displaystyle R}{\overset{\vert}{\underset{\underset{\displaystyle H}{\vert}}{C}}}-OH$	$R-\overset{\displaystyle R}{\overset{\vert}{\underset{\underset{\displaystyle R}{\vert}}{C}}}-OH$
I° or 1°	II° or 2°	III° or 3°
$CH_3-\overset{\displaystyle H}{\overset{\vert}{\underset{\underset{\displaystyle H}{\vert}}{C}}}-OH$	$CH_3-\overset{\displaystyle CH_3}{\overset{\vert}{\underset{\underset{\displaystyle H}{\vert}}{C}}}-OH$	$CH_3-\overset{\displaystyle CH_3}{\overset{\vert}{\underset{\underset{\displaystyle CH_3}{\vert}}{C}}}-OH$
ethanol	propan-2-ol	2-methylpropan-2-ol

Fig 44
The classification of alcohols (R represents any alkyl group)

However, the three types of alcohol differ in their reactions with oxidising agents such as acidified potassium dichromate(VI).

Oxidation of alcohols

Primary alcohols are oxidised first to aldehydes, such as ethanal, as shown in Fig 45.

$$\underset{\text{ethanol}}{CH_3-\overset{\displaystyle H}{\overset{\vert}{\underset{\underset{\displaystyle H}{\vert}}{C}}}-OH} + [O] \longrightarrow \underset{\text{ethanal}}{CH_3-C\overset{\displaystyle /\!\!/ O}{\underset{\displaystyle \diagdown H}{}}} + H_2O$$

Fig 45
The oxidation of ethanol

The use of [O] to represent the oxidant is an allowed simplification in this and other equations showing the oxidation of organic compounds. The equations, however, must still balance.

An aldehyde still has one hydrogen atom attached to the carbonyl carbon, so it can be oxidised one step further to a carboxylic acid (see Fig 46).

$$\underset{\text{ethanal}}{CH_3-C\overset{\displaystyle /\!\!/ O}{\underset{\displaystyle \diagdown H}{}}} + [O] \longrightarrow \underset{\text{ethanoic acid}}{CH_3-C\overset{\displaystyle /\!\!/ O}{\underset{\displaystyle \diagdown OH}{}}}$$

Fig 46
The oxidation of ethanal

In practice, a primary alcohol such as ethanol is dripped into a warm solution of acidified potassium dichromate(VI). The aldehyde, ethanal, is formed and immediately distils off, thereby preventing further oxidation to ethanoic acid, because the boiling point of ethanal (23 °C) is much lower than that of either the original alcohol ethanol (78 °C) or of ethanoic acid (118 °C). Both the alcohol and the acid have higher boiling points because of hydrogen bonding.

If oxidation of ethanol to ethanoic acid is required, the reagents must be heated together under reflux to prevent escape of the aldehyde before it can be oxidised further.

Alcoholic drinks such as wine and beer that are left exposed to air go sour. This is because the ethanol present is oxidised to ethanoic acid by oxygen using enzymes from bacteria. However, the bacteria cannot tolerate ethanol concentrations greater than about 20%, so the ethanol in high alcohol drinks is not oxidised. This applies to all spirits and also to fortified wines such as sherry and port.

Secondary alcohols are oxidised to ketones (see Fig 47). These have no hydrogen atoms attached to the carbonyl carbon and so cannot easily be oxidised further.

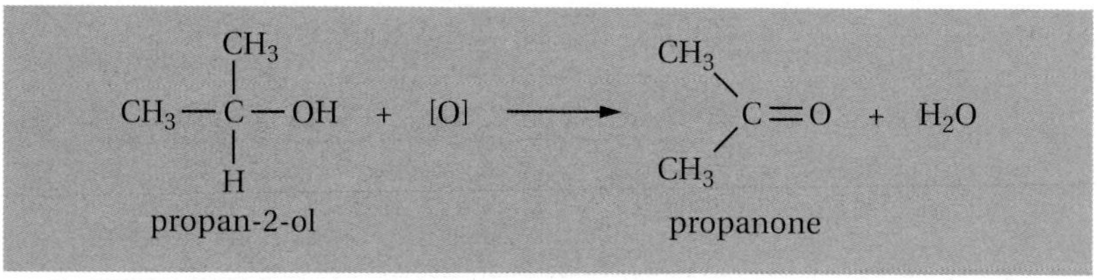

Fig 47
The oxidation of propan-2-ol

$$CH_3-\underset{\displaystyle H}{\overset{\displaystyle CH_3}{C}}-OH + [O] \longrightarrow \begin{matrix} CH_3 \\ CH_3 \end{matrix}\!\!>C{=}O + H_2O$$

propan-2-ol propanone

When orange potassium dichromate(VI) acts as an oxidising agent, it is reduced to green chromium(III) ions. Primary and secondary alcohols both turn acidified dichromate(VI) solution from orange to green when they are oxidised, and this colour change can be used to distinguish them from tertiary alcohols. Tertiary alcohols are not oxidised by acidified dichromate(VI) ions, so they have no effect on its colour, which remains orange.

Tertiary alcohols are not easily oxidised.

Distinguishing between aldehydes and ketones

The reaction with acidified potassium dichromate(VI) distinguishes between tertiary alcohols on the one hand and primary and secondary alcohols on the other, but cannot distinguish between primary and secondary alcohols. However, these can be differentiated by further tests on their oxidation products. Aldehydes formed from primary alcohols are easily oxidised to carboxylic acids, but ketones formed from secondary alcohols are not easily oxidised. Observing whether or not further oxidation occurs can therefore be used to differentiate between them.

Although ketones are not easily oxidised, they react with powerful oxidising agents causing carbon–carbon bonds to break and forming mixtures of carboxylic acids. Any distinguishing reaction must therefore use *mild* oxidising agents to prevent this bond-breaking from occurring. The following are two such examples.

Tollens' reagent contains the complex ion $[Ag(NH_3)_2]^+$ and is prepared by adding an excess of aqueous ammonia to silver nitrate solution. When gently warmed, aldehydes reduce this complex ion and produce a silver mirror on the walls of a test tube; ketones do not form a silver mirror.

Fehling's solution contains a deep blue copper(II) complex ion which on warming is reduced by aldehydes, but not by ketones, to form a red precipitate of copper(I) oxide, Cu_2O.

When aldehydes act as reducing agents in these reactions, they are oxidised to carboxylic acids as shown by the equation:

$$RCHO + [O] \longrightarrow RCOOH$$

Reduction of carbonyl compounds

Aldehydes can be reduced to primary alcohols and ketones to secondary alcohols. This conversion, which is the reverse of the oxidation reactions discussed above, can be achieved in two ways.

1. Catalytic hydrogenation

This involves reacting the carbonyl compound with hydrogen in the presence of a nickel or platinum catalyst. The carbon–oxygen double bond is saturated in exactly the same way that carbon–carbon double bonds are saturated when alkenes are hydrogenated. A molecule containing both carbon–oxygen and carbon–carbon double bonds will become completely saturated when catalytically hydrogenated, for example:

$$CH_3COCH_2CH_3 + H_2 \longrightarrow CH_3CH(OH)CH_2CH_3$$

butanone → butan-2-ol

$$CH_2{=}CHCHO + 2H_2 \longrightarrow CH_3CH_2CH_2OH$$

prop-2-enal → propan-1-ol

2. Reaction with sodium tetrahydridoborate(III) ($NaBH_4$) in methanol

This adds hydrogen via nucleophilic attack on the carbonyl group (see *Further Physical and Organic Chemistry*, section 13.5.1). Acidification of the intermediate ion produces the alcohol. Since the electron cloud of an alkene repels nucleophiles, $NaBH_4$ has no effect on alkenes. It can, however, be used for selective reduction of aldehydes and ketones, e.g. ethanal is reduced to ethanol as shown in Fig 48.

E: The use of [H] to represent the reductant is an allowed simplification in this and other equations showing the reduction of organic compounds. The equations, however, must still balance.

$$CH_3(H)C{=}O + 2[H] \longrightarrow CH_3{-}CH_2{-}OH$$

ethanal → ethanol

Fig 48
The reduction of ethanal

Propanone is reduced to propan-2-ol (see Fig 49).

$$(CH_3)_2C{=}O + 2[H] \longrightarrow CH_3{-}CH(CH_3){-}OH$$

propanone → propan-2-ol

Fig 49
The reduction of propanone

Prop-2-enal is reduced to prop-2-en-1-ol by selective reduction of the carbonyl group (see Fig 50).

$$H_2C{=}CH{-}CHO + 2[H] \longrightarrow H_2C{=}CH{-}CH_2{-}OH$$

prop-2-enal → prop-2-en-1-ol

Fig 50
The reduction of prop-2-enal

12.5.3 *Elimination*

Alcohols with a hydrogen atom on the carbon next to the OH group can be dehydrated to alkenes when heated to about 180 °C with concentrated sulphuric or concentrated phosphoric acid. The mechanism is an elimination. H^+ ions are used in the first step but released in the final step, i.e. the reaction is acid catalysed.

E: Some alcohols may form more than one alkene on dehydration, e.g. butan-2-ol can form but-1-ene and but-2-ene (See also elimination from haloalkanes in section 12.4.2).

Consider the dehydration of propan-2-ol to propene.

In the first step the alcohol is protonated as shown in Fig 51.

Fig 51
Protonation

$$CH_3-CH(OH)-CH_3 + H^+ \longrightarrow CH_3-CH(\overset{+}{O}H_2)-CH_3$$

The protonated alcohol then loses water to form a carbocation as shown in Fig 52.

Fig 52
Loss of water

$$CH_3-CH(\overset{+}{O}H_2)-CH_2 \longrightarrow H-CH_2-\overset{+}{C}H-CH_3 + H_2O$$

This ion then loses a proton to produce an alkene as shown in Fig 53.

Fig 53
Loss of a proton

$$H-CH_2-\overset{+}{C}H-CH_3 \longrightarrow H_2C{=}CH-CH_3 + H^+$$

The overall reaction is shown in Fig 54.

Fig 54
Overall dehydration of propan-2-ol

$$CH_3-CH(OH)-CH_3 \longrightarrow H_2C{=}CH-CH_3 + H_2O$$

propan-2-ol → propene

Similarly, cyclohexanol forms cyclohexene and ethanol forms ethene as shown in Figs 55 and 56.

Fig 55
Dehydration of cyclohexanol

cyclohexanol (OH) → cyclohexene + H_2O

Fig 56
Dehydration of ethanol

$$CH_3-CH_2OH \longrightarrow H_2C{=}CH_2 + H_2O$$

ethanol → ethene

AS 3 Foundation Chemistry 3 Sample module test

1 (a) List, in order of boiling point, any three fractions produced by the fractional distillation of crude oil, starting with the lowest boiling point.

Fraction 1 ....

Fraction 2 ....

Fraction 3 .... (4)

(b) (i) Octane, C_8H_{18}, is a member of the homologous series of alkanes.

Explain the term *homologous series*.

....

....

(ii) The combustion of octane in car engines is sometimes incomplete. Write an equation for the combustion of octane to form carbon monoxide and water only.

.... (2)

(c) Write an equation for the reaction between carbon monoxide and nitrogen monoxide which occurs in a catalytic converter. State the role of carbon monoxide in this reaction.

Equation ....

Role of carbon monoxide .... (2)

(d) An alkane with 14 carbon atoms undergoes thermal cracking to form only octane and propene.

(i) Give the molecular formula of the alkane.

....

(ii) Write an equation for the cracking reaction, name the type of mechanism involved in thermal cracking and give a use for the propene formed.

Equation ....

Type of mechanism ....

Use for propene.. (4)

(12)

2 (a) (i) There are three types of mechanistic step in the reaction of chlorine with methane. Name each type of step and write an equation to illustrate each one.

Name of first type ....

Equation ....

Name of second type ....

Equation ....

Name of third type ....

Equation ....

(ii) State why an excess of methane is used in this reaction when chloromethane is the required product.

....

.... *(7)*

(b) Predict the major organic product of the reaction between prop-1-ene and hydrogen bromide. Explain the basis of your prediction.

Major product ....

Explanation ....

....

.... *(4)*

(c) Three hydrocarbons, **D**, **E** and **F**, all have the molecular formula C_6H_{12}.

D decolourises an aqueous solution of bromine and shows geometric isomerism.

E also decolourises an aqueous solution of bromine but does not show geometric isomerism.

F does not decolourise an aqueous solution of bromine.

Draw one possible structure each for **D**, **E** and **F**.

Structure of **D** Structure of **E** Structure of **F**

(3)

(14)

3 (a) Name the initial oxidation product formed when propan-1-ol is warmed with acidified potassium dichromate(VI).

.... *(1)*

(b) Write an equation for the oxidation of propan-2-ol by acidified potassium dichromate(VI) showing clearly the structure of the organic product. You may use the symbol [O] for the oxidising agent.

.... *(2)*

(c) By stating a reagent and the observation with each compound, show how a simple test can distinguish between the products formed in (a) and (b).

Reagent..

Observation with product formed in part (a)

....

Observation with product formed in part (b)

.... *(3)*

(d) Give the reagent and conditions, and outline a mechanism for the formation of propene by the dehydration of propan-2-ol.

Reagent ....

Conditions...

Mechanism

(6)

(12)

4 Formation of epoxyethane by the partial oxidation of ethene by air, in the presence of a catalyst, is an exothermic process ($\Delta H = -210 \text{ kJ mol}^{-1}$).

(a) Write an equation for the reaction, name the catalyst and suggest a hazard associated with the process.

Equation

Catalyst ....

Hazardous feature .... *(4)*

(b) (i) Name the type of reaction which takes place between epoxyethane and water. Write an equation for the reaction between one mole of epoxyethane and one mole of water.

Type of reaction ....

Equation

(ii) Give the structure of the compound formed when the product in part (b)(i) undergoes a further reaction with two more moles of epoxyethane.

(3)

(7)

5 (a) (i) Ethanol can be produced industrially either from ethene or from a sugar such as glucose, $C_6H_{12}O_6$. For each route, name the method used, write an equation for the reaction and give one necessary condition. Suggest, with a reason, which route is likely to become the major method of production in the future. *(8)*

(ii) Ethanol can also be produced in a reaction involving reduction. Give a suitable reducing agent and write an equation for the reaction. You may use [H] to indicate the reductant in your equation. *(2)*

(b) Ethene can be converted into two different saturated hydrocarbons, one of which is a gas and the other a solid of high relative molecular mass.
Give the structure of each product and state the types of reaction involved. For the gaseous product only, give the conditions necessary for its formation and give an industrial use of this type of reaction. *(6)*

(16)

....

....

....

....

....

....

....

....

....

....

....

....

....

....

....

....

....

....

....

6 In aqueous ethanolic alkali, 2-bromo-2-methylbutane undergoes either substitution or elimination reactions to produce an alcohol or a mixture of two alkenes, respectively. Give the structures and names of these three compounds. Account for the formation of the various products by reference to the mechanisms of the reactions involved.

(14)

Module test answers

1 (a) Any three from: LPG / gasoline (petrol) / naphtha / kerosine (petrol) / gas oil (diesel) / mineral oil (lubricating oil) / fuel oil / wax, grease / bitumen

in correct order — 4

(b) (i) compounds with the same general formula
(ii) $C_8H_{18} + 8\frac{1}{2}O_2 \rightarrow 8CO + 9H_2O$ — 2

(c) $2CO + 2NO \rightarrow 2CO_2 + N_2$
reducing agent — 2

(d) (i) $C_{14}H_{30}$
(ii) $C_{14}H_{30} \rightarrow C_8H_{18} + 2C_3H_6$
radical
making plastics — 4

2 (a) (i) initiation
$Cl_2 \rightarrow 2Cl\cdot$
propagation
e.g. $CH_4 + Cl\cdot \rightarrow CH_3\cdot + HCl$
termination
e.g. $CH_3\cdot + Cl\cdot \rightarrow CH_3Cl$
(ii) to minimise further reaction — 7

(b) $CH_3CHBrCH_3$ or 2-bromopropane
secondary carbocation or $CH_3\overset{+}{C}HCH_3$
more stable
than primary carbocation or $CH_3CH_2\overset{+}{C}H_2$ — 4

(c)
D $CH_3CH{=}CHCH_2CH_2CH_3$ (and others)

E $(CH_3)_2C{=}C(CH_3)_2$ (and others)

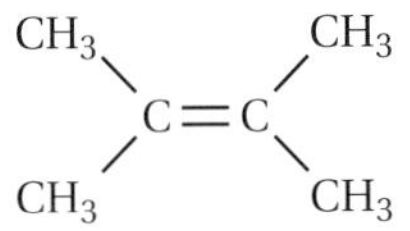

F cyclohexane (hexagon) (and others)

3

3 (a) propanal — 1

(b)
$$CH_3{-}CH(OH){-}CH_3 + [O] \rightarrow CH_3{-}C(=O){-}CH_3 + H_2O$$ 2

(c) Tollens' reagent (or Fehling's solution)
silver mirror (or red precipitate)
no reaction — 3

(d) concentrated H_2SO_4
180 °C

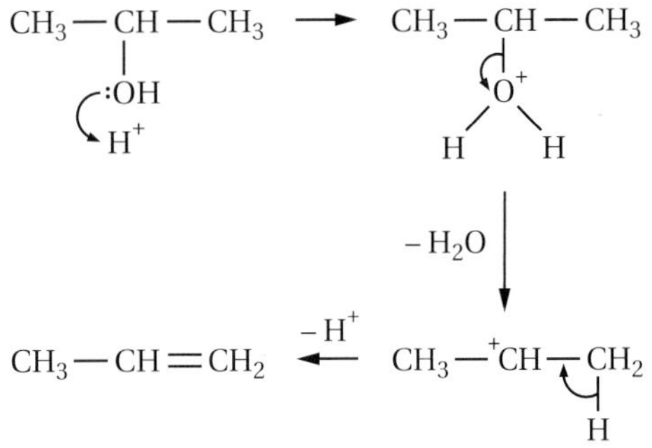

$CH_3{-}CH(:OH){-}CH_3$ + H^+ $\rightarrow$ $CH_3{-}CH(\overset{+}{O}H_2){-}CH_3$

$-H_2O$ $\downarrow$

$CH_3{-}\overset{+}{C}H{-}CH_2{-}H$ $\xrightarrow{-H^+}$ $CH_3{-}CH{=}CH_2$ — 6

4 (a)
$$2CH_2{=}CH_2 + O_2 \rightarrow 2CH_2{-}CH_2 \text{ (epoxide, bridged by O)}$$
silver
toxic or flammable or explosive — 4

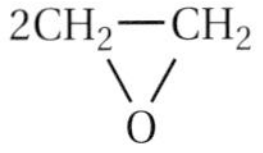

(b) (i) hydrolysis or hydration 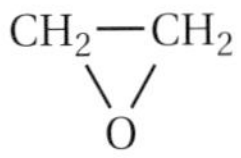
$$CH_2{-}CH_2 \text{ (epoxide, O)} + H_2O \rightarrow HOCH_2CH_2OH$$
(ii) $HO(CH_2CH_2O)_3H$ — 3

5 (a) (i) hydration
$CH_2{=}CH_2 + H_2O \rightarrow CH_3CH_2OH$
acid catalyst
fermentation
$C_6H_{12}O_6 \rightarrow 2C_2H_5OH + 2CO_2$
yeast/warm/anaerobic
fermentation in long term
sugars are renewable resource / oil will run out — 8
(ii) $NaBH_4$
$CH_3CHO + 2[H] \rightarrow CH_3CH_2OH$ — 2

(b) gas CH_3CH_3
hydrogenation
Ni or Pt catalyst
reaction used to make more saturated molecules
for margarine preparation
solid $\mathrm{+\!\!\!\!\!-}CH_2{-}CH_2\mathrm{+\!\!\!\!\!-}_n$
polymerisation — 6

6

3 products

$CH_3CH_2C(OH)(CH_3)_2$ (1)

$CH_3CH{=}C(CH_3)_2$ (1)

$CH_3CH_2C{=}CH_2$ (1), with CH_3 on the C (i.e. $CH_3CH_2C(CH_3){=}CH_2$)

2-methylbutan-2-ol (1)

2-methylbut-2-ene (1)

2-methylbut-1-ene (1)

Formation of alcohol: OH^- acts as nucleophile (1)

mechanism:

Br (1)

$CH_3CH_2-C-CH_3$ (1)

$H\bar{O}$: → C, CH_3

(1)

Formation of alkenes: OH^- acts as base (1)

mechanism:

Br (1)

$CH_3CH-C-CH_3 \longrightarrow CH_3CH{=}C(CH_3)_2$

H (1) CH_3

$H\bar{O}$: (1)

or via carbocation:

Br (1)

$CH_3CH_2-C-CH_3 \xrightarrow{OH^-} CH_3CH_2-\overset{+}{C}(CH_3)_2$ (1)

CH_3

then loss of H^+

either:

$CH_3-CH-\overset{+}{C}(CH_3)(CH_3) \longrightarrow CH_3CH{=}C(CH_3)_2$

H (1)

$H\bar{O}$:

or:

$CH_3-CH_2-\overset{+}{C}(CH_3)-CH_2-H \longrightarrow CH_3CH_2C(CH_3){=}CH_2$

H

$H\bar{O}$: (1)

max 14

NOTES

The Periodic Table of the Elements

Key:

atomic no
symbol
name
relative atomic mass

Period \ Group	1	2											3	4	5	6	7	0
1	1 **H** hydrogen 1.0																	2 **He** helium 4.0
2	3 **Li** lithium 6.9	4 **Be** beryllium 9.0											5 **B** boron 10.8	6 **C** carbon 12.0	7 **N** nitrogen 14.0	8 **O** oxygen 16.0	9 **F** fluorine 19.0	10 **Ne** neon 20.2
3	11 **Na** sodium 23.0	12 **Mg** magnesium 24.3											13 **Al** aluminium 26.9	14 **Si** silicon 28.1	15 **P** phosphorus 31.0	16 **S** sulphur 32.1	17 **Cl** chlorine 35.5	18 **Ar** argon 39.9
4	19 **K** potassium 39.1	20 **Ca** calcium 40.1	21 **Sc** scandium 45.0	22 **Ti** titanium 47.8	23 **V** vanadium 50.9	24 **Cr** chromium 52.0	25 **Mn** manganese 54.9	26 **Fe** iron 55.9	27 **Co** cobalt 58.9	28 **Ni** nickel 58.7	29 **Cu** copper 63.5	30 **Zn** zinc 65.4	31 **Ga** gallium 69.7	32 **Ge** germanium 72.6	33 **As** arsenic 74.9	34 **Se** selenium 79.0	35 **Br** bromine 79.9	36 **Kr** krypton 83.8
5	37 **Rb** rubidium 85.5	38 **Sr** strontium 87.6	39 **Y** yttrium 88.9	40 **Zr** zirconium 91.2	41 **Nb** niobium 92.9	42 **Mo** molybdenum 95.9	43 **Tc** technetium (98)	44 **Ru** ruthenium 101.1	45 **Rh** rhodium 102.9	46 **Pd** palladium 106.4	47 **Ag** silver 107.9	48 **Cd** cadmium 112.4	49 **In** indium 114.8	50 **Sn** tin 118.7	51 **Sb** antimony 121.8	52 **Te** tellurium 127.6	53 **I** iodine 126.9	54 **Xe** xenon 131.3
6	55 **Cs** caesium 132.9	56 **Ba** barium 137.3	57 **La** lanthanum 138.9	72 **Hf** hafnium 178.5	73 **Ta** tantalum 181.0	74 **W** tungsten 183.9	75 **Re** rhenium 186.2	76 **Os** osmium 190.2	77 **Ir** iridium 192.2	78 **Pt** platinum 195.1	79 **Au** gold 197.0	80 **Hg** mercury 200.6	81 **Tl** thallium 204.4	82 **Pb** lead 207.2	83 **Bi** bismuth 209.0	84 **Po** polonium (209)	85 **At** astatine (210)	86 **Rn** radon (222)
7	87 **Fr** francium (223)	88 **Ra** radium (226)	89 **Ac** actinium (227)	104 **Unq** unnilquadium (261)	105 **Unp** unnilpentium (262)	106 **Unh** unnilhexium (263)	107 **Uns** unnilseptium (262)	108 **Uno** unniloctium (265)	109 **Une** unnilennium (266)									

Lanthanides

58	59	60	61	62	63	64	65	66	67	68	69	70	71
Ce cerium 140.1	**Pr** praseodymium 140.9	**Nd** neodymium 144.2	**Pm** promethium (145)	**Sm** samarium 150.4	**Eu** europium 152.0	**Gd** gadolinium 157.3	**Tb** terbium 158.9	**Dy** dysprosium 162.5	**Ho** holmium 164.9	**Er** erbium 167.3	**Tm** thulium 168.9	**Yb** ytterbium 173.0	**Lu** lutetium 175.5

Actinides

90	91	92	93	94	95	96	97	98	99	100	101	102	103
Th thorium 232.0	**Pa** protactinium (231)	**U** uranium 238.1	**Np** neptunium (237)	**Pu** plutonium (244)	**Am** americium (243)	**Cm** curium (247)	**Bk** berkelium (247)	**Cf** californium (251)	**Es** einsteinium (254)	**Fm** fermium (253)	**Md** mendelevium (256)	**No** nobelium (254)	**Lr** lawrencium (257)

Collins Support Materials for AQA – ORDER FORM

This booklet covers one module from the AQA chemistry course at A-level.
If you would like to order further copies from the series, please send a completed copy of this page to Collins by telephone, fax or post.

Title	ISBN	Price	Evaluation copy	Order quantity
1 Atomic Structure, Bonding and Periodicity	000327701 1	£4.99		
2 Foundation Physical and Inorganic Chemistry	000327702 X	£4.99		
3 Introduction to Organic Chemistry	000327703 8	£4.99		
4 Further Physical and Organic Chemistry	000327704 6	£6.25		
5 Thermodynamics and Further Inorganic Chemistry	000327705 4	£6.99		
TOTAL ORDER VALUE				

Also available:

Collins Advanced Modular Sciences – Chemistry

comprehensive textbooks to support the AQA specification

Title	ISBN	Price	Evaluation copy	Order quantity
Chemistry AS	000327753 4	£19.99		
Chemistry A2	000327754 2	£19.99		
TOTAL ORDER VALUE				

Details of other A-level titles in this series are available on our website:

www.**Collins**Education.com

Please fill in your details and send your order to the address below:

Name Address	Tel: 0870 0100 442 Fax: 0141 306 3750 Post: Collins Educational HarperCollins Publishers FREEPOST GW2446 GLASGOW G64 1BR

The *illustrated* lecture series

GW00983880

Cardiology

The *illustrated* lecture series

Neurology and Psychiatry	P. N. Plowman
Endocrinology and Metabolic Diseases	P. N. Plowman
Nephrology, Electrolyte Pathophysiology and Poisoning	P. N. Plowman
Haematology and Immunology	P. N. Plowman
Cardiology	T. J. Phillips, P. N. Plowman
Respiratory Medicine	P. N. Plowman
Alimentary Medicine and Tropical Diseases	P. N. Plowman, T. J. Phillips, S. J. Rose

In Preparation

Surgery

Anatomy

Obstetrics and Gynaecology

Pathology

Physiology

Ophthalmology

The *illustrated* lecture series

Cardiology

Tania J. Phillips BSc MBBS MRCP
Research Fellow
Boston University School of Medicine
Massachusetts USA

P. N. Plowman MA MD (Cantab) MRCP FRCR
Consultant Physician
St Bartholomew's Hospital and Medical College
London EC1 UK

A Wiley-Phoenix Publication

JOHN WILEY & SONS
Chichester · New York · Brisbane · Toronto · Singapore

Distributors in United States and Canada:
Medical Examination Publishing Company,
A Division of Elsevier Science Publishing Co. Inc., New York

British Library Cataloguing in Publication Data:

Plowman, P. N.
Cardiology.—(The illustrated lecture
series).
1. Heart—Diseases
I. Title II. Plowman, P. N. III. Series
616.1'2 RC681

ISBN 0 471 91452 5

Printed and bound in Great Britain

Contents

Preface	vii
Publisher's Note	ix
Anatomy	3
Examination of the Cardiovascular System	6
Investigations in Cardiology—The Electrocardiogram	24
Abnormalities of Heart Rate and Rhythm	34
Other Investigations in Cardiology	43
Coronary Artery Disease	50
Myocardial Infarction	60
Hypertension	82
Heart Failure	97
Left Ventricular Failure	100
Right Ventricular Failure	103
Rheumatic Heart Disease	106
Mitral Valve Disease	113
Tricuspid Valve Disease	124
Aortic Valve Disease	127
Congenital Heart Disease	136
Acyanotic Heart Disease	136

Cyanotic Heart Disease 145

Infective Endocarditis 148

Heart Muscle Disease 155

Cardiomyopathies 156

Pericarditis 158

Venous Thrombosis and Pulmonary Embolism 164

Index **175**

Preface

Modern cardiology is an extremely complicated subject and for the new student coming onto a cardiac unit it must seem a daunting task even to know where to begin to learn. However, it is certain that the clinical fundamentals—pulse, blood pressure, peripheral signs of failure, apex beat position and heart sounds—should precede the more sophisticated investigations such as chest X-ray, electrocardiogram, echocardiogram and other imaging procedures. Only when the analysis takes this course will the correct diagnosis be most easily reached.

In this book, cardiology has been approached in just this way. The first quarter of the book is devoted to the clinical examination and the appropriate investigations; the major diseases of the heart are then explained using the knowledge gained from the earlier pages.

The major bonus for students using this Illustrated Lecture Series is the abundance of illustrations. Fortunately, cardiology lends itself well to illustration, and diagrams reinforce and commit to memory all the key statements in the text in a manner never before accomplished. The student who reads the text and then studies the adjacent figures will build up a knowledge of this complex subject in the easiest and most painless way.

P. N. Plowman

Publisher's Note

Modern medicine is now more interesting and challenging than ever before. Unfortunately, the subject is now so large that it presents a formidable task for students to encompass and for the practising physician to maintain an up-to-date knowledge.

This series of illustrated books represents an entirely new concept which we believe will open up a new method of medical teaching, adding an extra dimension which will keep the reader's interest alive and active throughout the whole syllabus of general medicine.

The most important feature of the book is the linkage and locking of prose with figures in such a way that illustrations (with repeated key phrases) reinforce the comprehension of the text at all stages as one proceeds through the pages. The content of the series also differs from many standard works in that not only does it bring in new sections on subjects such as coma, brain death, blood transfusion reactions, etc. omitted in older texts, but it also recognises that certain diseases (e.g. tertiary syphilis) no longer merit extensive description whilst other subjects (e.g. current successes in oncology) merit a more generous coverage.

This series, when completed and collected together, should comprise a uniquely illustrated textbook for the entire medical curriculum. Although primarily intended for the undergraduate student, these books should also prove substantially helpful to nurses, paramedics and social workers who are academically inclined, and offer a refresher course to the busy practitioner.

We have tried to make academic life for the student easier. We shall welcome criticisms, comments and suggestions from academics, students and other readers since we feel sure that these will help us to improve future editions.

Cardiology

ANATOMY

The heart is a pump. It consists of four chambers; the right atrium and ventricle, which receive venous blood and pump it into the pulmonary circulation, and the left atrium and ventricle, which pump oxygenated blood from the pulmonary veins into the systemic circulation.

The right atrium receives blood from the superior and inferior venae cavae and the coronary sinus. It is separated from the right ventricle by the tricuspid valve. Venous blood passes from the right ventricle into the pulmonary arteries through the pulmonary valve and becomes oxygenated in the lungs. The oxygenated blood passes via the pulmonary veins into the left atrium, and thence through the mitral valve into the left ventricle. The left ventricle has a thick muscular wall, and pumps blood into the systemic circulation through the aortic valve.

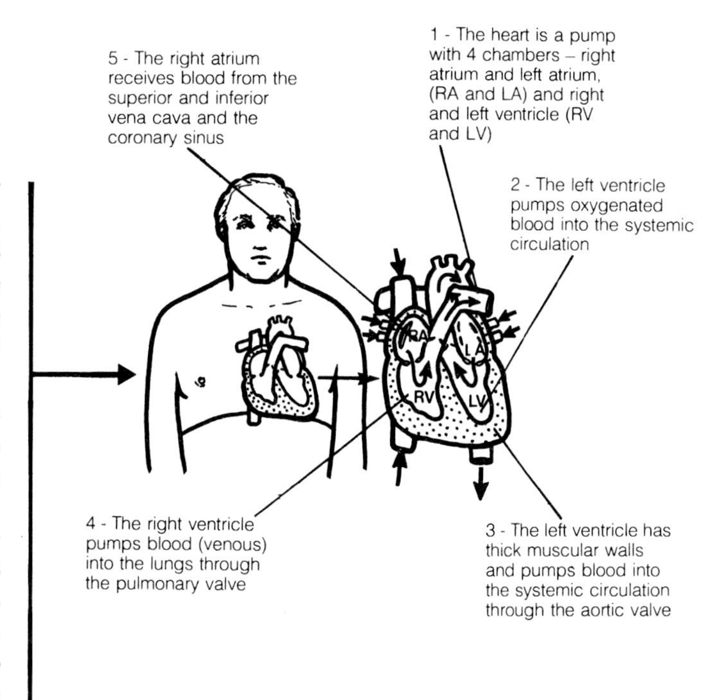
5 - The right atrium receives blood from the superior and inferior vena cava and the coronary sinus
1 - The heart is a pump with 4 chambers – right atrium and left atrium, (RA and LA) and right and left ventricle (RV and LV)
2 - The left ventricle pumps oxygenated blood into the systemic circulation
RA
LA
RV
LV
4 - The right ventricle pumps blood (venous) into the lungs through the pulmonary valve
3 - The left ventricle has thick muscular walls and pumps blood into the systemic circulation through the aortic valve

Blood supply of the heart

The heart is supplied by the right and left coronary arteries. The right coronary artery arises from the anterior part of the aortic sinus. It descends in the atrio-ventricular groove, and gives off marginal and interventricular branches.

The left coronary artery arises from the posterior part of the aortic sinus. It passes laterally around the left border of the heart in the atrio-ventricular groove. It gives off the important left anterior descending artery, which passes in the anterior interventricular groove to supply the anterior part of both ventricles.

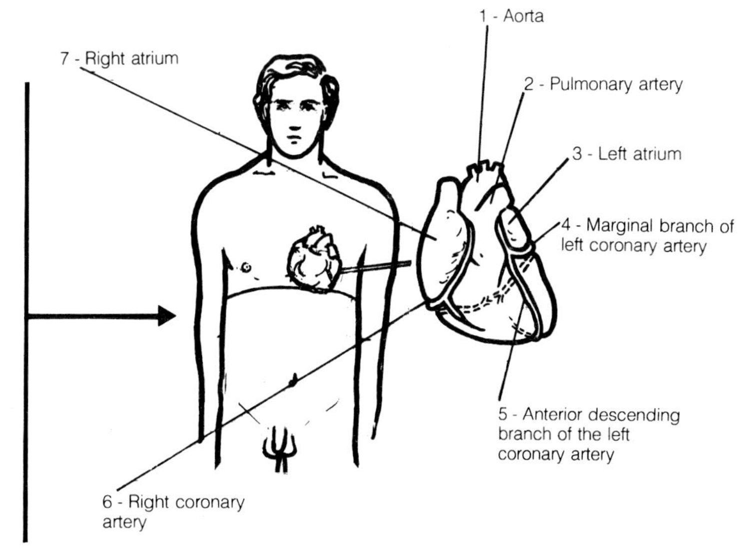
1 - Aorta
7 - Right atrium
2 - Pulmonary artery
3 - Left atrium
4 - Marginal branch of left coronary artery
5 - Anterior descending branch of the left coronary artery
6 - Right coronary artery

Physiology: The Starling Curve

The healthy heart is able to transfer the blood it receives from the venous to the arterial side of the circulation. If there is increased venous return, ventricular muscle fibres are stretched, and will contract more powerfully in systole to transfer blood to the arterial circulation.

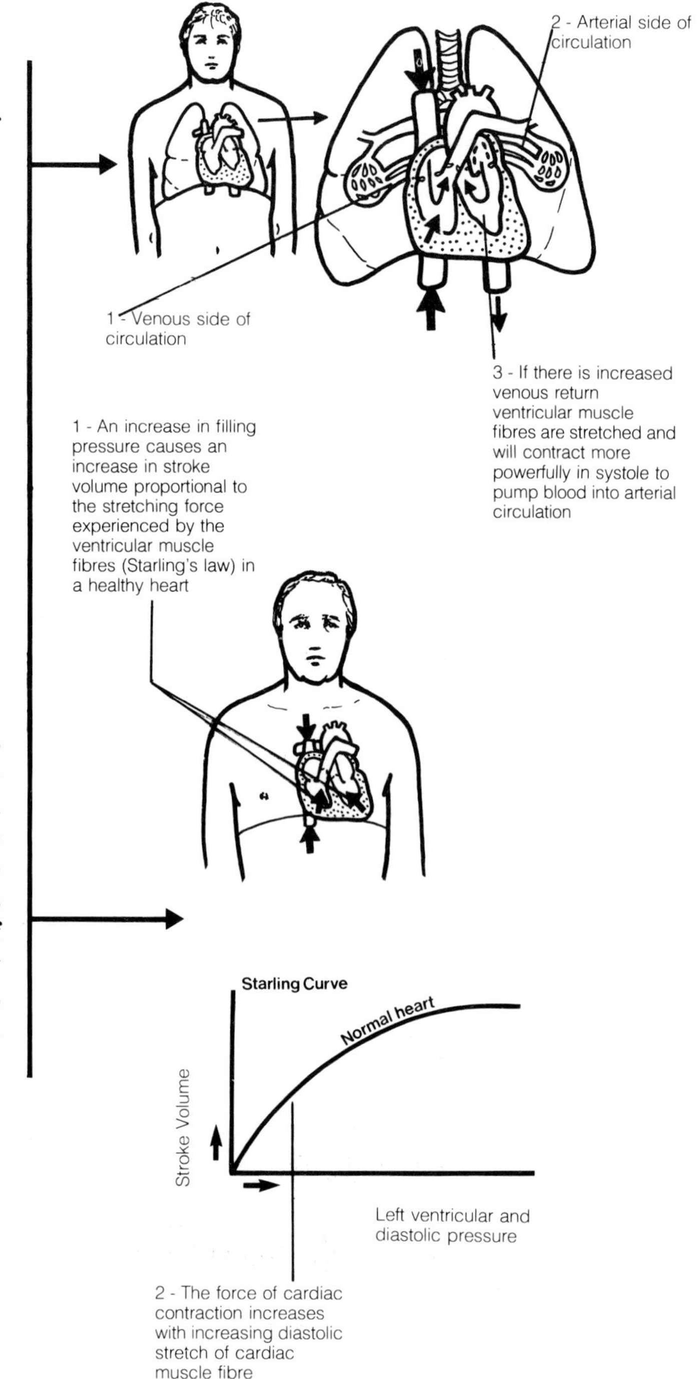
2 - Arterial side of circulation
1 - Venous side of circulation
3 - If there is increased venous return ventricular muscle fibres are stretched and will contract more powerfully in systole to pump blood into arterial circulation
1 - An increase in filling pressure causes an increase in stroke volume proportional to the stretching force experienced by the ventricular muscle fibres (Starling's law) in a healthy heart
Starling Curve
Normal heart
Stroke Volume
Left ventricular and diastolic pressure
2 - The force of cardiac contraction increases with increasing diastolic stretch of cardiac muscle fibre

Starling's law of the heart states that an increase in filling pressure (end diastolic pressure) produces an increase in stroke volume proportional to the degree of ventricular muscle fibre stretch, ie **the force of cardiac contraction increases with increasing diastolic stretch of cardiac muscle fibres** in the healthy heart. By this means the cardiac output always matches the venous return – in a normal person.

The conducting system of the heart and cardiac cycle

Each heartbeat originates in the sino-atrial (SA) node or pacemaker of the heart, which is found in the wall of the right atrium. Impulses from the SA node initiate atrial contraction, and then travel through the atria to the AV node, lying just above the atrio-ventricular septum. The electrical impulses then move down the bundle of His, which divides into right and left branches, and ventricular contraction results. Ventricular contraction is called **systole**, while the ventricular relaxation phase is known as **diastole**.

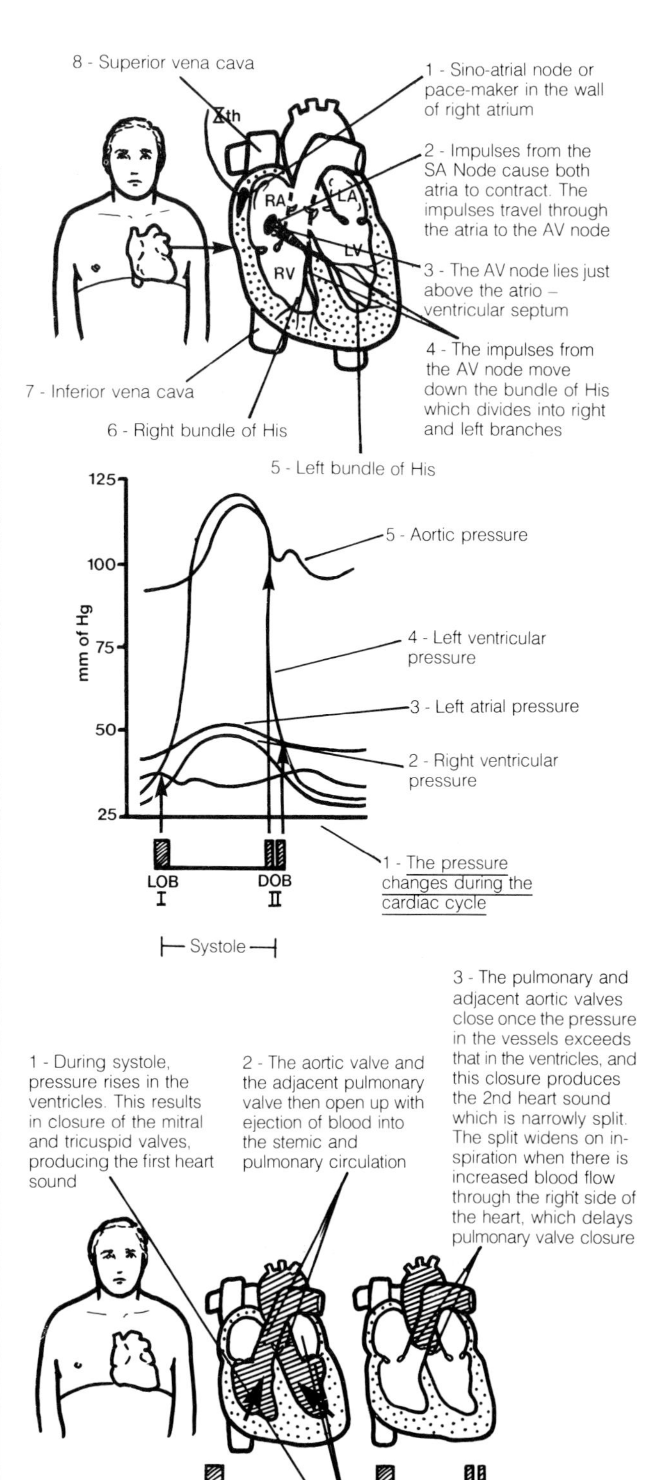
8 - Superior vena cava
1 - Sino-atrial node or pace-maker in the wall of right atrium
2 - Impulses from the SA Node cause both atria to contract. The impulses travel through the atria to the AV node
3 - The AV node lies just above the atrio – ventricular septum
4 - The impulses from the AV node move down the bundle of His which divides into right and left branches
5 - Left bundle of His
6 - Right bundle of His
7 - Inferior vena cava
RA
LA
RV
LV
125
100
75
50
25
mm of Hg
5 - Aortic pressure
4 - Left ventricular pressure
3 - Left atrial pressure
2 - Right ventricular pressure
1 - The pressure changes during the cardiac cycle
LOB
I
DOB
II
Systole
1 - During systole, pressure rises in the ventricles. This results in closure of the mitral and tricuspid valves, producing the first heart sound
2 - The aortic valve and the adjacent pulmonary valve then open up with ejection of blood into the stemic and pulmonary circulation
3 - The pulmonary and adjacent aortic valves close once the pressure in the vessels exceeds that in the ventricles, and this closure produces the 2nd heart sound which is narrowly split. The split widens on inspiration when there is increased blood flow through the right side of the heart, which delays pulmonary valve closure
I
I
II

The heart sounds

During systole, pressure falls in the atria and rises in the ventricles. This results in closure of the mitral and tricuspid valves, producing the first heart sound. This is usually single or closely split.

After closure of the mitral and tricuspid valves, the ventricles being to contract, with a rise in intraventricular pressure (isovolumetric contraction). The pulmonary and aortic valves then open, and blood is ejected into the systemic or pulmonary circulation. Once the pressure in the vessels exceeds that in the ventricles, the pulmonary and aortic valves close to produce the second heart sound. The sound is narrowly split, the aortic valve closure occuring just before pulmonary valve closure; the split widens in inspiration when there is increased blood flow through the right side of the heart, which delays pulmonary valve closure.

EXAMINATION OF THE CARDIOVASCULAR SYSTEM

In clinical examination, the rule is 'inspection, palpation, percussion and auscultation', in that order.

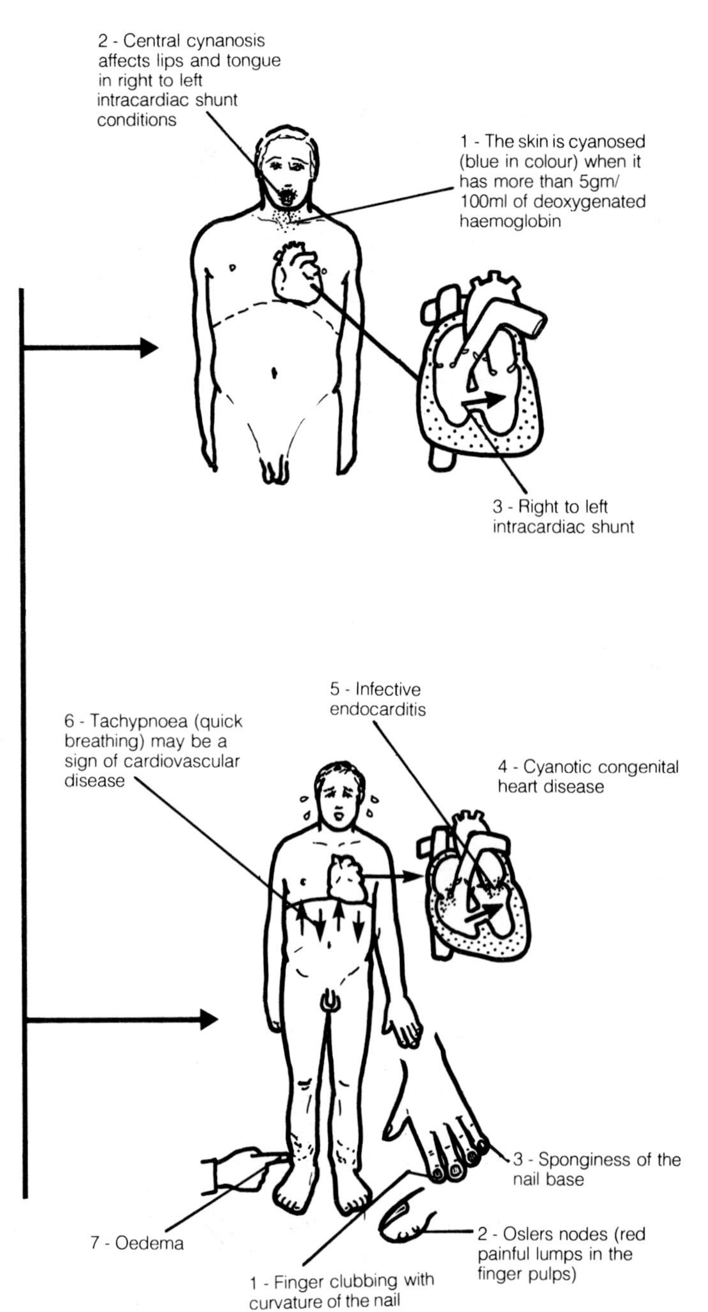
2 - Central cynanosis affects lips and tongue in right to left intracardiac shunt conditions
1 - The skin is cyanosed (blue in colour) when it has more than 5gm/100ml of deoxygenated haemoglobin
3 - Right to left intracardiac shunt
5 - Infective endocarditis
6 - Tachypnoea (quick breathing) may be a sign of cardiovascular disease
4 - Cyanotic congenital heart disease
3 - Sponginess of the nail base
7 - Oedema
2 - Oslers nodes (red painful lumps in the finger pulps)
1 - Finger clubbing with curvature of the nail

Inspection

The patient may be **cyanosed.** In cyanosis, the skin looks blue in colour because the blood perfusing the skin contains more than 5gm per 100ml of deoxygenated haemoglobin. In cardiovascular disease, central cyanosis, affecting the lips and tongue, is usually due to a right to left intracardiac shunt.

Finger clubbing may be present, with longitudinal curvature of the nail, sponginess at the nail and loss of angle between the nail and nail bed. This can occur in cyanotic congenital heart disease and infective endocarditis. If the latter is present, **splinter haemorrhages** may be present in the nails, or **Osler's nodes** (reddened painful lumps) in the finger pulps.

Tachypnoea and peripheral oedema may also be signs of cardiovascular disease.

Palpation of the pulse is most important. The rate and rhythm can be assessed at the wrist, while the character of the pulse is better assessed at the carotid artery in the neck.

Rate: The pulse rate at the wrist should be counted for at least half a minute. The normal rate can vary widely from 50-120 beats per minute. The rate is increased (tachycardia) by anxiety, after exercise, in fevers and in thyrotoxicosis. A slow heart rate can occur in athletes and in myxoedema. In complete heart block, the rate may be very slow (20-40 beats per minute).

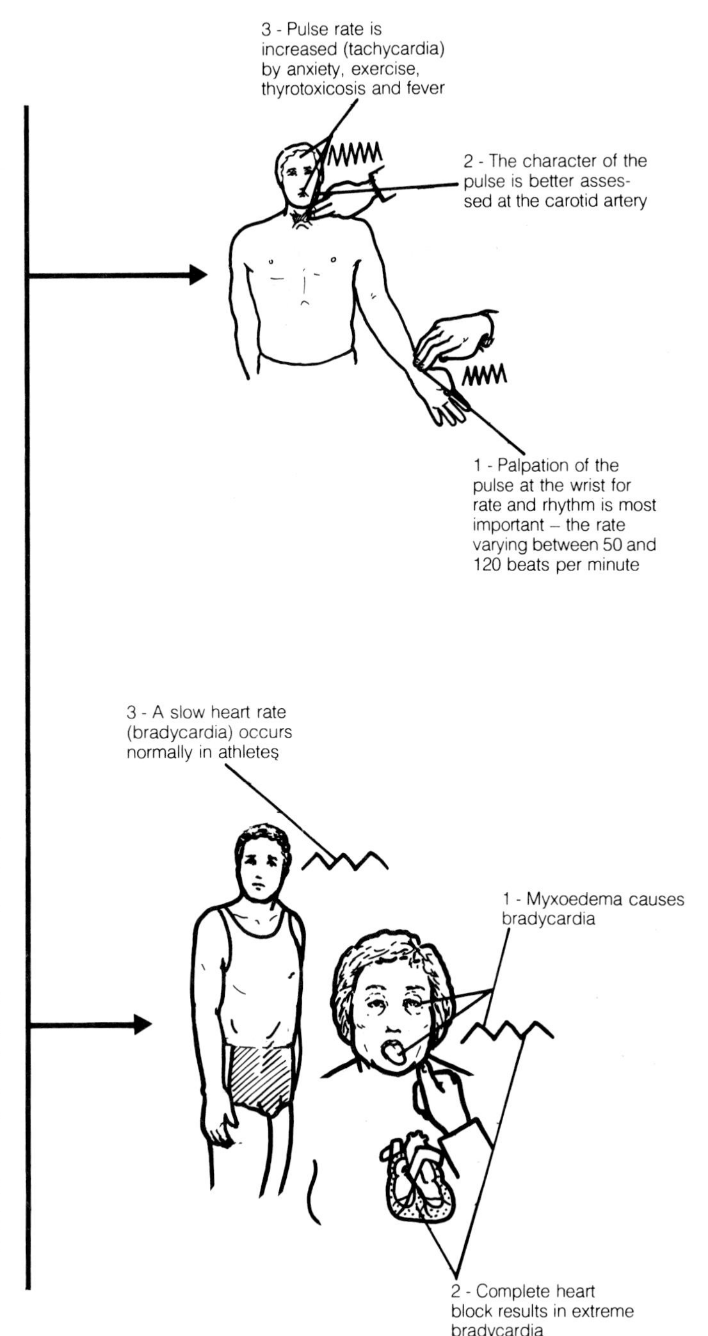
3 - Pulse rate is increased (tachycardia) by anxiety, exercise, thyrotoxicosis and fever
2 - The character of the pulse is better assessed at the carotid artery
1 - Palpation of the pulse at the wrist for rate and rhythm is most important – the rate varying between 50 and 120 beats per minute
3 - A slow heart rate (bradycardia) occurs normally in athletes
1 - Myxoedema causes bradycardia
2 - Complete heart block results in extreme bradycardia

Rhythm: The pulse may be regular or irregular. If it is completely irregular, the patient is likely to be in atrial fibrillation, and the rate recorded at the apex may be higher than that recorded at the radial pulse. If the rhythm is occasionally irregular, the patient may be having extrasystoles. They are usually abolished by exercise, whereas atrial fibrillation usually becomes more pronounced after exertion. (An extrasystole is an extra, conducted electrical impulse arising from a pacemaking focus outside the SA node).

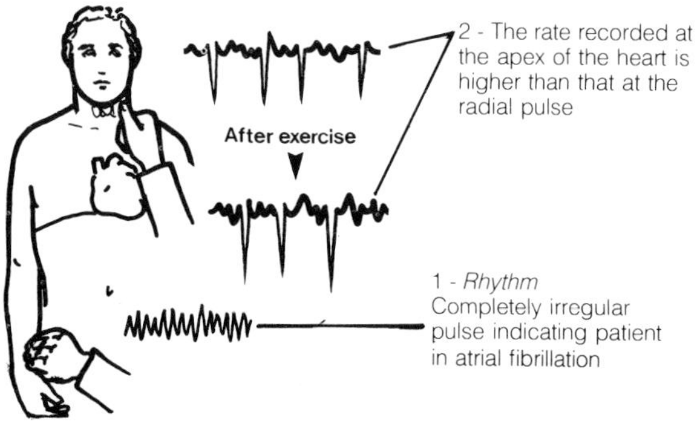

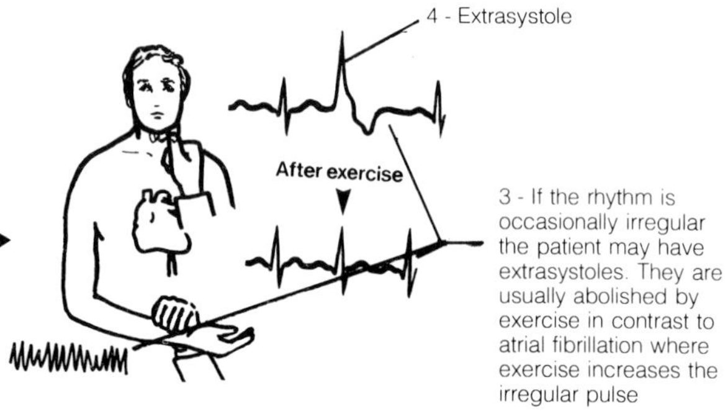

2 - The rate recorded at the apex of the heart is higher than that at the radial pulse

After exercise

1 - *Rhythm*
Completely irregular pulse indicating patient in atrial fibrillation

4 - Extrasystole

After exercise

3 - If the rhythm is occasionally irregular the patient may have extrasystoles. They are usually abolished by exercise in contrast to atrial fibrillation where exercise increases the irregular pulse

Character: The character of the pulse is best felt with the thumb at the carotid artery, at the level of the thyroid cartilage.

1 - The character of the pulse is best felt with the thumb on the carotid artery at the level of the thyroid cartilage

p t d p t d

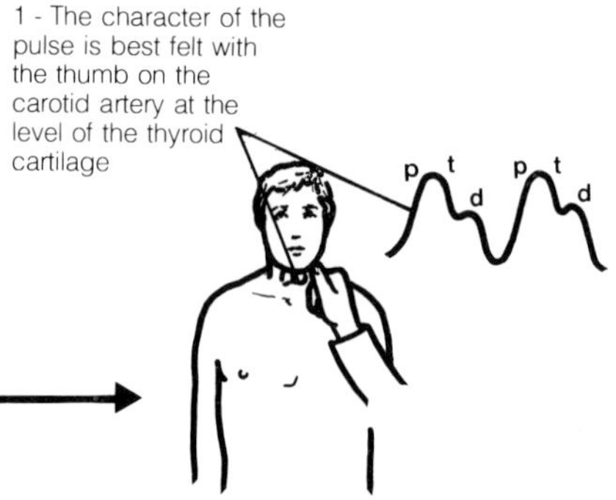

Slow rising pulse. In aortic stenosis, the pulse is slow rising.

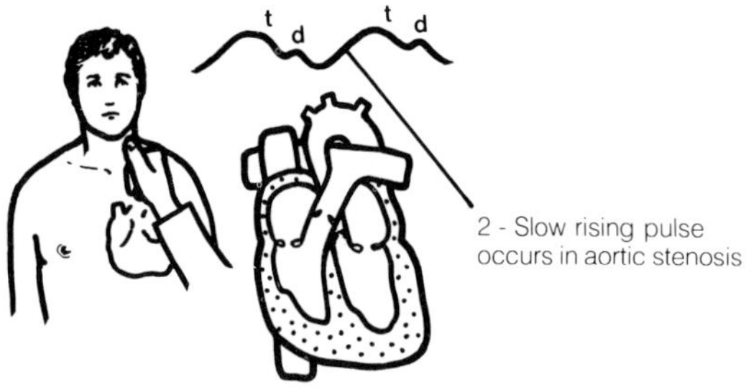

t d t d

2 - Slow rising pulse occurs in aortic stenosis

Collapsing (Water Hammer) Pulse. Here there is a sharp upstroke to the pulse, which is sometimes well felt with the patient's arm well elevated and the examiner's fingertips pressed flat against the radial pulse. This type of pulse occurs where there is an abnormal leak from the arterial system, with no systemic vascular resistance, for example in aortic regurgitation, arteriovenous fistula or patent ductus arteriosus.

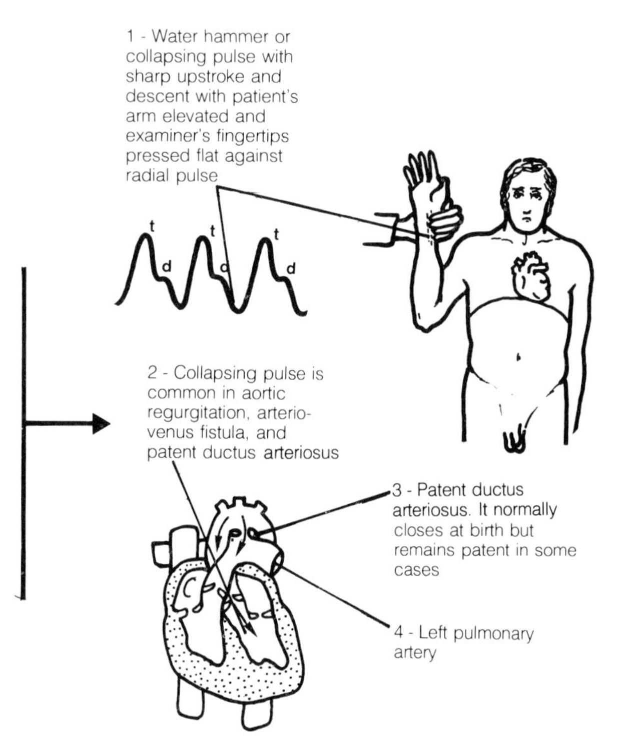
1 - Water hammer or collapsing pulse with sharp upstroke and descent with patient's arm elevated and examiner's fingertips pressed flat against radial pulse
t
d
t
d
t
d
2 - Collapsing pulse is common in aortic regurgitation, arterio-venus fistula, and patent ductus arteriosus
3 - Patent ductus arteriosus. It normally closes at birth but remains patent in some cases
4 - Left pulmonary artery

Bisferiens pulse. This pulse has two systolic peaks and occurs with mixed aortic valve disease (stenosis and regurgitation combined).

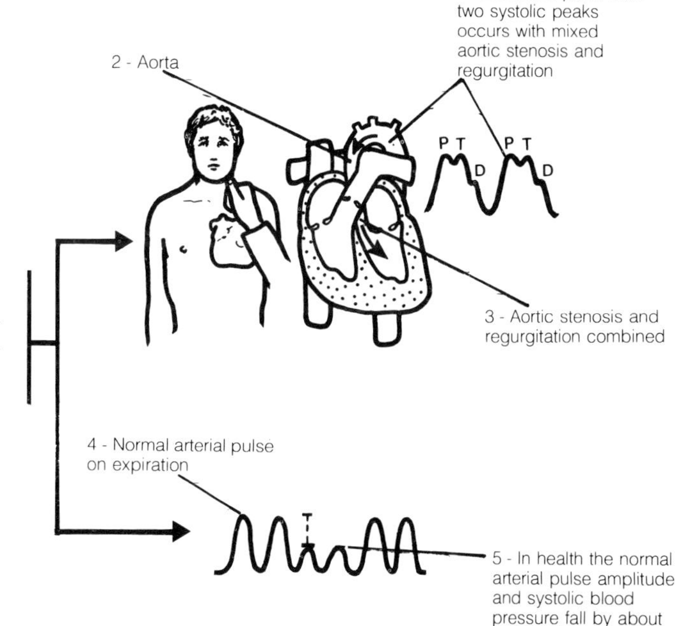
1 - Bisferiens pulse with two systolic peaks occurs with mixed aortic stenosis and regurgitation
2 - Aorta
P T
D
P T
D
3 - Aortic stenosis and regurgitation combined
4 - Normal arterial pulse on expiration
5 - In health the normal arterial pulse amplitude and systolic blood pressure fall by about 10mm Hg during inspiration due to the negative intrathoracic pressure on the vessels

The Jugular Venous Pulse

The neck veins should be examined with the patient lying at an angle of 45° degrees to the horizontal. In healthy subjects, the venous pressure at this angle will be at the level of the clavicle. In heart failure, the venous pressure is raised, and is measured above the level of the manubrium sterni.

The **venous pulse** has three positive waves **a**, **c** and **v**, and two negative waves, **x** and **y**. The **a** wave is caused by right atrial contraction and coincides with the first heart sound. The **c** wave is due to transmitted carotid pulsation.

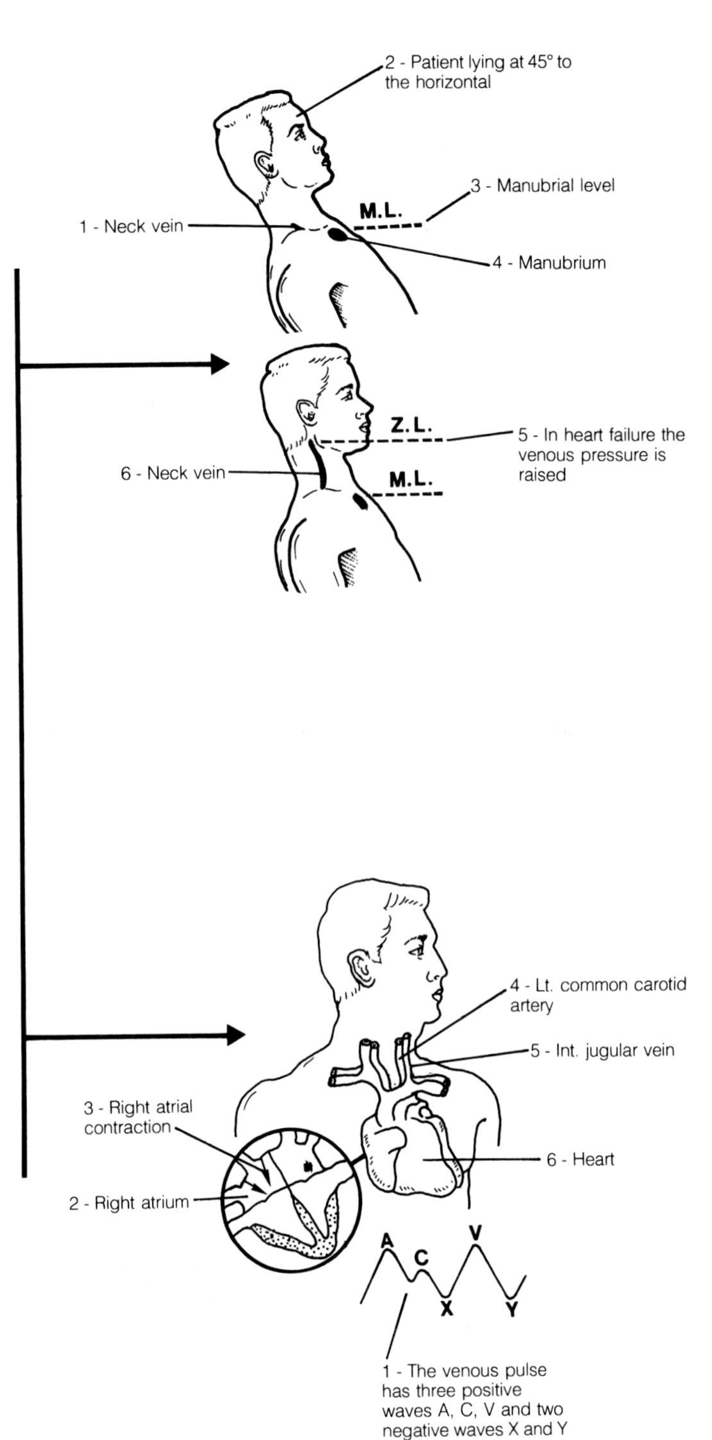
2 - Patient lying at 45° to the horizontal
3 - Manubrial level
M.L.
1 - Neck vein
4 - Manubrium
Z.L.
5 - In heart failure the venous pressure is raised
6 - Neck vein
M.L.
4 - Lt. common carotid artery
5 - Int. jugular vein
3 - Right atrial contraction
6 - Heart
2 - Right atrium
A
C
V
X
Y
1 - The venous pulse has three positive waves A, C, V and two negative waves X and Y

The **x** descent occurs during atrial relaxation, and the **v** wave occurs during right ventricular filling while the tricuspid valve is closed.

The tricuspid valve then opens and blood rapidly enters the right ventricle, causing the **y** descent.

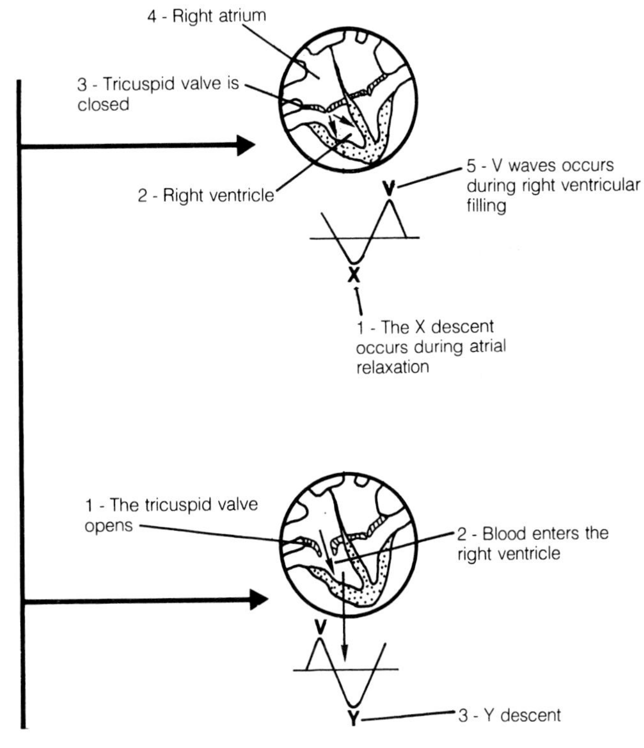
4 - Right atrium
3 - Tricuspid valve is closed
2 - Right ventricle
5 - V waves occurs during right ventricular filling
V
X
1 - The X descent occurs during atrial relaxation
1 - The tricuspid valve opens
2 - Blood enters the right ventricle
V
Y
3 - Y descent

In **atrial fibrillation**, the 'a' waves disappear because there is no co-ordinated atrial contraction, while in tricuspid stenosis the 'a' waves are prominent.

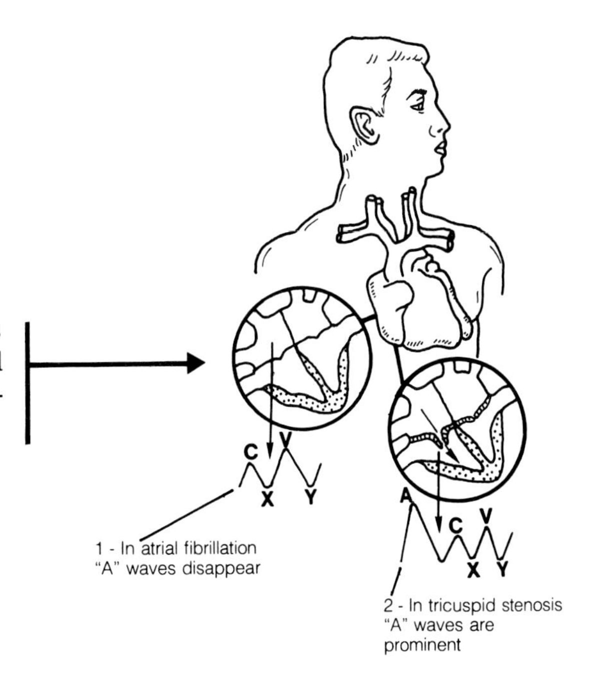
C
V
X
Y
1 - In atrial fibrillation "A" waves disappear
A
C
V
X
Y
2 - In tricuspid stenosis "A" waves are prominent

In **tricuspid regurgitation**, prominent systolic waves appear in the venous pulse, due to regurgitation of blood through the tricuspid valve during ventricular contraction.

2 - Very large "**a**" waves appear in the neck

A C V X Y

1 - Complete heart block

3 - Right atrium

4 - Right ventricle

In complete heart block, when atrial and ventricular contractions are dissociated, very large 'a' waves known as **cannon waves** appear in the neck.

In constrictive pericarditis, a rapid **y** descent occurs.

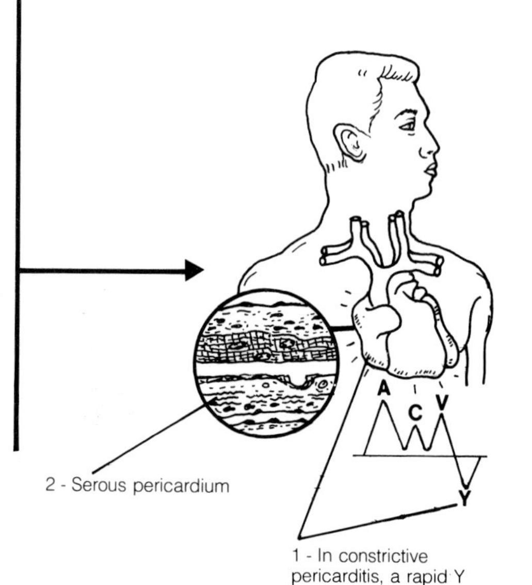
A
C
V
Y
2 - Serous pericardium
1 - In constrictive pericarditis, a rapid Y descent occurs

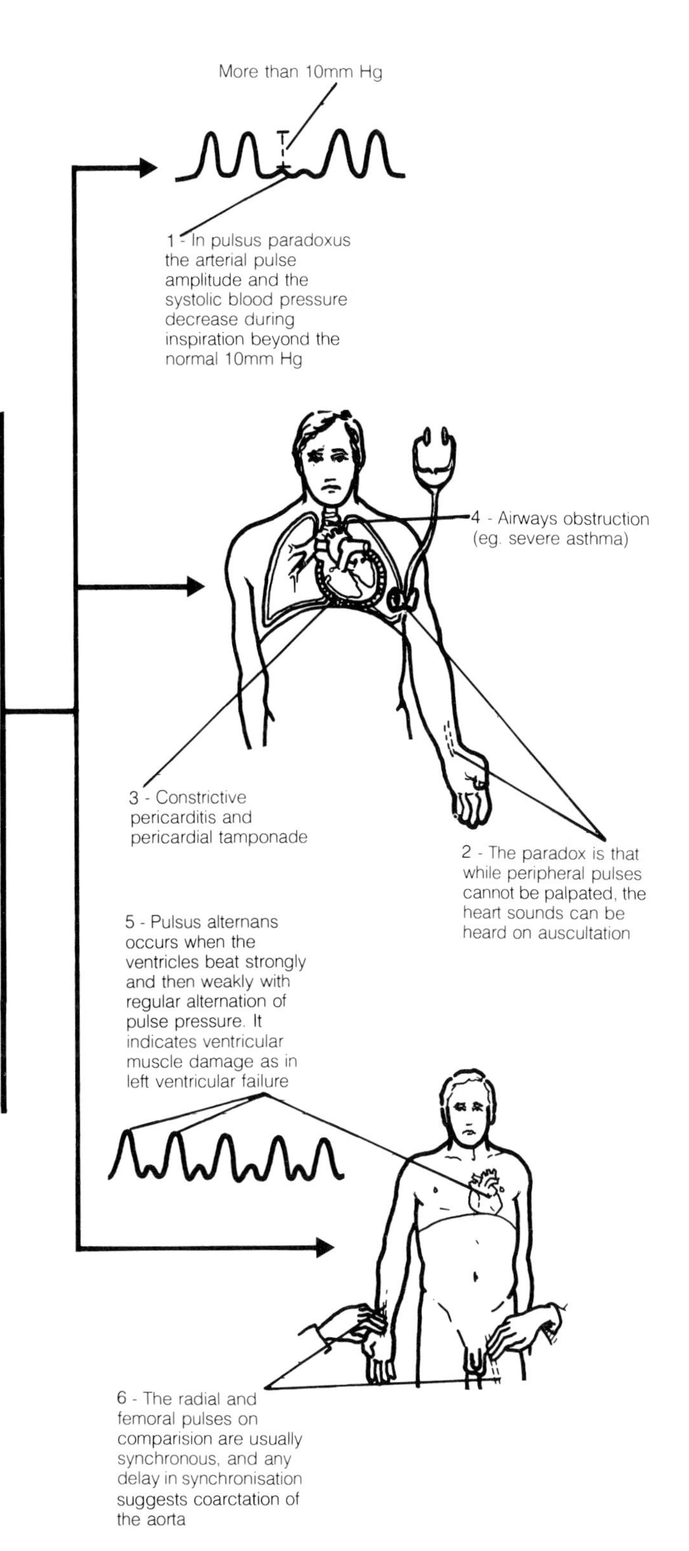
More than 10mm Hg
1 - In pulsus paradoxus the arterial pulse amplitude and the systolic blood pressure decrease during inspiration beyond the normal 10mm Hg
4 - Airways obstruction (eg. severe asthma)
3 - Constrictive pericarditis and pericardial tamponade
2 - The paradox is that while peripheral pulses cannot be palpated, the heart sounds can be heard on auscultation
5 - Pulsus alternans occurs when the ventricles beat strongly and then weakly with regular alternation of pulse pressure. It indicates ventricular muscle damage as in left ventricular failure
6 - The radial and femoral pulses on comparision are usually synchronous, and any delay in synchronisation suggests coarctation of the aorta

Pulsus paradoxus. In health, the amplitude of the arterial pulse and the systolic blood pressure fall by up to 10 mmHg during inspiration. In pulsus paradoxus, this decrease is accentuated. (The paradox is that while the peripheral pulses cannot be palpated, the heart sounds can be heard on auscultation). Pulsus paradoxus occurs in constrictive pericarditis, pericardial tamponade and severe airways obstruction.

In **Pulsus alternans** the ventricle beats strongly and then weakly, so that there is regular alternation of pulse pressure. This usually indicates ventricular muscle damage, as in left ventricular failure.

The radial and femoral pulses should always be compared. They are usually synchronous, and if there is any delay coarctation of the aorta should be suspected.

The Blood Pressure

This is a measurement of the pressure within arteries during the cardiac cycle. The highest pressure recorded is the systolic pressure and the lowest reading is the diastolic pressure. The difference between these pressures is the pulse pressure. Blood pressure is measured using a sphygmomanometer. The patient should be resting, and the cuff is firmly placed round the upper arm. A 12-14 cm width cuff is generally used in adults, but in obese patients this may result in a falsely high reading and a larger cuff should be used. Conversely in children or patients with thin arms, a narrower cuff should be used. The patient must be relaxed, as anxiety can raise blood pressure. The brachial artery is palpated while the cuff is inflated to a pressure of 30 mm above the level at which pulsation can no longer be felt. The stethoscope is then applied over the brachial artery, and then the cuff slowly deflated. The pressure at which sounds appear is the systolic blood pressure. The point at which sounds have become muffled is Korotkov phase four. The sounds disappear completely at Korotkov phase five, which is now taken as the diastolic blood pressure. (It is less liable to produce an error and corresponds more accurately to intra-arterial blood pressure than phase four). The normal blood pressure is between 100-140 mmHg systolic and between 60-90 mmHg diastolic. This reading increases with age.

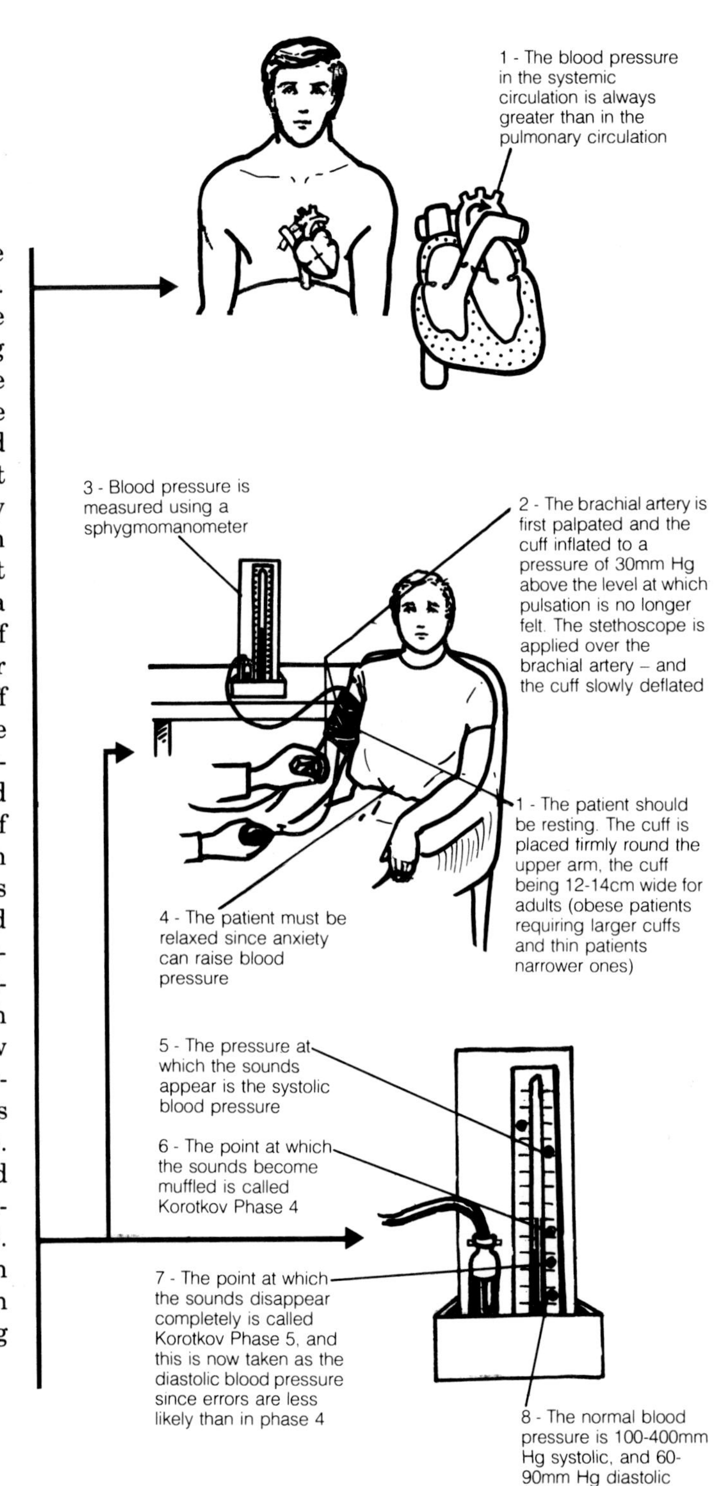
1 - The blood pressure in the systemic circulation is always greater than in the pulmonary circulation
3 - Blood pressure is measured using a sphygmomanometer
2 - The brachial artery is first palpated and the cuff inflated to a pressure of 30mm Hg above the level at which pulsation is no longer felt. The stethoscope is applied over the brachial artery – and the cuff slowly deflated
1 - The patient should be resting. The cuff is placed firmly round the upper arm, the cuff being 12-14cm wide for adults (obese patients requiring larger cuffs and thin patients narrower ones)
4 - The patient must be relaxed since anxiety can raise blood pressure
5 - The pressure at which the sounds appear is the systolic blood pressure
6 - The point at which the sounds become muffled is called Korotkov Phase 4
7 - The point at which the sounds disappear completely is called Korotkov Phase 5, and this is now taken as the diastolic blood pressure since errors are less likely than in phase 4
8 - The normal blood pressure is 100-400mm Hg systolic, and 60-90mm Hg diastolic though these values increase with age

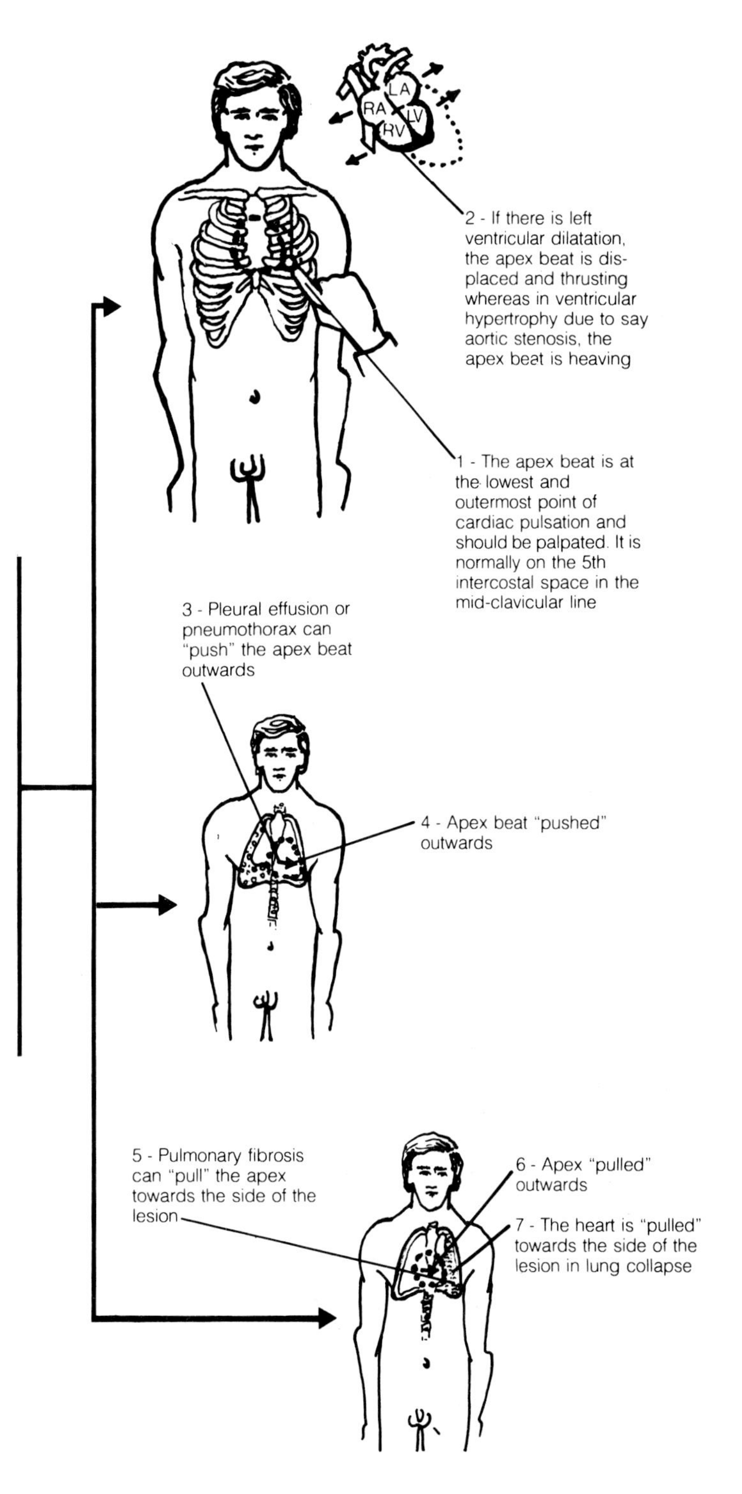
LA
RA
LV
RV
2 - If there is left ventricular dilatation, the apex beat is displaced and thrusting whereas in ventricular hypertrophy due to say aortic stenosis, the apex beat is heaving
1 - The apex beat is at the lowest and outermost point of cardiac pulsation and should be palpated. It is normally on the 5th intercostal space in the mid-clavicular line
3 - Pleural effusion or pneumothorax can "push" the apex beat outwards
4 - Apex beat "pushed" outwards
5 - Pulmonary fibrosis can "pull" the apex towards the side of the lesion
6 - Apex "pulled" outwards
7 - The heart is "pulled" towards the side of the lesion in lung collapse

The Heart

The apex beat, which is the lowest and outermost point of cardiac pulsation, should be palpated. This is usually in the fifth intercostal space in the mid-clavicular line. When there is left ventricular dilatation, the apex beat is often displaced and thrusting, whereas in hypertrophy due to outflow tract obstruction (for example aortic stenosis) the apex beat is heaving in nature. The position of the apex beat may be displaced due to chest deformity or lung disorders; for example, pneumothorax or pleural effusion will "push" the apex beat, while pulmonary fibrosis or lung collapse will "pull" it towards the side of the lesion. In right ventricular hypertrophy, forceful right ventricular contraction can be detected by placing the flat of the hand over the third and fourth intercostal spaces at the left sternal edge, when a "heave" can be felt. If valvular disease is present, the heart sounds may be accentuated to produce a palpable impulse; for example in mitral stenosis the **"tapping"** apex beat is due to a palpable loud first heart sound.

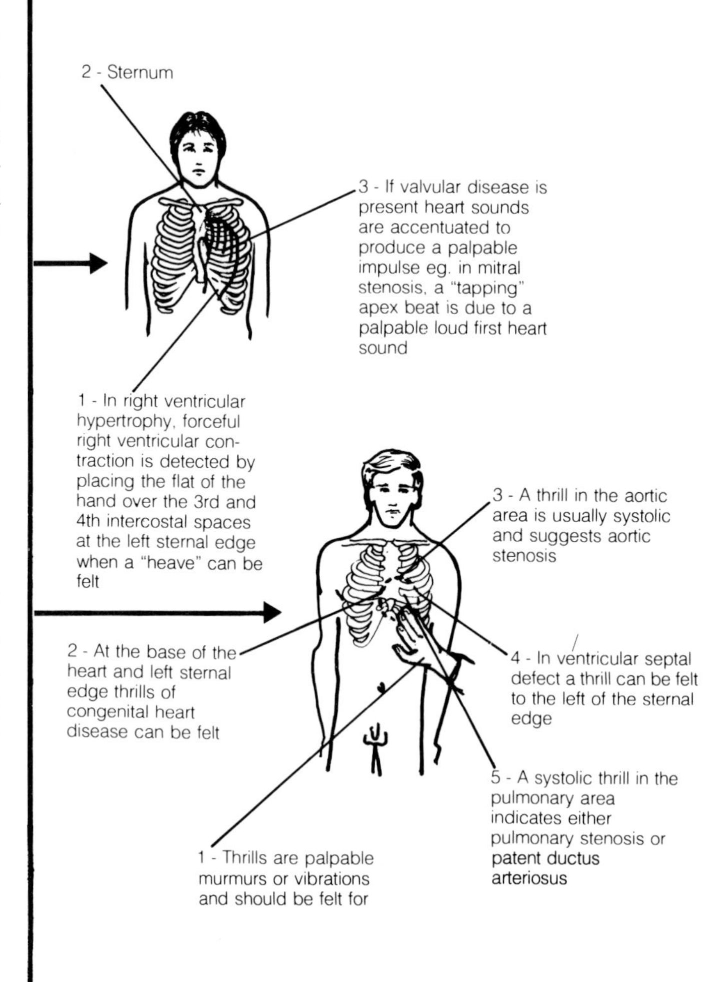
2 - Sternum
3 - If valvular disease is present heart sounds are accentuated to produce a palpable impulse eg. in mitral stenosis, a "tapping" apex beat is due to a palpable loud first heart sound
1 - In right ventricular hypertrophy, forceful right ventricular contraction is detected by placing the flat of the hand over the 3rd and 4th intercostal spaces at the left sternal edge when a "heave" can be felt
3 - A thrill in the aortic area is usually systolic and suggests aortic stenosis
2 - At the base of the heart and left sternal edge thrills of congenital heart disease can be felt
4 - In ventricular septal defect a thrill can be felt to the left of the sternal edge
5 - A systolic thrill in the pulmonary area indicates either pulmonary stenosis or patent ductus arteriosus
1 - Thrills are palpable murmurs or vibrations and should be felt for

Thrills are palpable murmurs, and should be felt for. Systolic thrills can sometimes be felt in mitral incompetence, aortic stenosis and ventricular septal defects. Diastolic thrills can occur in mitral stenosis.

Percussion of the heart is rarely useful.

Auscultation is usually carried out in certain areas, though the whole of the precordium should be carefully examined. The **mitral area** is in the fifth intercostal space, mid-axillary line, and mitral murmurs may radiate to the axilla. The **aortic area** is in the second intercostal space, to the right of the sternum, though aortic murmurs are often well heard at the lower left sternal edge and may radiate up into the neck. The **pulmonary area** is in the second intercostal space to the left of the sternum. The **tricuspid area** is at the lower end of the sternum. Low pitched sounds, such as mitral murmurs, are best heard with the bell of the stethoscope, while high pitched murmurs, such as that heard in aortic regurgitation, are best heard with the diaphragm.

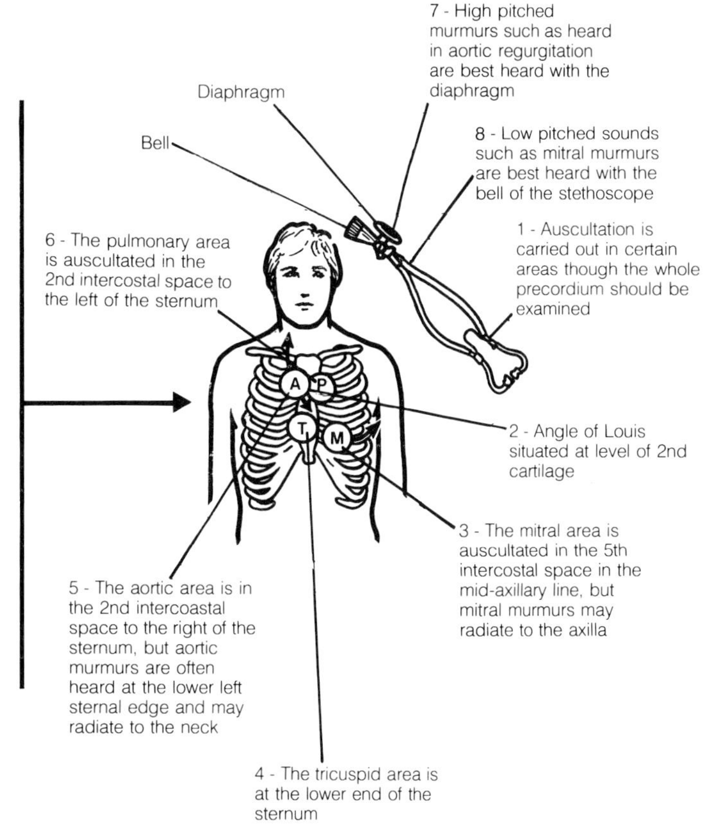
7 - High pitched murmurs such as heard in aortic regurgitation are best heard with the diaphragm
Diaphragm
Bell
8 - Low pitched sounds such as mitral murmurs are best heard with the bell of the stethoscope
1 - Auscultation is carried out in certain areas though the whole precordium should be examined
6 - The pulmonary area is auscultated in the 2nd intercostal space to the left of the sternum
A
P
T
M
2 - Angle of Louis situated at level of 2nd cartilage
3 - The mitral area is auscultated in the 5th intercostal space in the mid-axillary line, but mitral murmurs may radiate to the axilla
5 - The aortic area is in the 2nd intercoastal space to the right of the sternum, but aortic murmurs are often heard at the lower left sternal edge and may radiate to the neck
4 - The tricuspid area is at the lower end of the sternum

Heart Sounds: The **first sound** is caused by closure of the mitral and tricuspid valves. The sound is usually single or closely split. The tricuspid component may be delayed in right bundle branch block, when there is delayed onset of right or ventricular systole. If the PR interval is prolonged or ventricular contraction is delayed, the first heart sound is soft. In

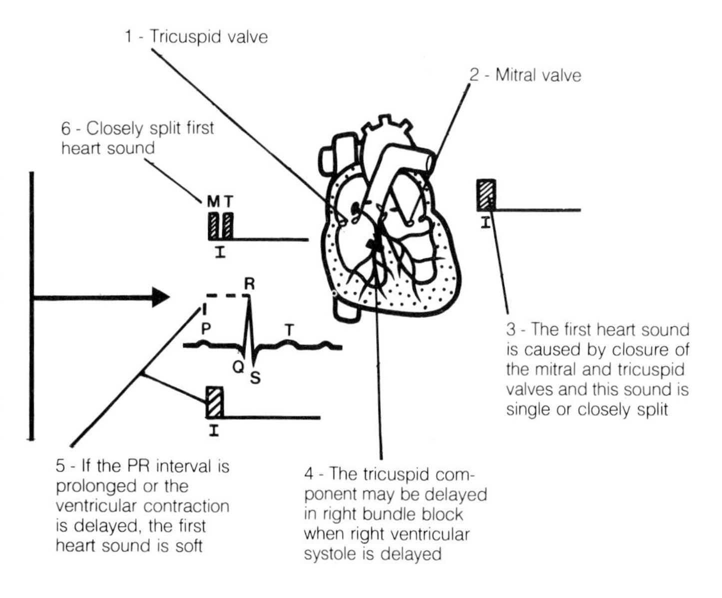
1 - Tricuspid valve
2 - Mitral valve
6 - Closely split first heart sound
M T
I
R
P
T
Q
S
I
I
3 - The first heart sound is caused by closure of the mitral and tricuspid valves and this sound is single or closely split
5 - If the PR interval is prolonged or the ventricular contraction is delayed, the first heart sound is soft
4 - The tricuspid component may be delayed in right bundle block when right ventricular systole is delayed

mitral valve disease, the anterior cusp of the mitral valve is usually stiff, and closure from an open position produces a loud first sound. The **second heart sound** is produced by the closure of aortic and pulmonary valves. The sound is split during inspiration, when there is increased blood flow through the right side of the heart, with delayed closure of the pulmonary valve. In right bundle branch block, delay of the pulmonary component occurs, with widened splitting of the second sound. In atrial septal defect, there is fixed splitting of the second sound; this is because there is free communication between the left and right atria.

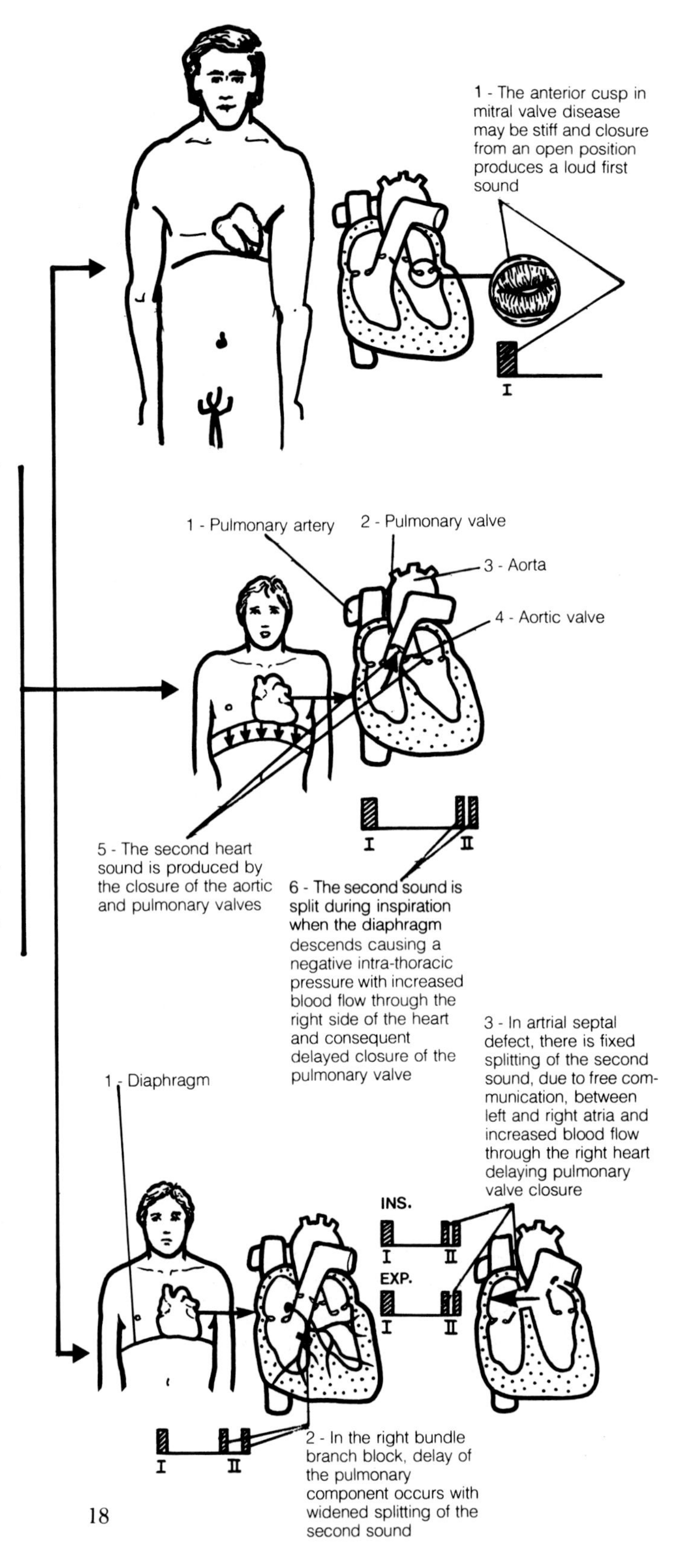
1 - The anterior cusp in mitral valve disease may be stiff and closure from an open position produces a loud first sound
I
1 - Pulmonary artery
2 - Pulmonary valve
3 - Aorta
4 - Aortic valve
I
II
5 - The second heart sound is produced by the closure of the aortic and pulmonary valves
6 - The second sound is split during inspiration when the diaphragm descends causing a negative intra-thoracic pressure with increased blood flow through the right side of the heart and consequent delayed closure of the pulmonary valve
3 - In artrial septal defect, there is fixed splitting of the second sound, due to free communication, between left and right atria and increased blood flow through the right heart delaying pulmonary valve closure
1 - Diaphragm
INS.
I
II
EXP.
I
II
I
II
2 - In the right bundle branch block, delay of the pulmonary component occurs with widened splitting of the second sound

The **third heart sound** is produced by rapid ventricular filling, and can be heard normally in youth. In older people, it is associated with left ventricular failure. The **fourth heart sound** is produced by atrial contraction, and is not usually heard in healthy people. It often occurs in left ventricular overload. When there is severe heart failure, the third and fourth heart sounds may combine to produce a so-called **"gallop rhythm"**.

The opening snap This is the sound produced by a mobile mitral valve where there is mitral stenosis. It is a sharp sound occurring just after the second heart sound, and is best heard at the left sternal edge.

Ejection clicks. These are sharp, high pitched sounds occurring in early systole and are sometimes heard in aortic or pulmonary stenosis when the valve is still mobile. The sound is produced by sudden distension of the valve.

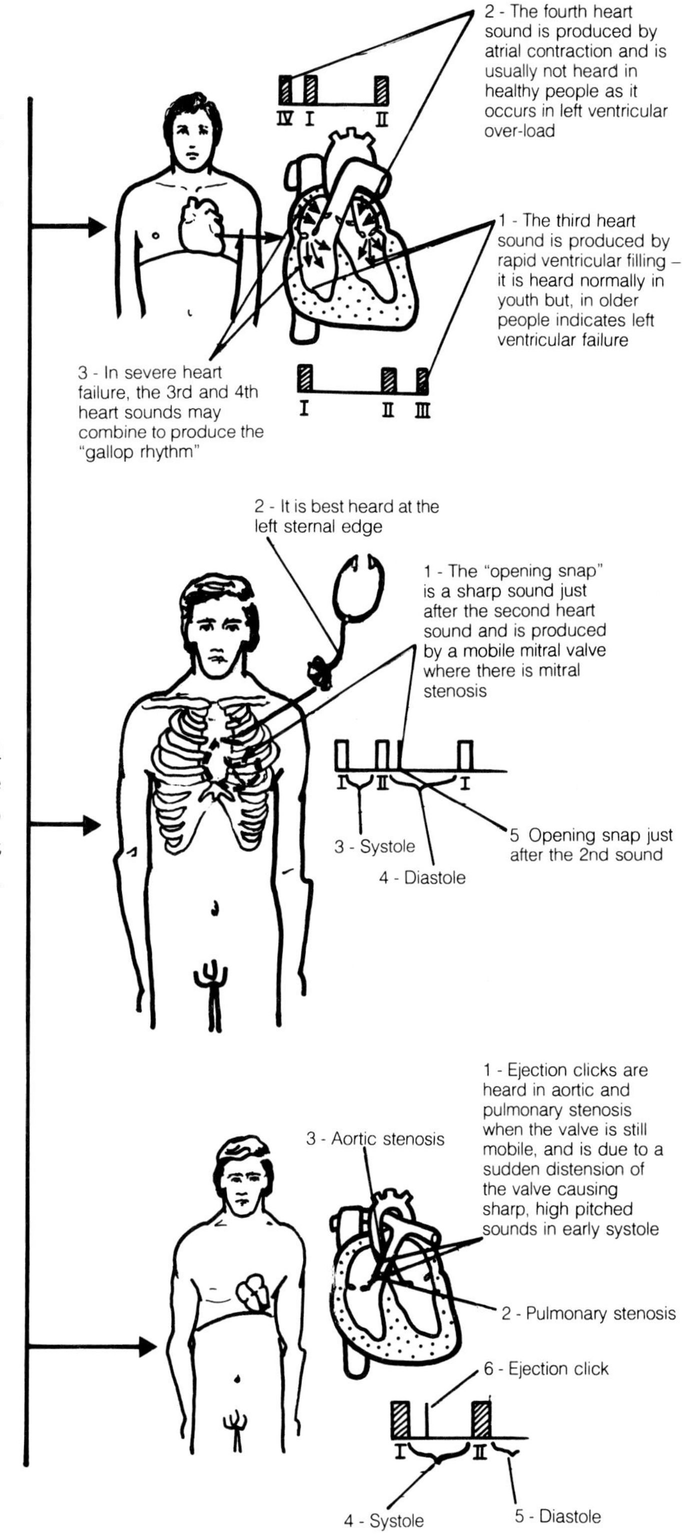
2 - The fourth heart sound is produced by atrial contraction and is usually not heard in healthy people as it occurs in left ventricular over-load
IV I II
1 - The third heart sound is produced by rapid ventricular filling – it is heard normally in youth but, in older people indicates left ventricular failure
3 - In severe heart failure, the 3rd and 4th heart sounds may combine to produce the "gallop rhythm"
I II III
2 - It is best heard at the left sternal edge
1 - The "opening snap" is a sharp sound just after the second heart sound and is produced by a mobile mitral valve where there is mitral stenosis
I II I
3 - Systole
4 - Diastole
5 Opening snap just after the 2nd sound
1 - Ejection clicks are heard in aortic and pulmonary stenosis when the valve is still mobile, and is due to a sudden distension of the valve causing sharp, high pitched sounds in early systole
3 - Aortic stenosis
2 - Pulmonary stenosis
6 - Ejection click
I II
4 - Systole
5 - Diastole

Murmurs are sounds produced by turbulent blood flow through valves. Murmurs are described according to their timing in the cardiac cycle. **Systolic murmurs** occur between the first and second heart sounds. **Diastolic murmurs** occur after the second heart sound. **Ejection systolic murmurs** commence after the first heart sound, build up to a peak in mid-systole and die away before the second heart sound. They occur in aortic stenosis and pulmonary stenosis, or may be innocent. **Pansystolic murmurs** last throughout systole, and tend to obscure the first and second sounds. They are heard in mitral and tricuspid regurgitation and in ventricular septal defects.

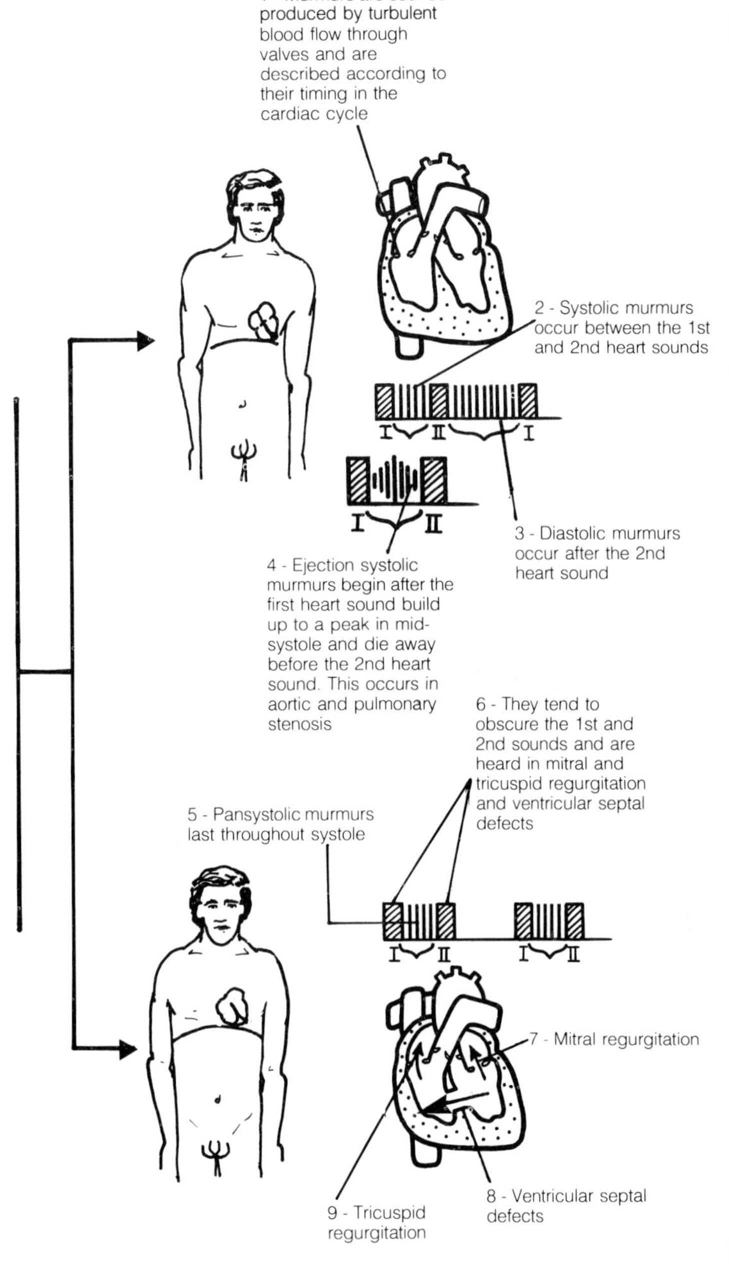
1 - Murmurs are sounds produced by turbulent blood flow through valves and are described according to their timing in the cardiac cycle
2 - Systolic murmurs occur between the 1st and 2nd heart sounds
I
II
I
I
II
3 - Diastolic murmurs occur after the 2nd heart sound
4 - Ejection systolic murmurs begin after the first heart sound build up to a peak in mid-systole and die away before the 2nd heart sound. This occurs in aortic and pulmonary stenosis
6 - They tend to obscure the 1st and 2nd sounds and are heard in mitral and tricuspid regurgitation and ventricular septal defects
5 - Pansystolic murmurs last throughout systole
I
II
I
II
7 - Mitral regurgitation
8 - Ventricular septal defects
9 - Tricuspid regurgitation

Late systolic murmurs. Some murmurs begin in mid-systole, extending to the second heart sound, and are often preceded by a click. This can occur in mitral valve prolapse, coarctation of the aorta, and hypertrophic cardiomyopathy.

Innocent systolic murmurs. Systolic murmurs are often heard which are of no pathological significance.They are usually associated with increased cardiac output, such as in anaemias, fevers, thyrotoxicosis and pregnancy. They are usually soft, mid-systolic, and not associated with any other signs of cardiac disease.

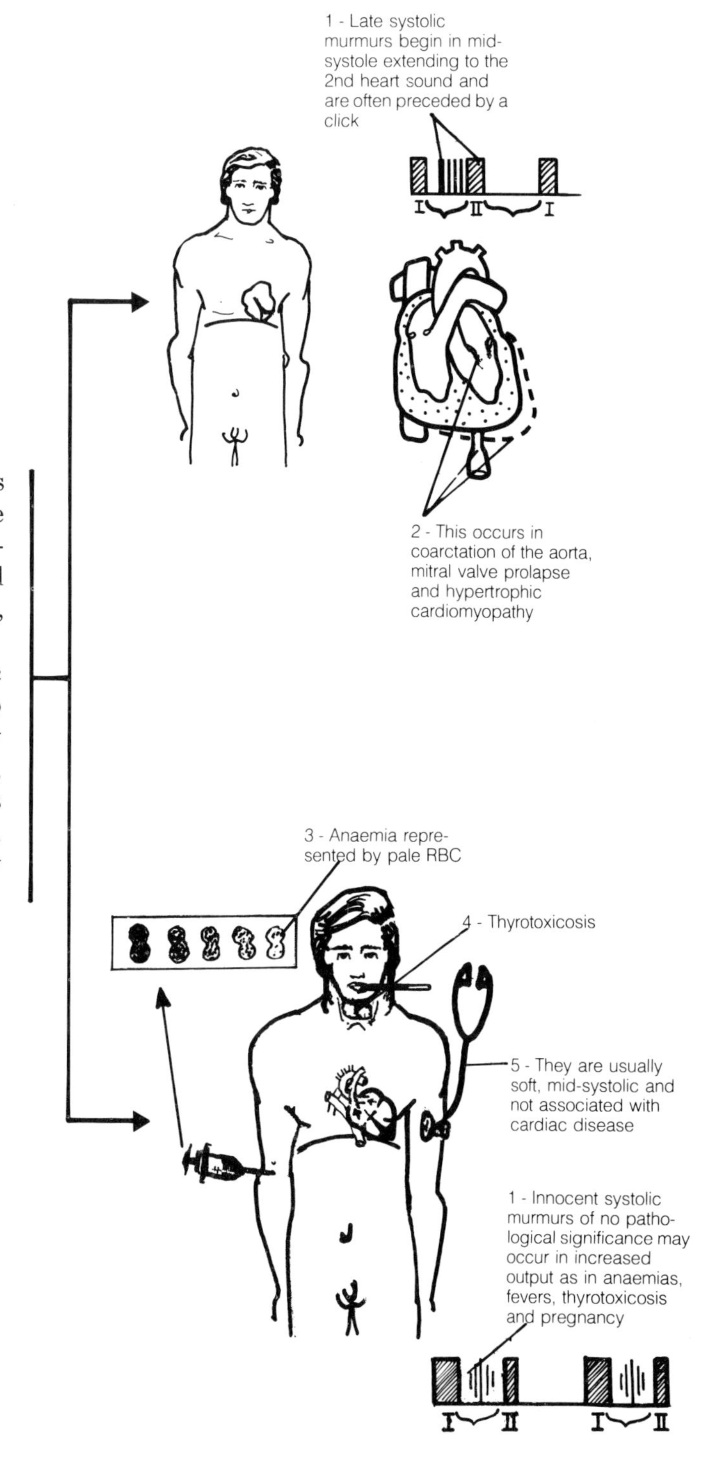
1 - Late systolic murmurs begin in mid-systole extending to the 2nd heart sound and are often preceded by a click
I
II
I
2 - This occurs in coarctation of the aorta, mitral valve prolapse and hypertrophic cardiomyopathy
3 - Anaemia represented by pale RBC
4 - Thyrotoxicosis
5 - They are usually soft, mid-systolic and not associated with cardiac disease
1 - Innocent systolic murmurs of no pathological significance may occur in increased output as in anaemias, fevers, thyrotoxicosis and pregnancy
I
II
I
II

Diastolic murmurs are often difficult to hear, and must be very carefully listened for.

1 - Diastolic murmurs are often difficult to hear

2 - Immediate diastolic murmurs arise from regurgitant aortic and pulmonary valves

3 - Diaphragm on expiration

Immediate diastolic murmurs usually arise from regurgitant aortic or pulmonary valves. They occur immediately after the second heart sound, and are best heard in expiration, along the left sternal edge, using the diaphragm of the stethoscope, pressed hard against the chest.

4 - Aortic or pulmonary diastolic murmurs occur immediately after the 2nd heart sound and are best heard on examination along the left sternal edge, using the diaphragm of the stethoscope

I II I II

5 - Systole

Mid-diastolic murmurs arise from stenosed mitral or tricuspid valves. They are rough, low pitched, and best heard with the bell of the stethoscope. There is a short gap between the second heart sound and the beginning of the murmur. Tricus-

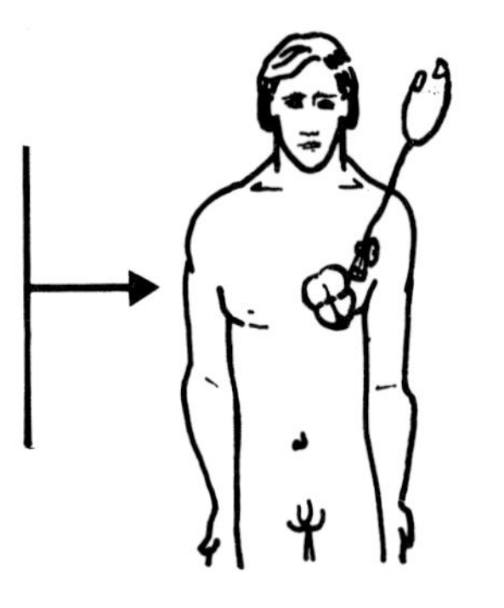

2 - There is a short gap between the 2nd heart sound and the beginning of the murmur

I II I

1 - Mid-diastolic murmurs arise from stenosed mitral or tricuspid valves and are rough, low-pitched and best heard with the bell stethoscope

4 - Stenosed mitral valve

3 - Stenosed tricuspid valve

pid murmurs become louder during inspiration. The **Austin Flint murmur** is a mid-diastolic murmur which occurs in aortic regurgitation. It is due to vibration of the anterior mitral valve leaflet by the aortic regurgitant jet.

Continuous (machinery) murmur. This is characteristically heard in patent ductus arteriosus, where the murmur is continuous throughout systole and diastole, with maximum intensity around the second heart sound.

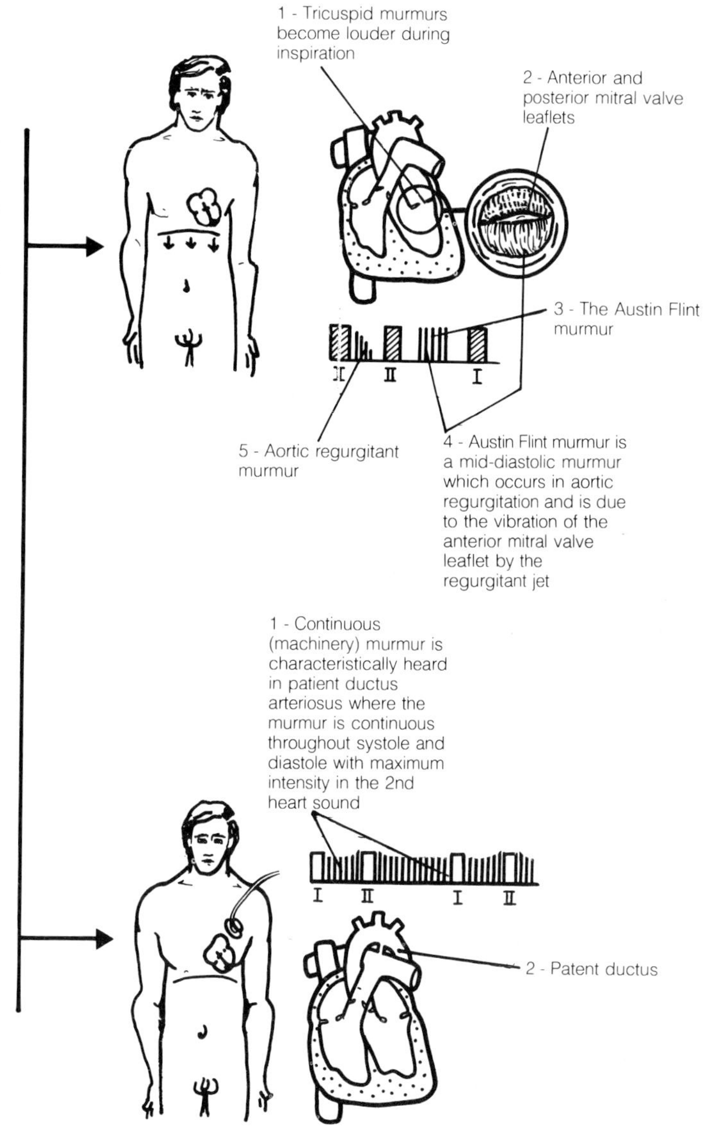
1 - Tricuspid murmurs become louder during inspiration
2 - Anterior and posterior mitral valve leaflets
3 - The Austin Flint murmur
4 - Austin Flint murmur is a mid-diastolic murmur which occurs in aortic regurgitation and is due to the vibration of the anterior mitral valve leaflet by the regurgitant jet
5 - Aortic regurgitant murmur
II
I
1 - Continuous (machinery) murmur is characteristically heard in patient ductus arteriosus where the murmur is continuous throughout systole and diastole with maximum intensity in the 2nd heart sound
I
II
I
II
2 - Patent ductus

INVESTIGATIONS IN CARDIOLOGY—THE ELECTROCARDIOGRAM

The electrocardiogram (ECG). Contraction of heart muscle is caused by electrical changes which can be detected by electrodes attached to the surface of the body; these electric potentials can be amplified and recorded onto moving paper to produce the electrocardiogram. The various waves on the ECG are named, P, Q, R, S, and T. The P wave reflects atrial activation and the QRS complex signifies contraction of the larger ventricular muscle mass. The T wave represents ventricular repolarisation.

ECG machines run at a standard rate of 25 mm/second. Each large square on the ECG paper represents 0.2 seconds and each small square is equivalent to 0.04 seconds. The PR interval is a measure of conduction time in the bundle of His and should not last longer than 0.2 seconds in the normal subject. The QRS complex lasts up to 0.12 seconds (3 small squares).

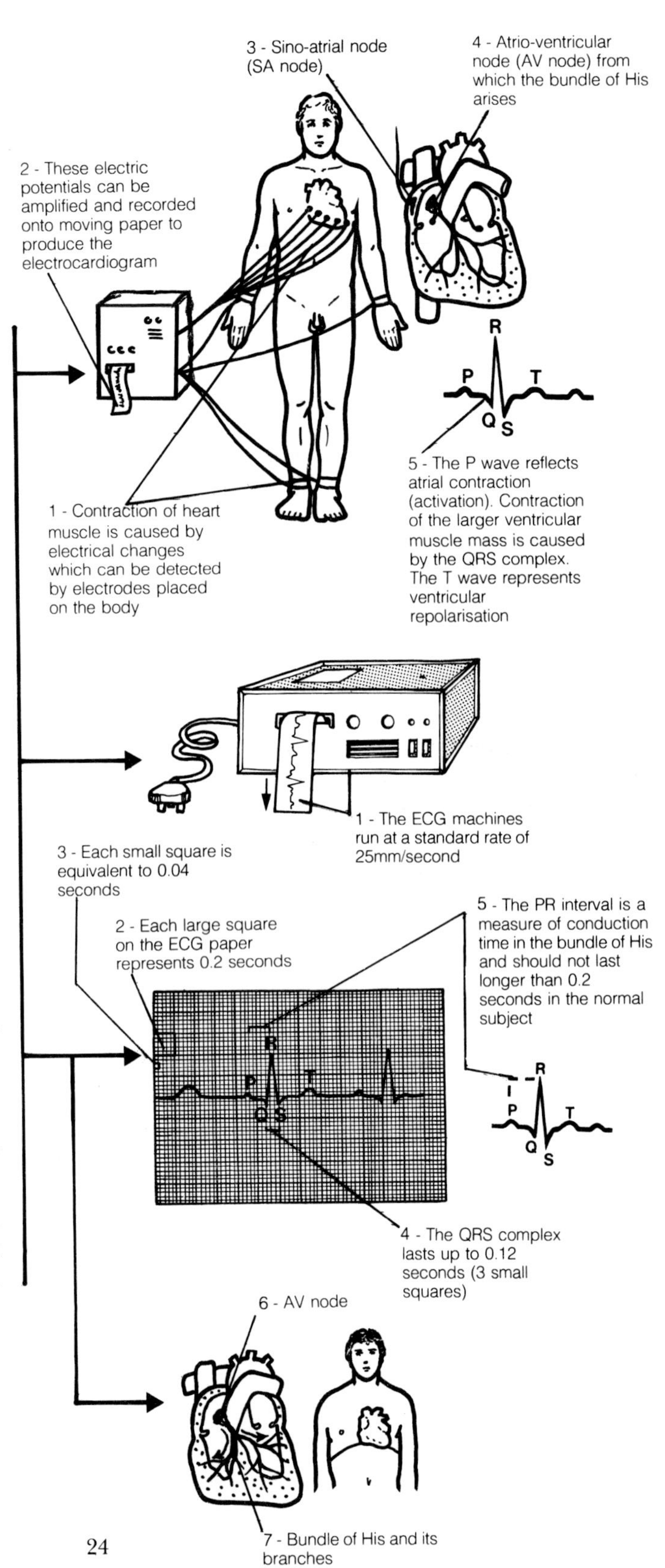
3 - Sino-atrial node (SA node)
4 - Atrio-ventricular node (AV node) from which the bundle of His arises
2 - These electric potentials can be amplified and recorded onto moving paper to produce the electrocardiogram
R
P
T
Q
S
1 - Contraction of heart muscle is caused by electrical changes which can be detected by electrodes placed on the body
5 - The P wave reflects atrial contraction (activation). Contraction of the larger ventricular muscle mass is caused by the QRS complex. The T wave represents ventricular repolarisation
1 - The ECG machines run at a standard rate of 25mm/second
3 - Each small square is equivalent to 0.04 seconds
2 - Each large square on the ECG paper represents 0.2 seconds
5 - The PR interval is a measure of conduction time in the bundle of His and should not last longer than 0.2 seconds in the normal subject
4 - The QRS complex lasts up to 0.12 seconds (3 small squares)
6 - AV node
7 - Bundle of His and its branches

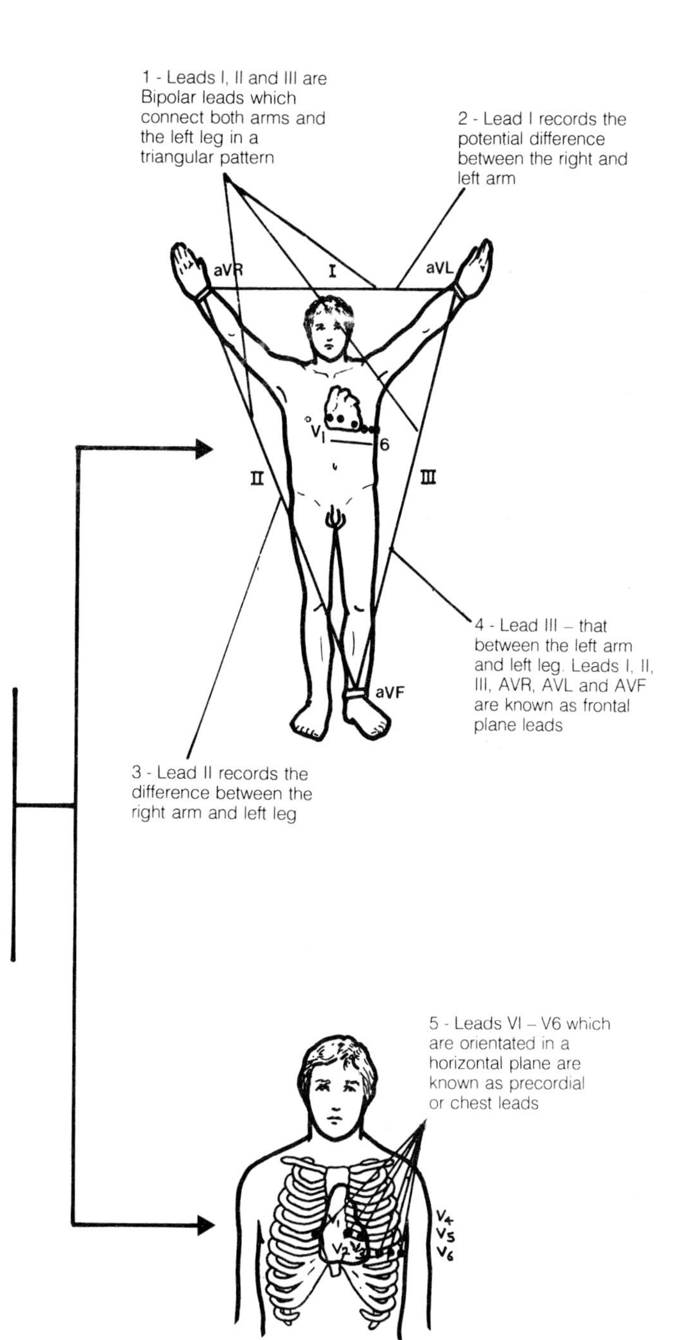

1 - Leads I, II and III are Bipolar leads which connect both arms and the left leg in a triangular pattern
2 - Lead I records the potential difference between the right and left arm
aVR
I
aVL
V1
6
II
III
aVF
4 - Lead III – that between the left arm and left leg. Leads I, II, III, AVR, AVL and AVF are known as frontal plane leads
3 - Lead II records the difference between the right arm and left leg
5 - Leads VI – V6 which are orientated in a horizontal plane are known as precordial or chest leads
V4
V5
V6

The electrode leads. There are twelve standard electrode leads which "look" at the heart from different angles. Leads, I, II and III are bipolar and leads aVR, aVL, aVF and V1 – V6 are unipolar. Leads, I, II, III, AVR, AVL and AVF are known as frontal plane leads, while leads V1 – V6 are orientated in the horizontal plane and are known as precordial or chest leads.

Leads I and AVL "look" at the left lateral surface of the heart. Leads II, III and aVF "look" at the inferior surface and aVR "looks" at the atria. Leads V1 and V2 "look" at the right ventricle, leads V3, and V4 "look" at the septum while leads V5 and V6 "look" at the lateral walls of the left ventricle.

1 - Leads I and aVL "look" at the left lateral surface of the heart

2 - Leads II, and III and aVF "look" at the inferior surface of the heart

3 - Lead aVR "looks" at the atria

4 - Leads V_1 and V_2 "look" at the right ventricle

5 - Lead V_3 and V_4 "look" at the interventricular septum

6 - Leads 5 and 6 "look" at the lateral walls of the left ventricle

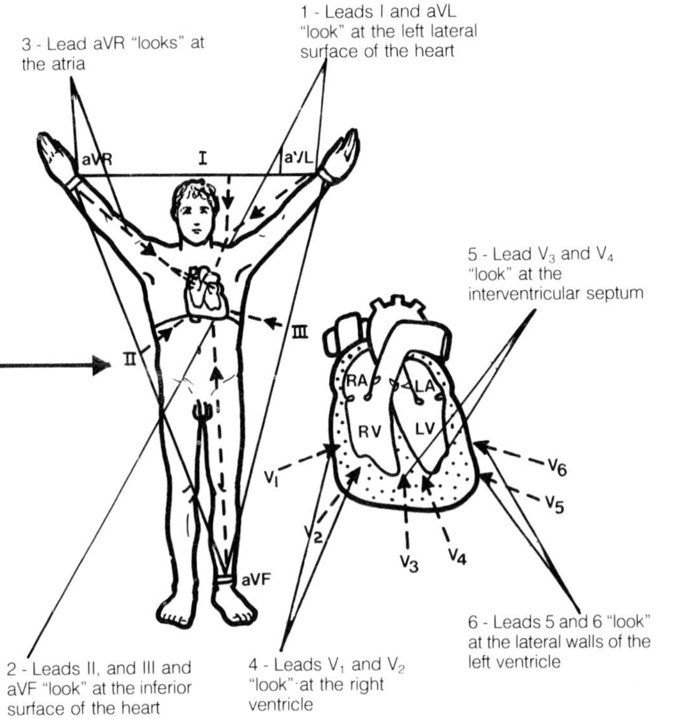

Reading the ECG. When reading the ECG, the heart rate and rhythm must be noted. Cardiac axis should be assessed and the width of the PR interval and QRS complexes should be measured. The P waves, QRS complexes, ST segments and T waves should all be examined.

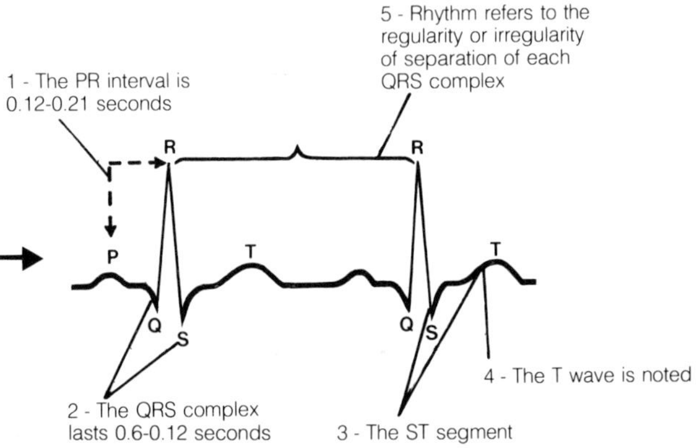

1 - The PR interval is 0.12-0.21 seconds

2 - The QRS complex lasts 0.6-0.12 seconds

3 - The ST segment

4 - The T wave is noted

5 - Rhythm refers to the regularity or irregularity of separation of each QRS complex

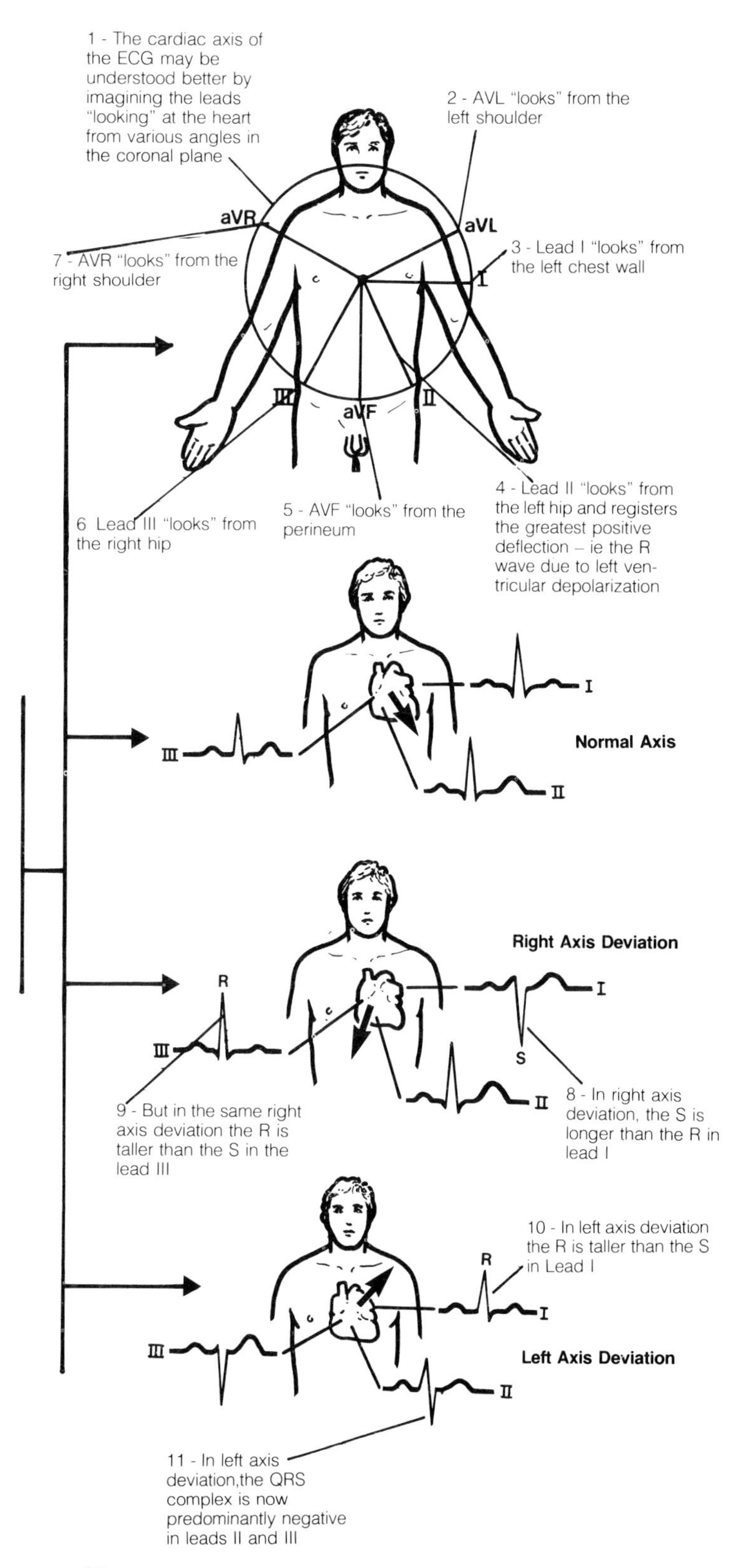
1 - The cardiac axis of the ECG may be understood better by imagining the leads "looking" at the heart from various angles in the coronal plane
2 - AVL "looks" from the left shoulder
aVR
aVL
7 - AVR "looks" from the right shoulder
3 - Lead I "looks" from the left chest wall
I
III
aVF
II
6 Lead III "looks" from the right hip
5 - AVF "looks" from the perineum
4 - Lead II "looks" from the left hip and registers the greatest positive deflection – ie the R wave due to left ventricular depolarization
I
III
Normal Axis
II
Right Axis Deviation
R
I
III
S
II
9 - But in the same right axis deviation the R is taller than the S in the lead III
8 - In right axis deviation, the S is longer than the R in lead I
10 - In left axis deviation the R is taller than the S in Lead I
R
I
III
Left Axis Deviation
II
11 - In left axis deviation,the QRS complex is now predominantly negative in leads II and III

Cardiac axis. The cardiac axis is the direction of spread of the depolarising wave. The normal axis is associated with a greater upward deflection in lead II than in leads I or III. In right axis deviation the deflection in lead I becomes negative while that in lead III becomes positive. In left axis deviation the QRS complex becomes predominantly negative in leads II and III and positive in lead I.

Ventricular hypertrophy. In left ventricular hypertrophy, there is increased muscle mass of the left ventricle, with increased electrical activity. This is reflected by tall R waves in the left ventricular leads (V5 and V6) and deep S waves in the right ventricular leads (V1 and V2). T wave inversion may also occur in the left ventricular leads; this is known as "left ventricular strain pattern". Left ventricular hypertrophy is diagnosed on the ECG when the sum of the S wave in V1 and the R wave in V5 or V6 exceeds 35mm.

1 - Increased muscle mass in left ventricular hypertrophy with increased electrical activity. Here the R waves are tall in leads V_5 and V_6. In these left ventricular leads, inversion of T wave may occur and are known as "left ventricular strain pattern."

2 - Deep S waves are seen in right ventricular leads of V_1 and V_2

3 - Left ventricular hypertrophy is diagnosed on the ECG when the sum of the S wave in V_1 and the R wave in V_5 or V_6 (as explained above) exceeds 35mm

1 - Right ventricular hypertrophy, R waves are tall in right ventricular leads V_1 and V_2 with an inversion of T wave ("strain pattern")

2 - In right ventricular hypertrophy deep S waves may be present in left chest leads such as V_5 and V_6

In right ventricular hypertrophy, tall R waves occur in the right ventricular leads (V1 and V2), associated with a strain pattern (T wave inversion). There may be deep S waves in the left chest leads.

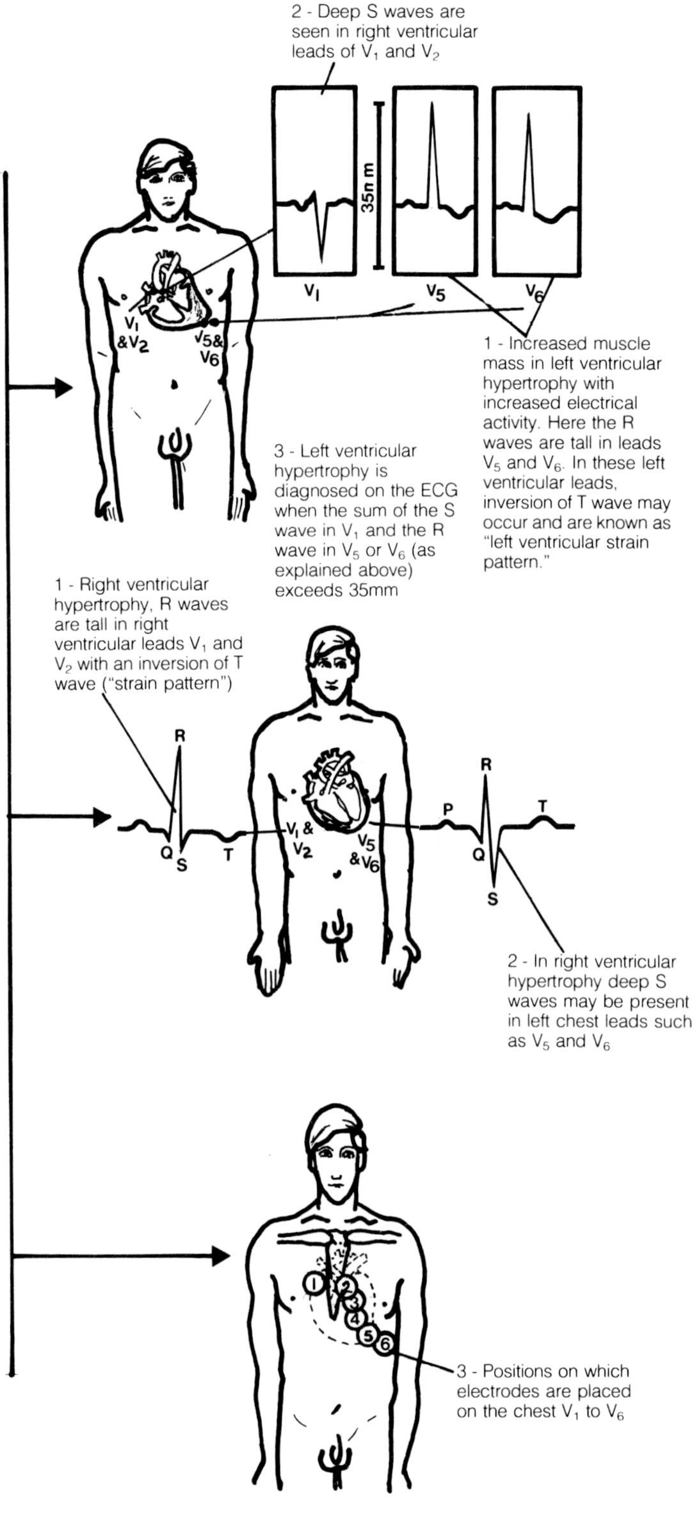

3 - Positions on which electrodes are placed on the chest V_1 to V_6

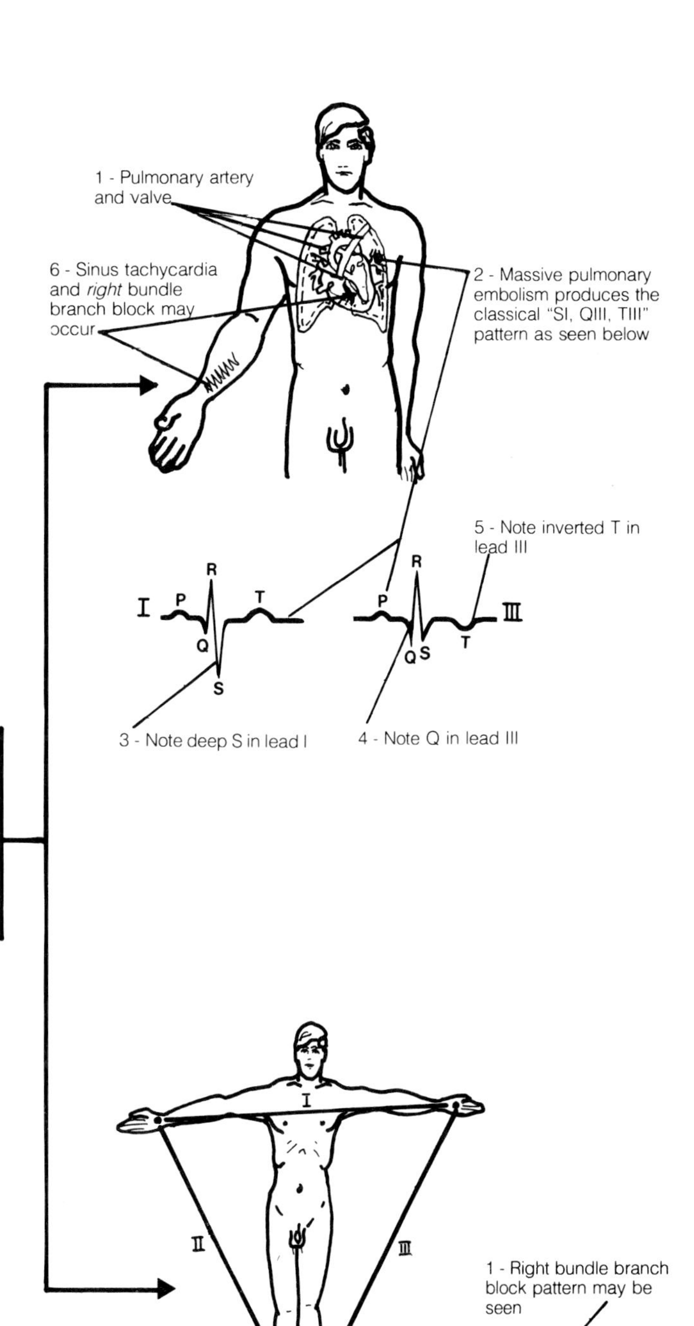
1 - Pulmonary artery and valve
6 - Sinus tachycardia and *right* bundle branch block may occur
2 - Massive pulmonary embolism produces the classical "SI, QIII, TIII" pattern as seen below
5 - Note inverted T in lead III
I
P
R
T
Q
S
P
R
Q
S
T
III
3 - Note deep S in lead I
4 - Note Q in lead III
I
II
III
1 - Right bundle branch block pattern may be seen
2 - Note classic "M" complex seen in lead V_1

Acute right ventricular strain occurs, for example, with massive pulmonary embolism, and produces the classical "SI, QIII, TIII" pattern with a large S wave in lead I and Q wave and inverted T waves in lead III. Sinus tachycardia and right bundle branch block may also occur.

Myocardial infarction

If myocardial infarction causes complete muscle death, an electrical "window" is created between the inside and the outside of the heart and an electrode looking at the heart over that window will record a Q wave. This is due to electrical activity spreading away from the electrode on the other side of the heart – "seen" through this window—and which is registered on the ECG as a negative deflection – the Q wave. The sequence of ECG changes is usually as follows: the T waves become tall and upright (hyperacute changes). The ST segments become elevated, Q waves appear and T waves become inverted. Over the next few days the ST segment gradually returns to the isoelectric line and the T waves become upright again. Q waves usually persist. The site of infarction can be inferred from the pattern of ECG changes. Infarction of the anterior part of the ventricular wall produces changes in leads I, aVL and V1-V6 while infarction of the inferior wall results in changes in leads II, III and aVF. In lateral infarction changes occur in leads I, aVL, V5 and V6. Sometimes only T wave changes occur and this is known as subendocardial infarction.

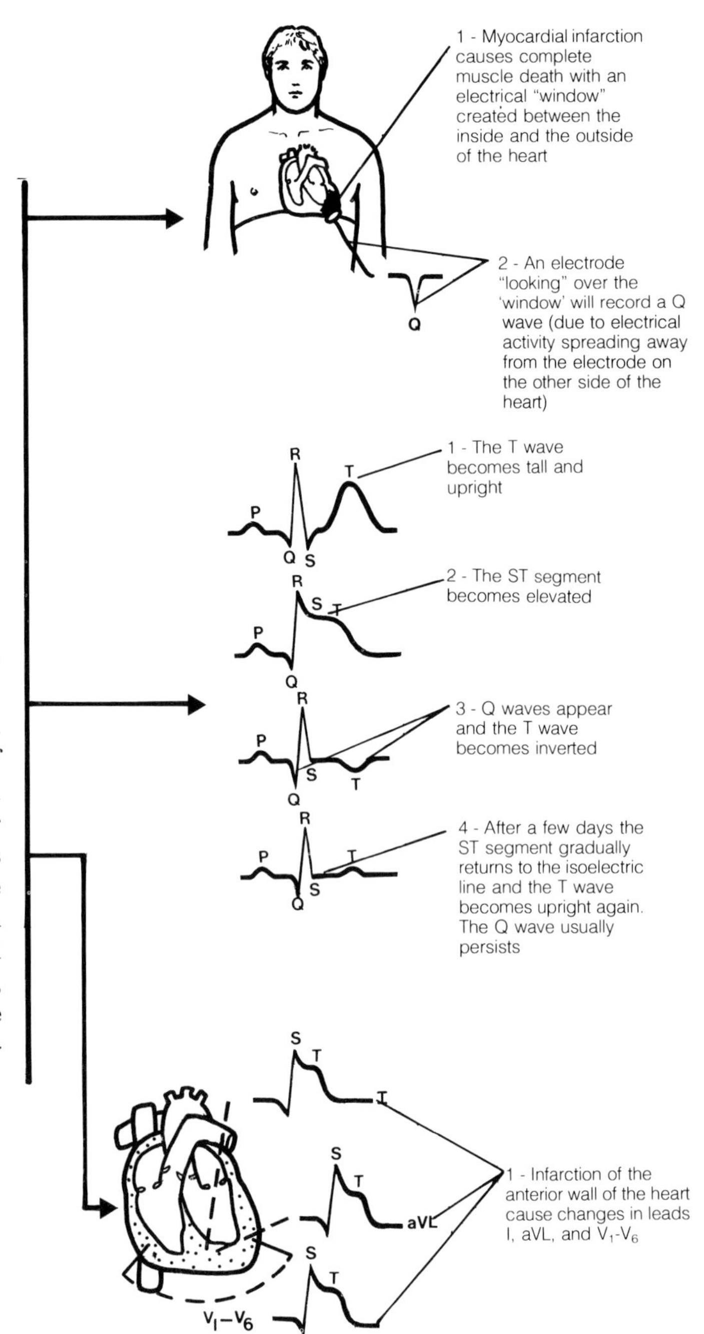
1 - Myocardial infarction causes complete muscle death with an electrical "window" created between the inside and the outside of the heart
2 - An electrode "looking" over the 'window' will record a Q wave (due to electrical activity spreading away from the electrode on the other side of the heart)
Q
1 - The T wave becomes tall and upright
2 - The ST segment becomes elevated
3 - Q waves appear and the T wave becomes inverted
4 - After a few days the ST segment gradually returns to the isoelectric line and the T wave becomes upright again. The Q wave usually persists
P
Q
R
S
T
aVL
V_1-V_6
1 - Infarction of the anterior wall of the heart cause changes in leads I, aVL, and V_1-V_6

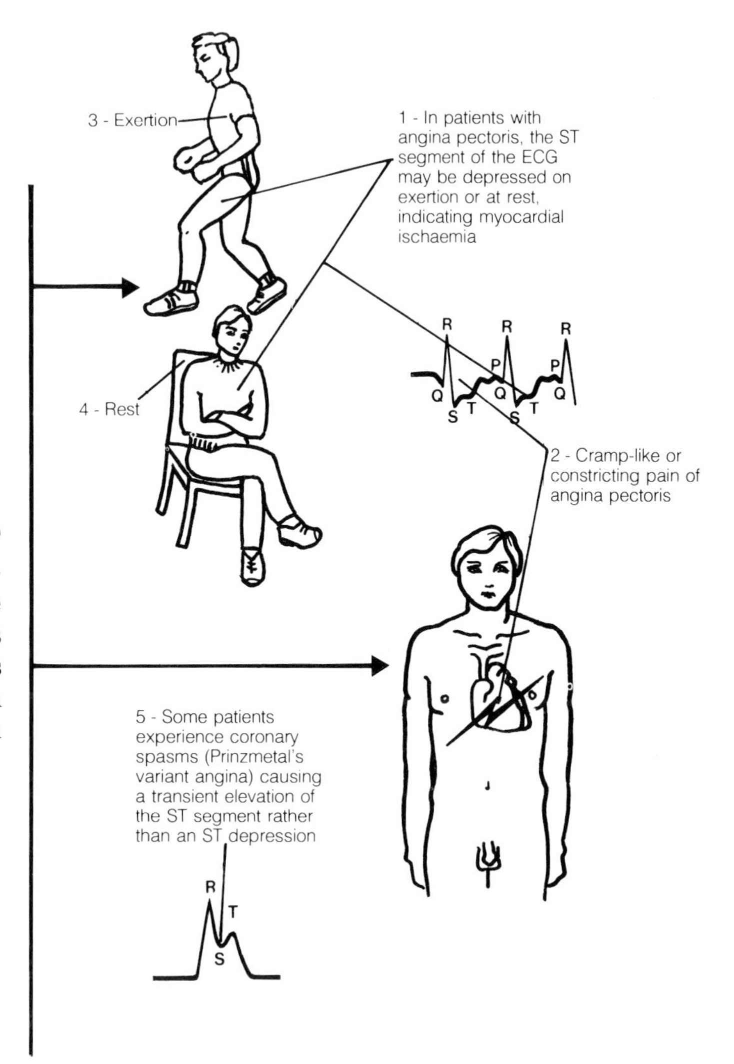
3 - Exertion
1 - In patients with angina pectoris, the ST segment of the ECG may be depressed on exertion or at rest, indicating myocardial ischaemia
R
R
R
P
P
Q
Q
Q
S
T
S
T
4 - Rest
2 - Cramp-like or constricting pain of angina pectoris
5 - Some patients experience coronary spasms (Prinzmetal's variant angina) causing a transient elevation of the ST segment rather than an ST depression
R
T
S

Myocardial ischaemia

In patients with angina pectoris, the ST segment of the ECG may become depressed at rest or on exertion. This is indicative of myocardial ischaemia. Some patients experience coronary spasm (**Prinzmetal's variant angina**). This is associated with transient elevation of the ST segment on the ECG rather than ST depression.

Bundle Branch Block The bundle of His divides into right, left anterior and left posterior divisions. Bundle branch block occurs when there is delayed conduction through one of these branches.

Right bundle branch block causes characteristic changes on the ECG. There is an M shaped QRS complex in the right ventricular leads (VI and V2). Standard leads I and AVL show a broad, slurred S wave. The QRS complex is widened to greater than 0.1 second.

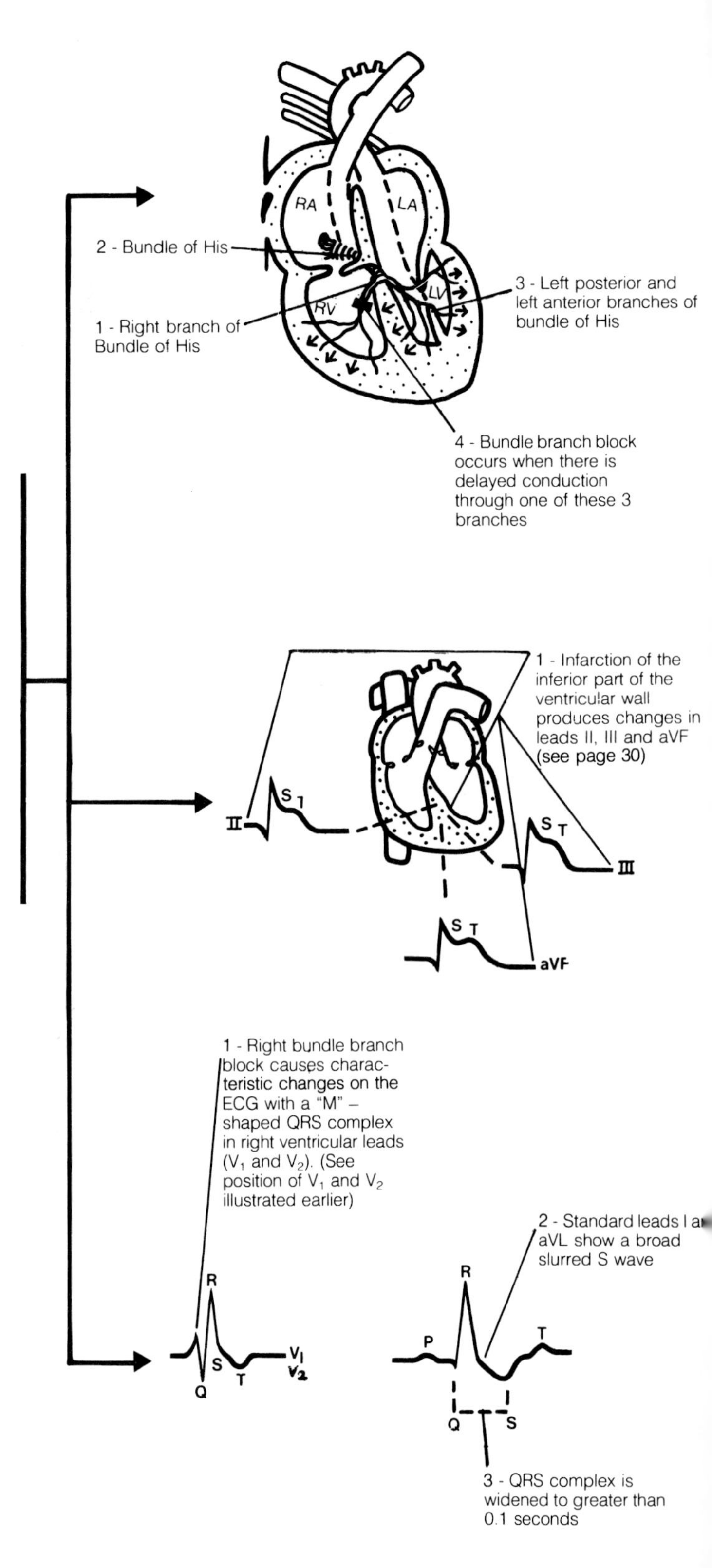

RA
LA
RV
LV
2 - Bundle of His
1 - Right branch of Bundle of His
3 - Left posterior and left anterior branches of bundle of His
4 - Bundle branch block occurs when there is delayed conduction through one of these 3 branches
1 - Infarction of the inferior part of the ventricular wall produces changes in leads II, III and aVF (see page 30)
II
III
aVF
S T
1 - Right bundle branch block causes characteristic changes on the ECG with a "M" – shaped QRS complex in right ventricular leads (V_1 and V_2). (See position of V_1 and V_2 illustrated earlier)
R
Q
S
T
V_1
V_2
2 - Standard leads I a aVL show a broad slurred S wave
P
3 - QRS complex is widened to greater than 0.1 seconds

Left bundle branch block Here, the QRS complex becomes widened and the R wave becomes wide and notched in leads I, AVL, V5 and V6.

Left anterior hemiblock Here the anterior division of the left bundle is blocked, resulting in left axis deviation of more than -60°.

Left posterior hemiblock The posterior division of the left bundle is blocked resulting in right axis deviation of more than 120°.

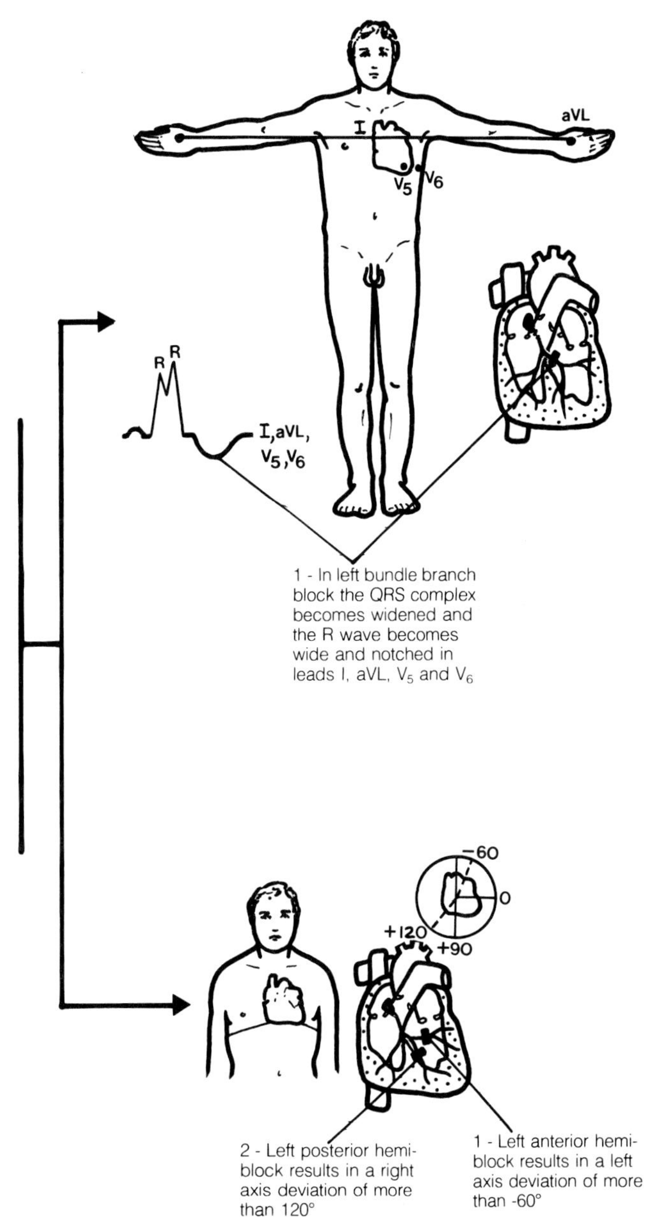
aVL
I
V_5
V_6
R
R
I,aVL,
V_5,V_6
-60
0
+120
+90

1 - In left bundle branch block the QRS complex becomes widened and the R wave becomes wide and notched in leads I, aVL, V_5 and V_6

2 - Left posterior hemiblock results in a right axis deviation of more than 120°

1 - Left anterior hemiblock results in a left axis deviation of more than -60°

ABNORMALITIES OF HEART RATE AND RHYTHM

Sinus tachycardia Here the heart rate is increased to over 100 beats/minute. This can occur in thyrotoxicosis, fevers, anaemia, pregnancy, exercise, haemorrhage and responses to fear or pain. Although the heart rate is fast the P, QRS and T complexes on the ECG are normal.

Sinus bradycardia The ECG is again normal in form but the heart rate is less than 60 beats/minute. This can be seen in very fit athletes and sometimes in myxoedema or hypothermia.

Sinus arrhythmia The heart rate tends to increase during inspiration and decrease during expiration. This is a normal finding, and the ECG is normal apart from variation in RR intervals with the respiratory cycle.

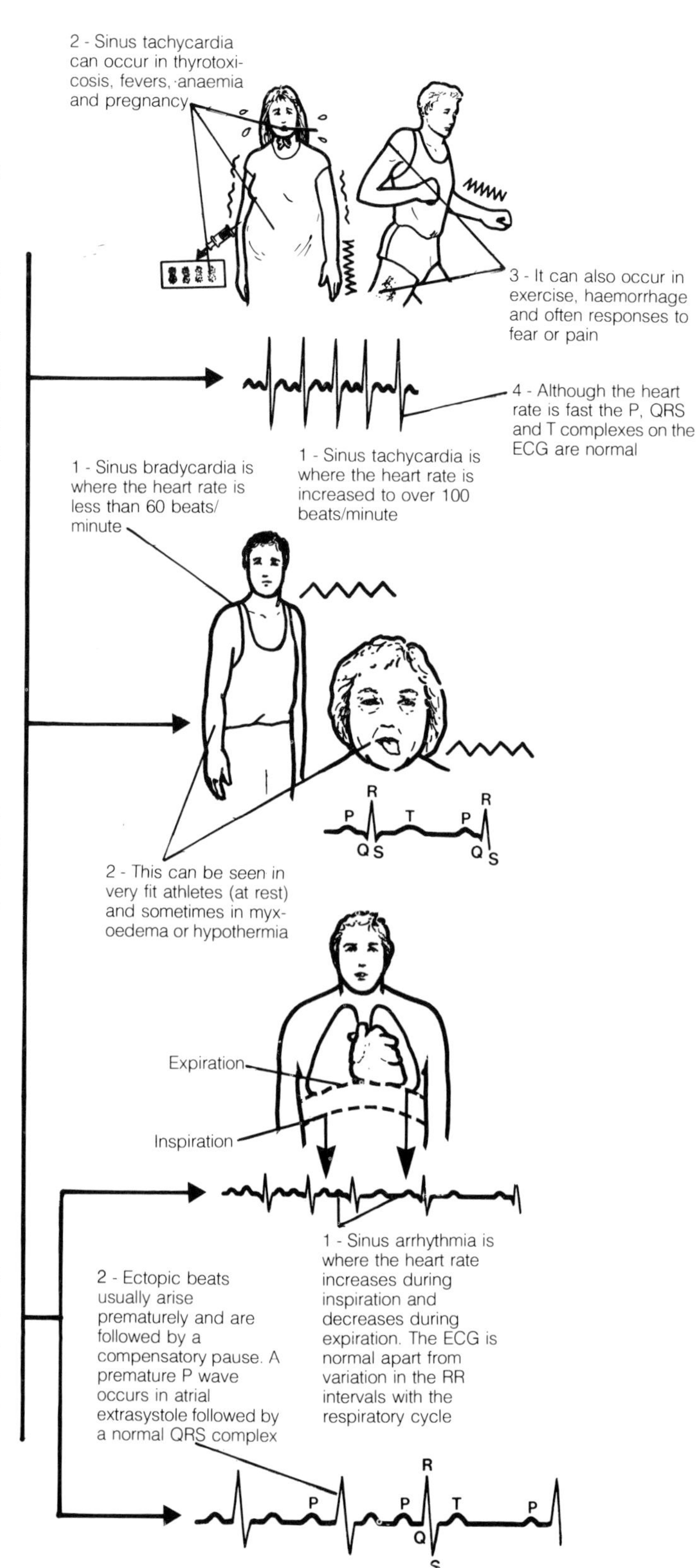

2 - Sinus tachycardia can occur in thyrotoxicosis, fevers, anaemia and pregnancy
3 - It can also occur in exercise, haemorrhage and often responses to fear or pain
4 - Although the heart rate is fast the P, QRS and T complexes on the ECG are normal
1 - Sinus tachycardia is where the heart rate is increased to over 100 beats/minute
1 - Sinus bradycardia is where the heart rate is less than 60 beats/minute
R
P
T
P
R
Q S
Q S
2 - This can be seen in very fit athletes (at rest) and sometimes in myxoedema or hypothermia
Expiration
Inspiration
1 - Sinus arrhythmia is where the heart rate increases during inspiration and decreases during expiration. The ECG is normal apart from variation in the RR intervals with the respiratory cycle
2 - Ectopic beats usually arise prematurely and are followed by a compensatory pause. A premature P wave occurs in atrial extrasystole followed by a normal QRS complex
P
P
R
T
P
Q
S

Extrasystoles (ectopic beats) Extrasystoles or ectopic beats occur when the cardiac impulse arises in an abnormal focus, which may be in the atria or the ventricles. Ectopic beats usually arise prematurely, and are followed by a compensatory pause. In atrial extrasystoles a premature P wave occurs, followed by a normal QRS complex. Ventricular ectopics are characterised by bizarre widened QRS complexes with absent P waves. Extrasystoles rarely require any treatment. Occasionally, following myocardial infarction, the R wave of a ventricular ectopic beat is superimposed on the T wave of the sinus beat (**R on T phenomenon**). This may herald ventricular fibrillation and requires treatment.

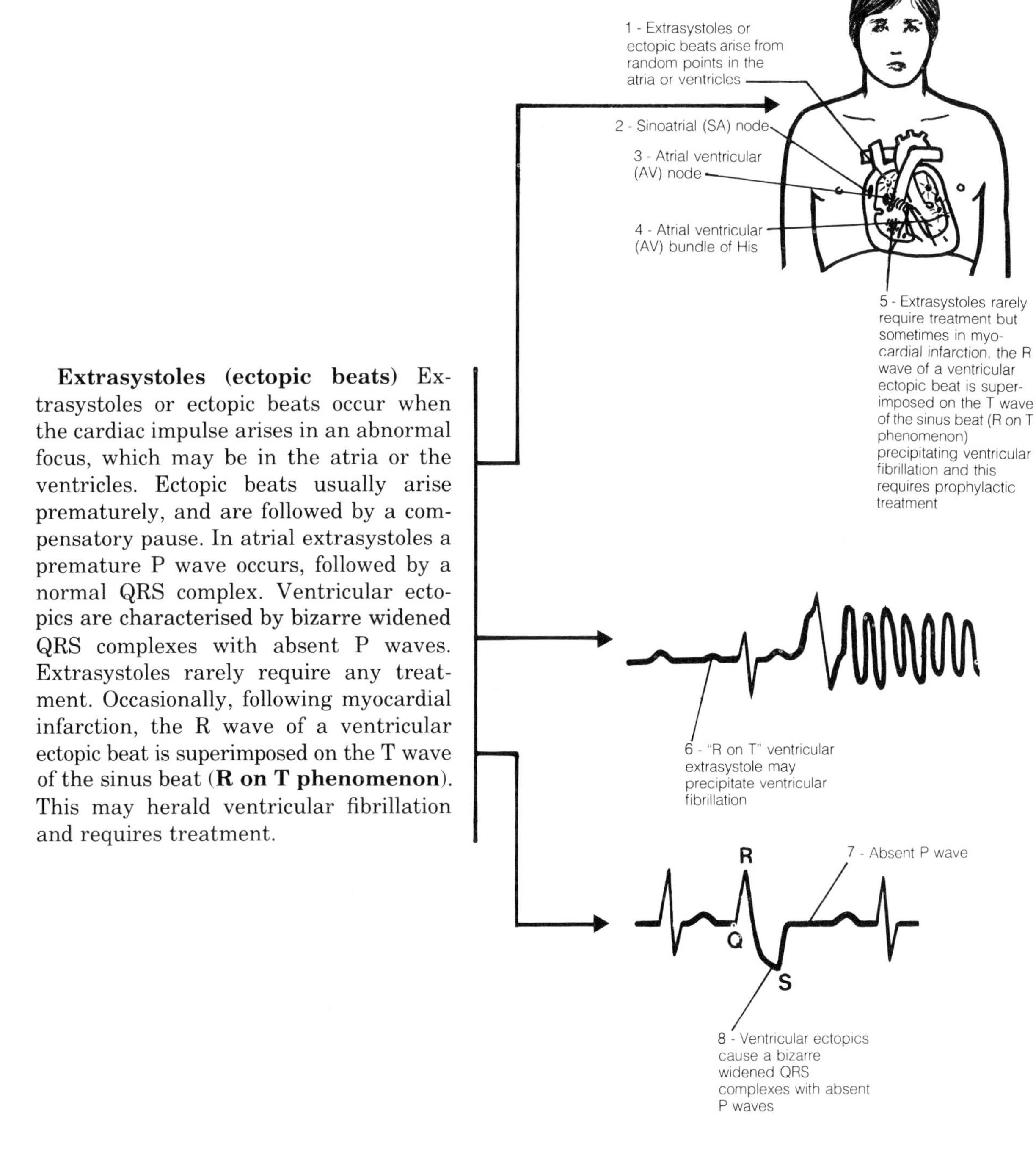
1 - Extrasystoles or ectopic beats arise from random points in the atria or ventricles
2 - Sinoatrial (SA) node
3 - Atrial ventricular (AV) node
4 - Atrial ventricular (AV) bundle of His
5 - Extrasystoles rarely require treatment but sometimes in myocardial infarction, the R wave of a ventricular ectopic beat is superimposed on the T wave of the sinus beat (R on T phenomenon) precipitating ventricular fibrillation and this requires prophylactic treatment
6 - "R on T" ventricular extrasystole may precipitate ventricular fibrillation
R
7 - Absent P wave
Q
S
8 - Ventricular ectopics cause a bizarre widened QRS complexes with absent P waves

Atrial tachycardias In atrial tachycardia (supraventricular tachycardia) and atrial flutter, there is an ectopic focus in the atrium which discharges rapidly. A tachycardia occurs; the P waves may be abnormal or difficult to identify, but the QRS complexes are normal and narrow. The heart rate is 150-180 beats/minute. At the fast rate, not all the atrial impulses are conducted to the ventricles and 3 : 1 or 4 : 1 block may develop.

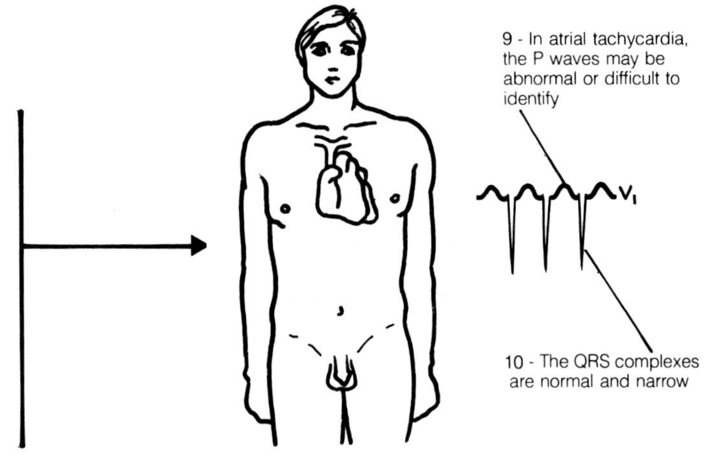
9 - In atrial tachycardia, the P waves may be abnormal or difficult to identify
V_1
10 - The QRS complexes are normal and narrow

Atrial fibrillation This is caused by multiple ectopic atrial foci, which discharge independently, at a rate of 360-400/minute. There is therefore no coordinated atrial activity and depolarisation spreads at irregular intervals down the bundle of His, to produce irregular ventricular contraction. The ECG shows fibrillation waves instead of P waves. The

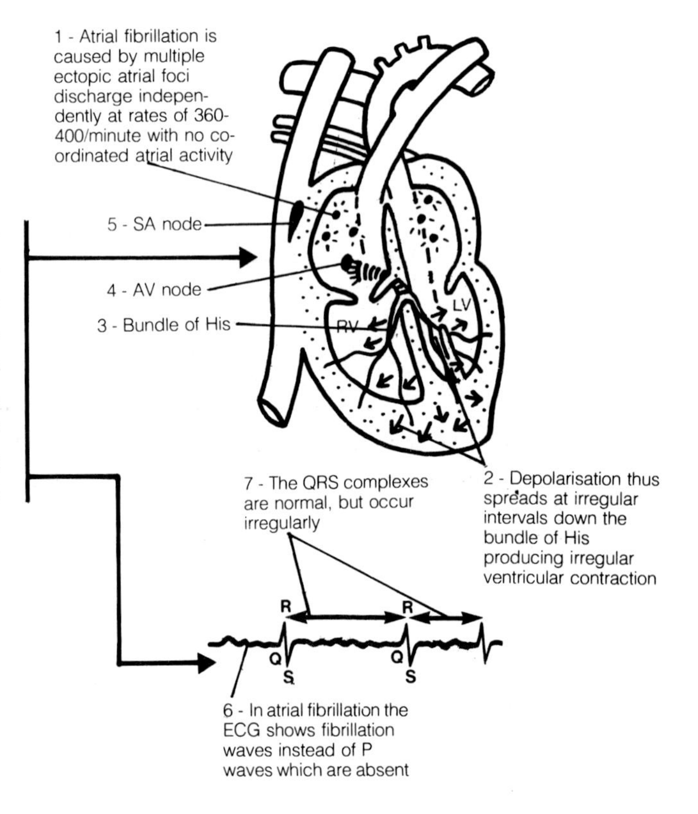
1 - Atrial fibrillation is caused by multiple ectopic atrial foci discharge independently at rates of 360-400/minute with no co-ordinated atrial activity
5 - SA node
4 - AV node
3 - Bundle of His
RV
LV
2 - Depolarisation thus spreads at irregular intervals down the bundle of His producing irregular ventricular contraction
7 - The QRS complexes are normal, but occur irregularly
R
R
Q
S
Q
S
6 - In atrial fibrillation the ECG shows fibrillation waves instead of P waves which are absent

QRS complexes are normal but occur irregularly. Atrial fibrillation is often associated with ischaemic heart disease, thyrotoxicosis, and mitral stenosis.

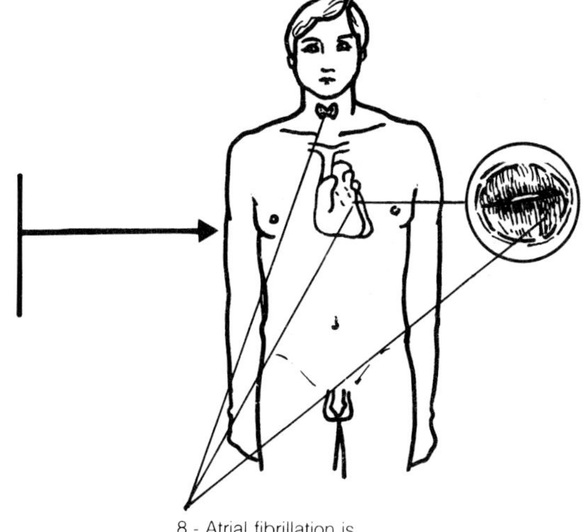
8 - Atrial fibrillation is often associated with ischaemic heart disease thyrotoxicosis and mitral stenosis

Wolff-Parkinson-White Syndrome The Wolff-Parkinson-White syndrome is characterised by a normal P wave, a short PR interval (0.11 seconds or less), a slurred upstroke to the QRS complex (delta wave) and increased QRS duration. There is a tendancy for atrial dysrhythmias to occur. The underlying pathology is thought to be due to an accesory conduction pathway which bypasses the AV node.

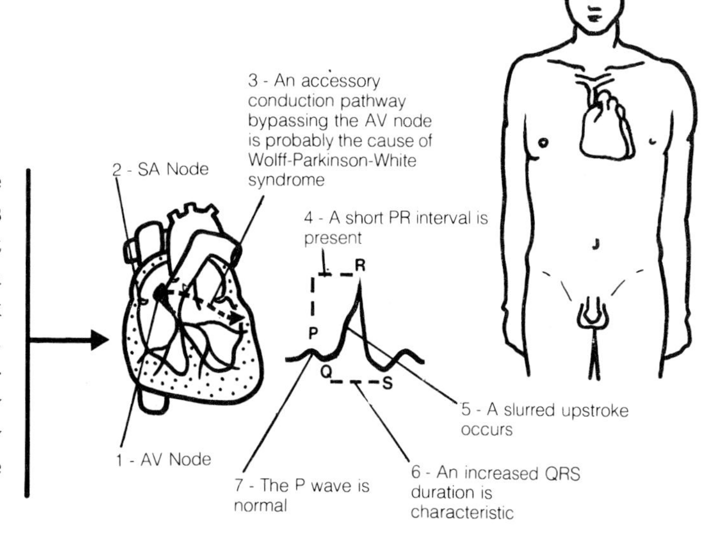
2 - SA Node
3 - An accessory conduction pathway bypassing the AV node is probably the cause of Wolff-Parkinson-White syndrome
4 - A short PR interval is present
R
P
Q
S
5 - A slurred upstroke occurs
1 - AV Node
7 - The P wave is normal
6 - An increased QRS duration is characteristic

Sick sinus syndrome This condition tends to occur in the elderly. Intermittent bradyarrhythmias and tachyarrhythmias may occur, causing dizziness and syncope.

Ventricular tachycardia This occurs when a ventricular focus depolarises at high frequency. The QRS complexes are wide and abnormal and no P waves are visible. Prompt treatment by electrical cardioversion or intravenous lignocaine is usually required (see myocardial infarction section).

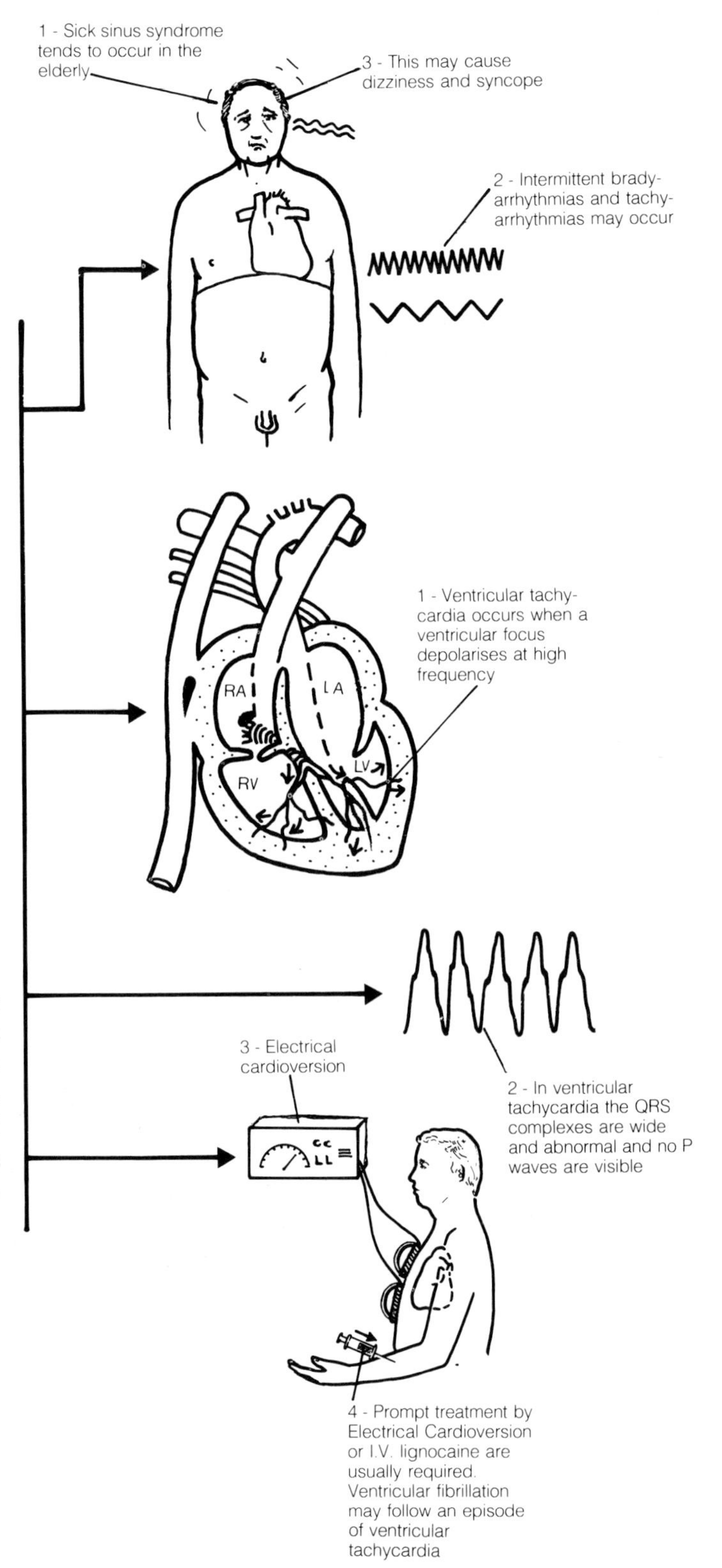
1 - Sick sinus syndrome tends to occur in the elderly
3 - This may cause dizziness and syncope
2 - Intermittent brady-arrhythmias and tachy-arrhythmias may occur
1 - Ventricular tachy-cardia occurs when a ventricular focus depolarises at high frequency
RA
LA
RV
LV
3 - Electrical cardioversion
2 - In ventricular tachycardia the QRS complexes are wide and abnormal and no P waves are visible
4 - Prompt treatment by Electrical Cardioversion or I.V. lignocaine are usually required. Ventricular fibrillation may follow an episode of ventricular tachycardia

Ventricular fibrillation This may follow an episode of ventricular tachycardia, or be initiated by an R on T ventricular ectopic beat. The ECG does not show any QRS complexes, but irregular fibrillary waves can be seen. This rhythm is fatal unless treated promptly by electrical cardioversion.

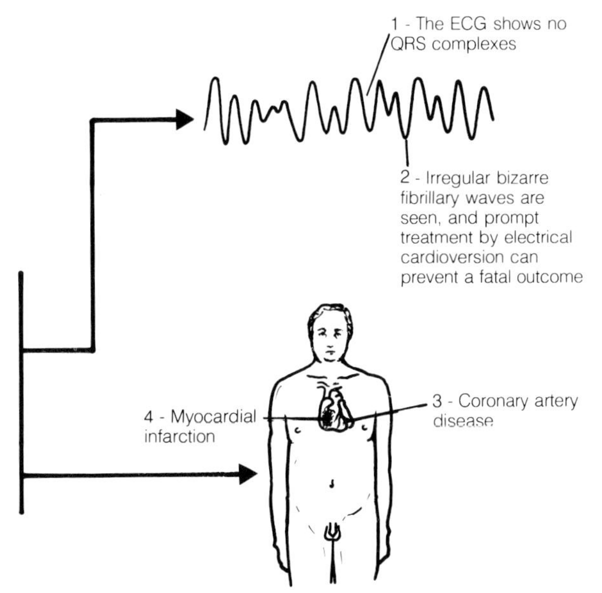
1 - The ECG shows no QRS complexes
2 - Irregular bizarre fibrillary waves are seen, and prompt treatment by electrical cardioversion can prevent a fatal outcome
3 - Coronary artery disease
4 - Myocardial infarction

Atrio-ventricular block (heart block) This is characterized by interrupted conduction through the AV node and bundle of His.

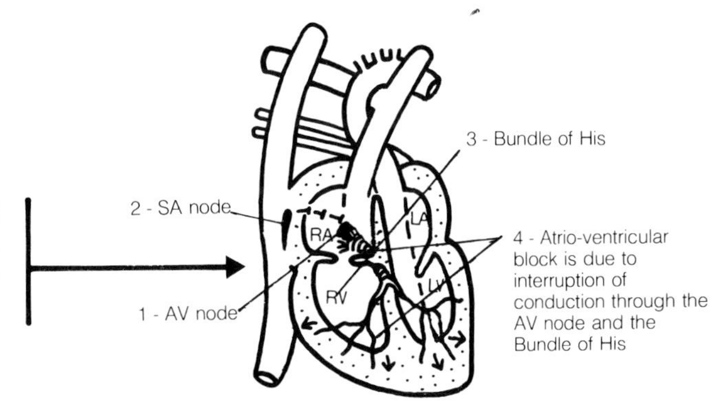
3 - Bundle of His
2 - SA node
RA
LA
RV
LV
4 - Atrio-ventricular block is due to interruption of conduction through the AV node and the Bundle of His
1 - AV node

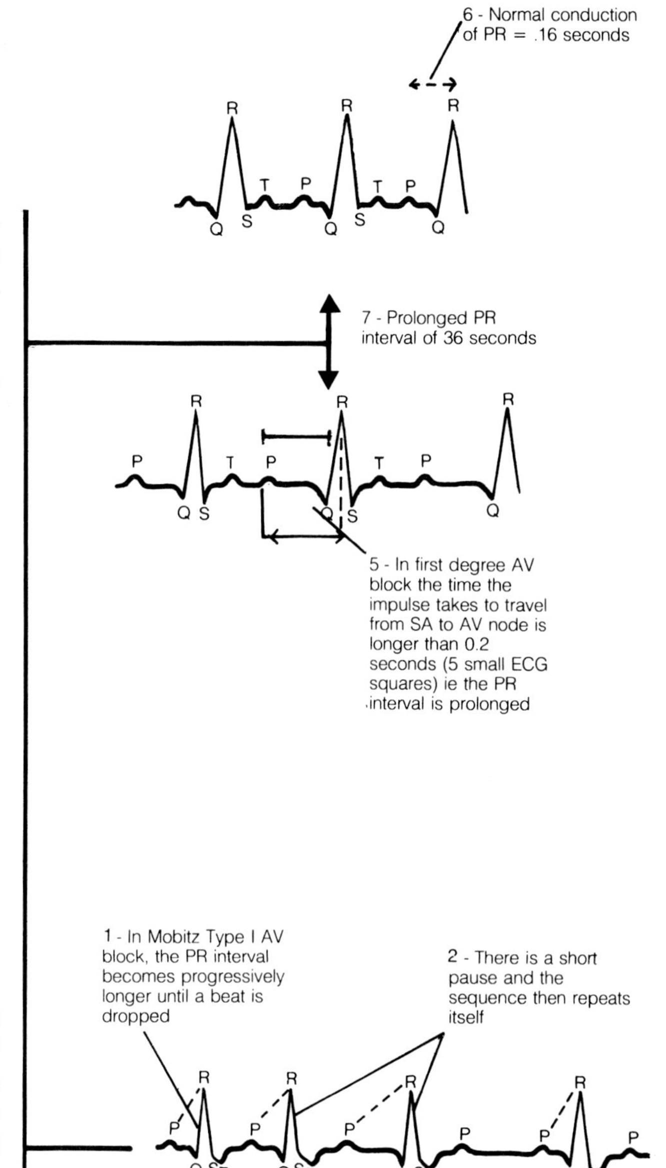
6 - Normal conduction of PR = .16 seconds
R
R
R
T
P
T
P
Q
S
Q
S
Q
7 - Prolonged PR interval of 36 seconds
R
R
R
P
T
P
T
P
Q S
S
Q
5 - In first degree AV block the time the impulse takes to travel from SA to AV node is longer than 0.2 seconds (5 small ECG squares) ie the PR interval is prolonged
1 - In Mobitz Type I AV block, the PR interval becomes progressively longer until a beat is dropped
2 - There is a short pause and the sequence then repeats itself
R
R
R
R
P
P
P
P
P
P
Q ST
Q ST
Q ST
Q ST

First degree heart block is characterised by a prolonged PR interval; the time taken for the impulse to travel from the SA to the AV node is longer than 0.2 seconds (five small squares on the ECG). First degree block can be found in acute rheumatic fever, patients taking digoxin, and coronary artery disease.

Second degree AV block Here, sinus impulses are blocked so that not every P wave is followed by a QRS complex. In **Type I AV block (Mobitz type I, Wenkebach phenomenon)** the PR interval becomes progressively longer until a beat is dropped. There is a short pause and the sequence then repeats itself.

Type II AV block (Mobitz type II)
Here the PR interval is constant, with intermittent dropped beats.

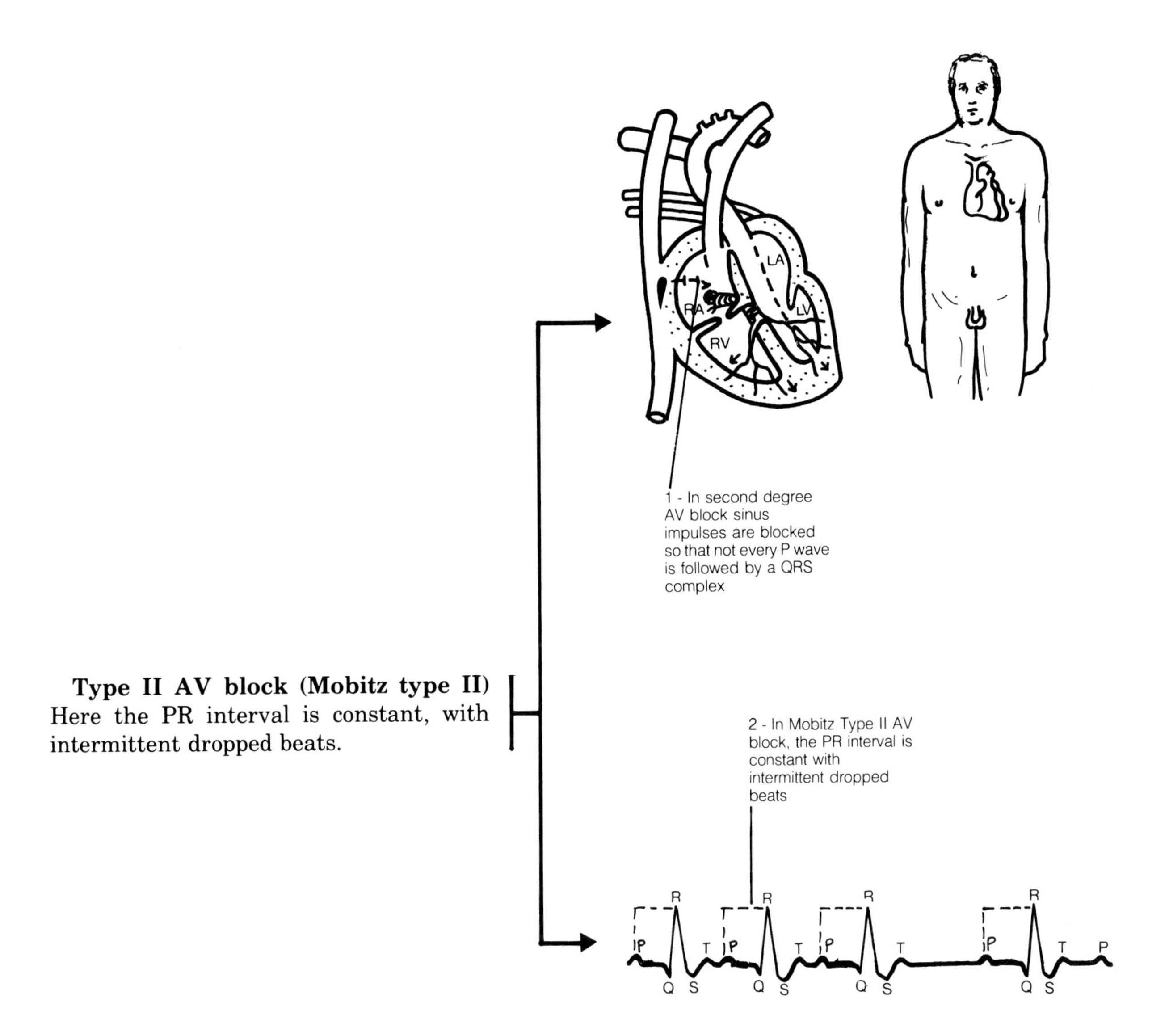

LA
RA
LV
RV
1 - In second degree AV block sinus impulses are blocked so that not every P wave is followed by a QRS complex
2 - In Mobitz Type II AV block, the PR interval is constant with intermittent dropped beats
R
P
T
Q
S

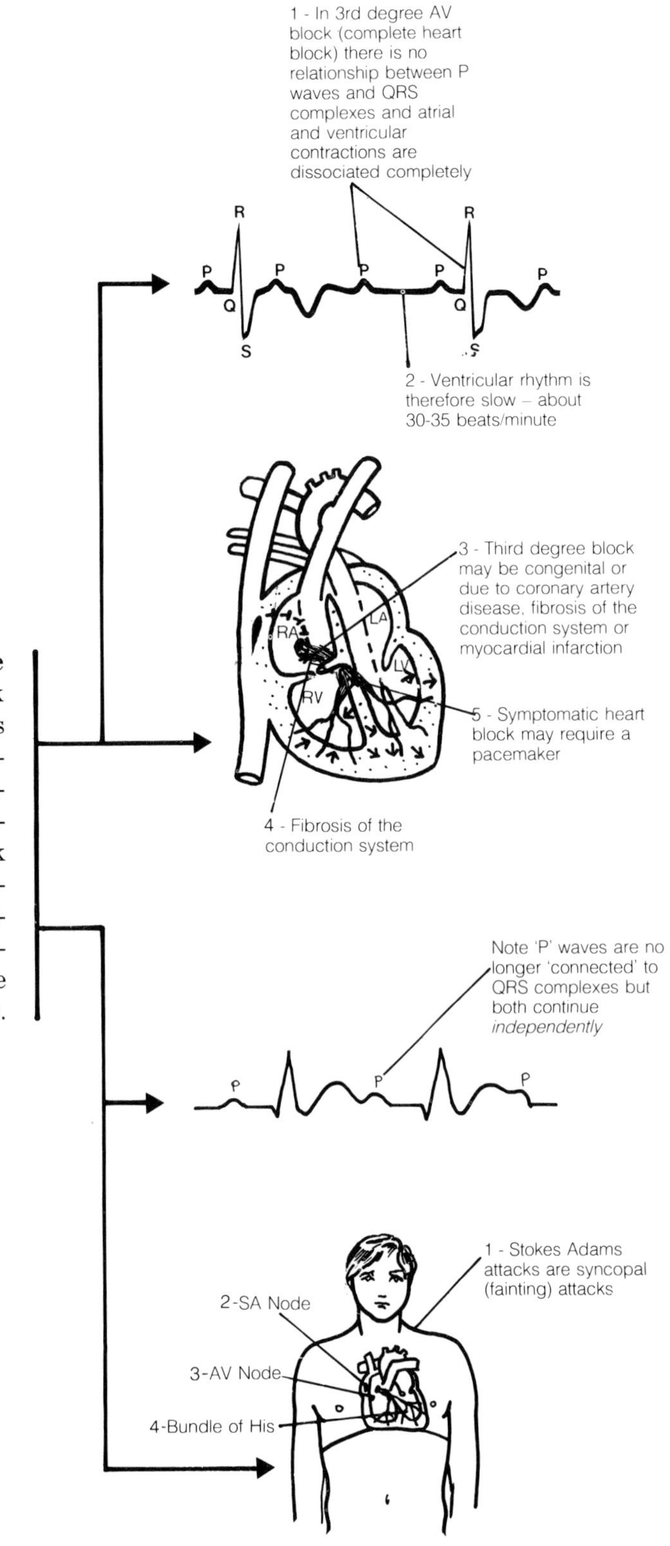

1 - In 3rd degree AV block (complete heart block) there is no relationship between P waves and QRS complexes and atrial and ventricular contractions are dissociated completely
R
P
Q
S
2 - Ventricular rhythm is therefore slow – about 30-35 beats/minute
3 - Third degree block may be congenital or due to coronary artery disease, fibrosis of the conduction system or myocardial infarction
RA
LA
RV
LV
5 - Symptomatic heart block may require a pacemaker
4 - Fibrosis of the conduction system
Note 'P' waves are no longer 'connected' to QRS complexes but both continue *independently*
1 - Stokes Adams attacks are syncopal (fainting) attacks
2-SA Node
3-AV Node
4-Bundle of His

Third degree AV block (complete heart block) In complete heart block there is no relationship between P waves and QRS complexes, and atrial and ventricular contraction are dissociated completely. The ventricular rhythm is usually slow (30-35 beats/minute). AV block may be congenital or secondary to coronary artery disease, fibrosis of the conducting system, or it may occur after myocardial infarction. A pacemaker may be required if the heart block is symptomatic.

Stokes-Adams attacks are syncopal attacks due to ventricular standstill. The underlying arrhythmia is ventricular fibrillation or ventricular tachycardia in 50% of cases, and asystole in the other 50%. Characteristically the patient falls to the ground during an attack and becomes pulseless. When cardiac output returns the patient flushes and regains consciousness.

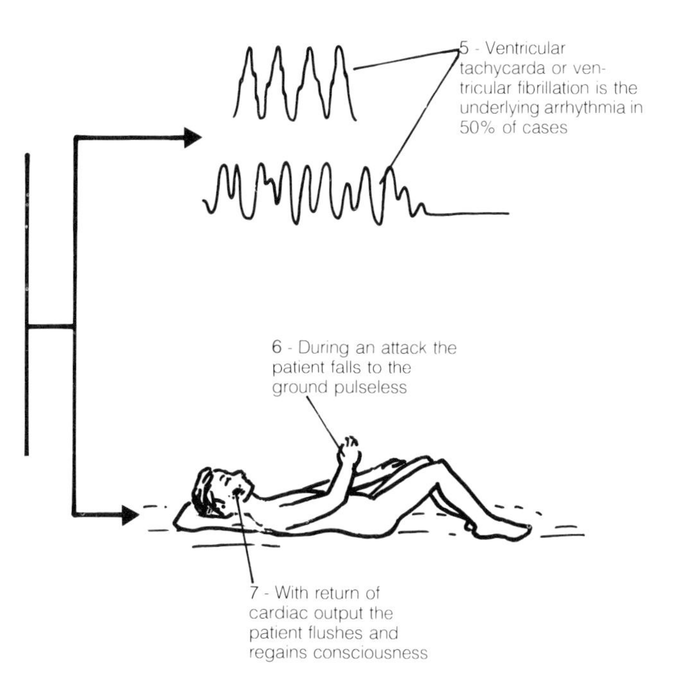
5 - Ventricular tachycarda or ventricular fibrillation is the underlying arrhythmia in 50% of cases
6 - During an attack the patient falls to the ground pulseless
7 - With return of cardiac output the patient flushes and regains consciousness

OTHER INVESTIGATIONS IN CARDIOLOGY

The Chest X-Ray

The chest X-ray can be very useful in assessing patients with cardiac disease. The cardiac silhouette is made up of the superior vena cava and right atrium on the right and the aortic knuckle, pulmonary artery and left ventricle on the left. Cardiac size can be assessed by measuring the cardiothoracic ratio, that is the maximal transverse diameter of the heart, divided by the maximal internal diameter of the thoracic cage. This ratio is

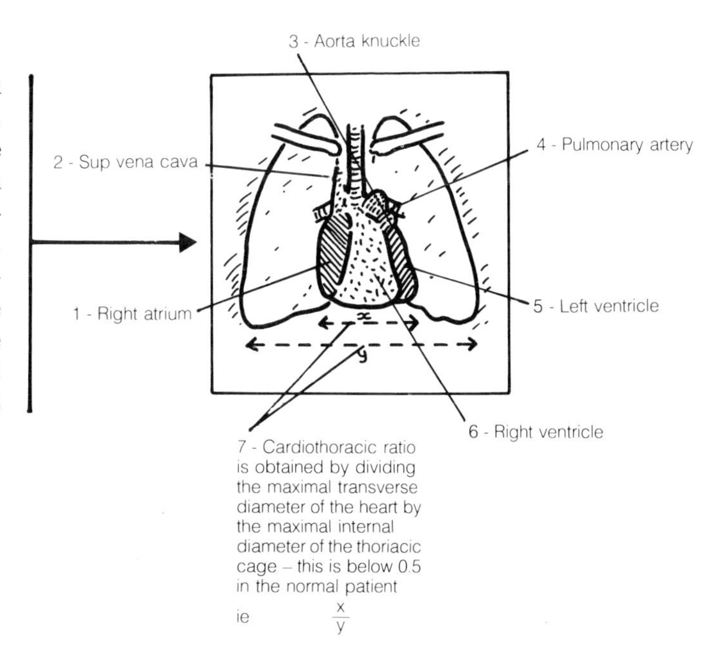
3 - Aorta knuckle
2 - Sup vena cava
4 - Pulmonary artery
1 - Right atrium
5 - Left ventricle
x
y
6 - Right ventricle
7 - Cardiothoracic ratio is obtained by dividing the maximal transverse diameter of the heart by the maximal internal diameter of the thoriacic cage – this is below 0.5 in the normal patient
ie $\frac{x}{y}$

below 0.5 in the normal patient and should not increase by more than 1.5cm between serial chest X-rays. Enlargement of the left atrium (eg. in mitral stenosis) causes splaying of the carinal angle, with elevation of the left main bronchus. A double contour may be seen along the right border of the heart, and the lateral chest X-ray may also show bulging of the left atrium posteriorly. It is not usually possible to distinguish between right and left ventricular enlargement on conventional radiography.

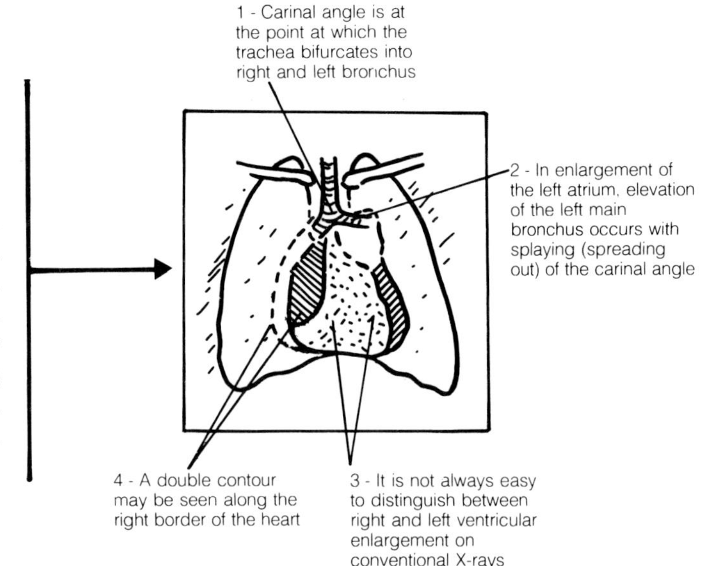
1 - Carinal angle is at the point at which the trachea bifurcates into right and left bronchus
2 - In enlargement of the left atrium, elevation of the left main bronchus occurs with splaying (spreading out) of the carinal angle
3 - It is not always easy to distinguish between right and left ventricular enlargement on conventional X-rays
4 - A double contour may be seen along the right border of the heart

The pulmonary outflow tract may be enlarged in pulmonary hypertension, while dilation of the ascending aorta can occur in aortic valve disease, aortic aneurysm, aortic coarctation and patent ductus

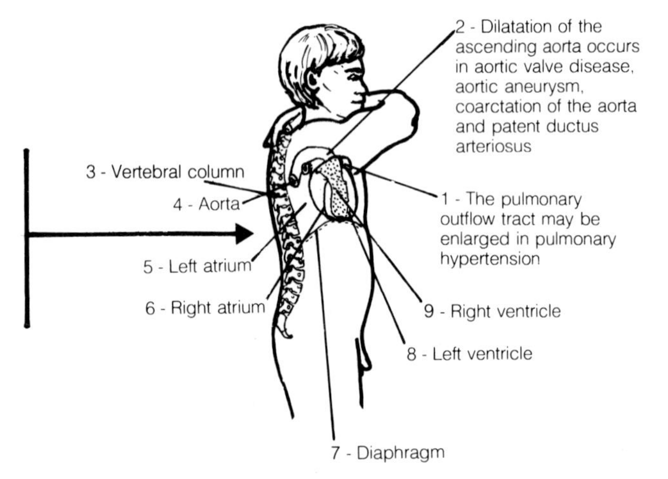
2 - Dilatation of the ascending aorta occurs in aortic valve disease, aortic aneurysm, coarctation of the aorta and patent ductus arteriosus
3 - Vertebral column
4 - Aorta
1 - The pulmonary outflow tract may be enlarged in pulmonary hypertension
5 - Left atrium
6 - Right atrium
9 - Right ventricle
8 - Left ventricle
7 - Diaphragm

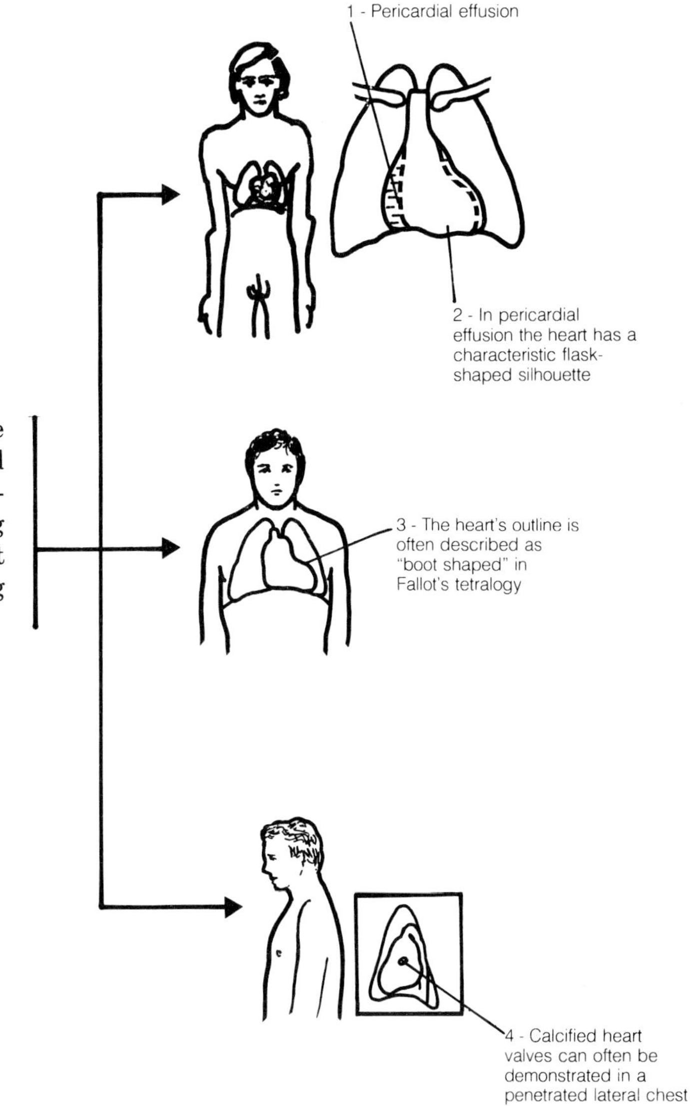
1 - Pericardial effusion
2 - In pericardial effusion the heart has a characteristic flask-shaped silhouette
3 - The heart's outline is often described as "boot shaped" in Fallot's tetralogy
4 - Calcified heart valves can often be demonstrated in a penetrated lateral chest X-ray

arteriosus. In pericardial effusion, the heart has a characteristic flask-shaped silhouette. In Fallot's tetralogy, the cardiac outline is often described as being "boot shaped"! A penetrated, lateral chest X-ray is often helpful in demonstrating calcified heart valves.

The Lung Fields in Heart Disease

The lung Fields may be congested in heart failure. The upper lobe blood vessels become engorged, and Kerley B lines become more apparent due to collection of fluid in the interlobular septa. These appear as short, parallel, horizontal lines at the costophrenic angles. The hila become congested to produce a characteristic "butterfly" shadow. In the left to right shunts the pulmonary vessels are larger than normal; this is called **pulmonary plethora**.

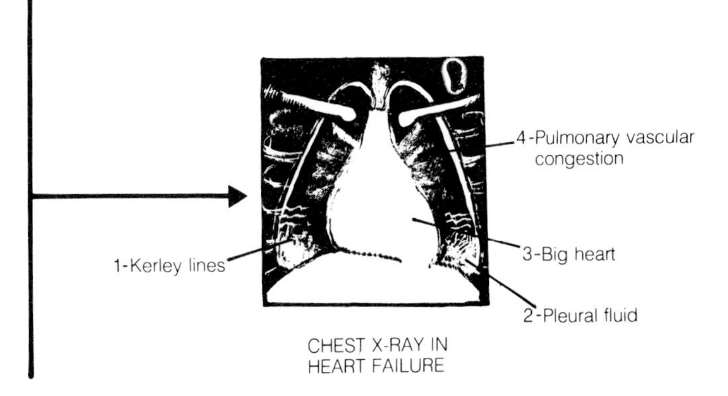
4-Pulmonary vascular congestion
3-Big heart
1-Kerley lines
2-Pleural fluid

CHEST X-RAY IN HEART FAILURE

Pulmonary oligaemia Here there are reduced pulmonary vascular markings, due to diminished pulmonary blood flow, as in pulmonary stenosis, tricuspid atresia and Fallot's tetralogy. Acquired causes include pulmonary embolism and emphysema.

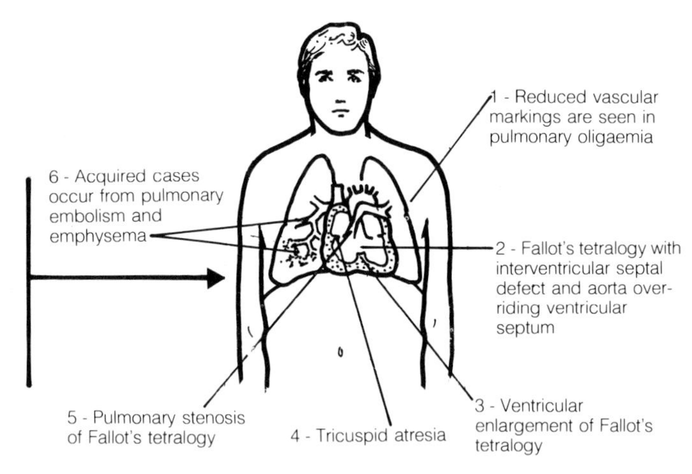
1 - Reduced vascular markings are seen in pulmonary oligaemia
2 - Fallot's tetralogy with interventricular septal defect and aorta over-riding ventricular septum
3 - Ventricular enlargement of Fallot's tetralogy
4 - Tricuspid atresia
5 - Pulmonary stenosis of Fallot's tetralogy
6 - Acquired cases occur from pulmonary embolism and emphysema

Echocardiography

Echocardiography is a harmless, non-invasive technique which depends on the reflection of high frequency sound echoes from intracardiac structures. The M-mode echocardiogram uses a single narrow ultrasound beam; the reflected ultrasound echoes are recorded on moving paper. The 2D echocardiogram uses multiple echo beams, and provides a cross-sectional moving image of the heart, which can be

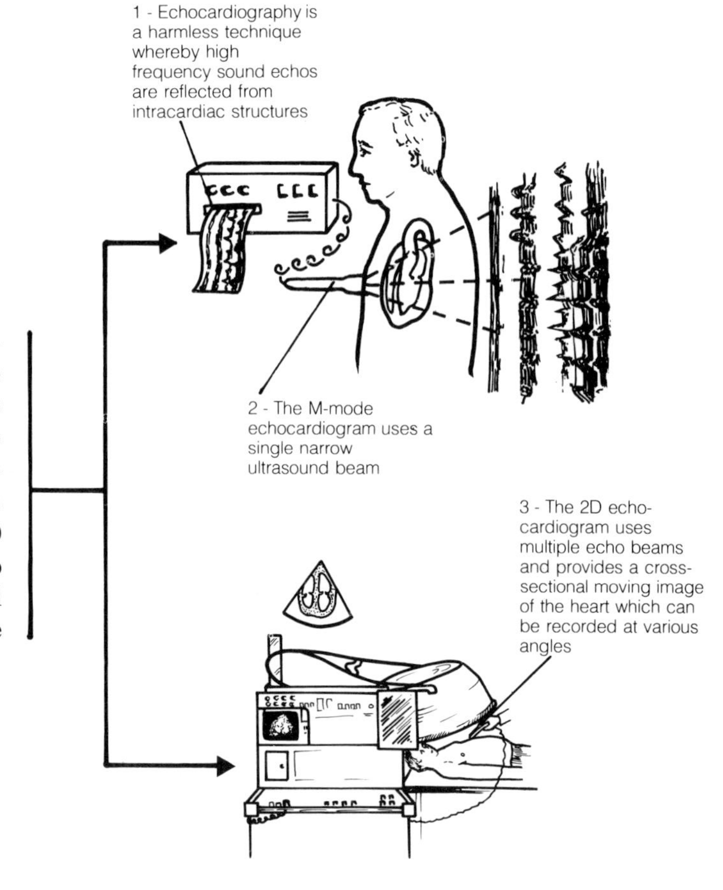
1 - Echocardiography is a harmless technique whereby high frequency sound echos are reflected from intracardiac structures
2 - The M-mode echocardiogram uses a single narrow ultrasound beam
3 - The 2D echo-cardiogram uses multiple echo beams and provides a cross-sectional moving image of the heart which can be recorded at various angles

recorded from several angles. The echocardiogram is particularly useful in the detection of pericardial effusions and valvular heart disease. Valvular thickening can be visualised, and in aortic regurgitation fluttering of the mitral valve caused by the aortic regurgitant jet is a very good indicator of aortic valvular incompetence. In patients with infective endocarditis, vegetations of the affected valve can often be well visualised and the technique will reliably identify atrial myxomas and hypertrophic cardiomyopathy. The echocardiogram is also useful in assessing size of the cardiac chambers as well as their function.

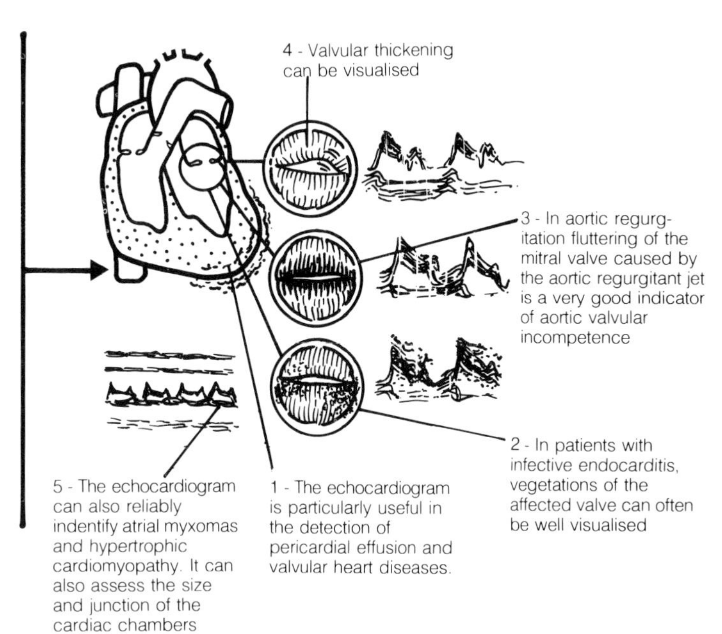
4 - Valvular thickening can be visualised
3 - In aortic regurgitation fluttering of the mitral valve caused by the aortic regurgitant jet is a very good indicator of aortic valvular incompetence
2 - In patients with infective endocarditis, vegetations of the affected valve can often be well visualised
5 - The echocardiogram can also reliably indentify atrial myxomas and hypertrophic cardiomyopathy. It can also assess the size and junction of the cardiac chambers
1 - The echocardiogram is particularly useful in the detection of pericardial effusion and valvular heart diseases.

The exercise ECG

Patients with coronary artery disease may have normal resting ECGs, but may develop ischaemic changes on exertion. An exercise ECG is often a useful test in these patients. The patient's ECG is continuously monitored, before, during and after exercise on a treadmill or bicycle. ST segment depression may occur during or after exercise and is very suggestive of myocardial ischaemia.

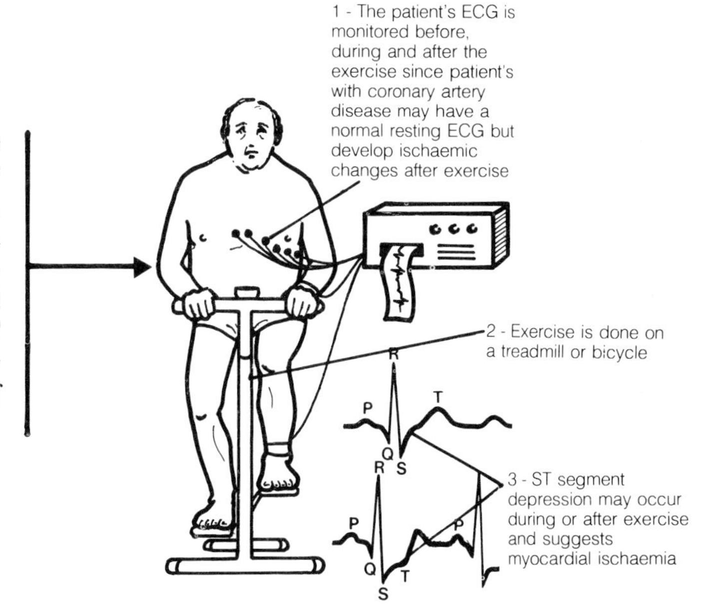
1 - The patient's ECG is monitored before, during and after the exercise since patient's with coronary artery disease may have a normal resting ECG but develop ischaemic changes after exercise
2 - Exercise is done on a treadmill or bicycle
R
P
T
Q
S
3 - ST segment depression may occur during or after exercise and suggests myocardial ischaemia

Radionuclide investigation of the heart

Radionuclide imaging is helpful in assessing patients with coronary artery disease. **Thallium 201** is taken up into the myocardium in proportion to local blood flow. The isotope is therefore useful in assessment of ischaemic heart disease, investigation of chest pain, and assessing patients prior to coronary artery bypass surgery. **Technetium 99M** can be used in evaluating ventricular function at rest and during stress. A blood sample is taken, the patient's red cells are labelled with technetium 99M and then reinjected. Imaging is performed using a gamma camera, and ejection fractions, regional wall movement and ventricular dilatation can be studied. The technique is useful in assessing ischaemic heart disease and ventricular function, the diagnosis of ventricular aneurysms and the assessment of the haemodynamic significance of valvular disease.

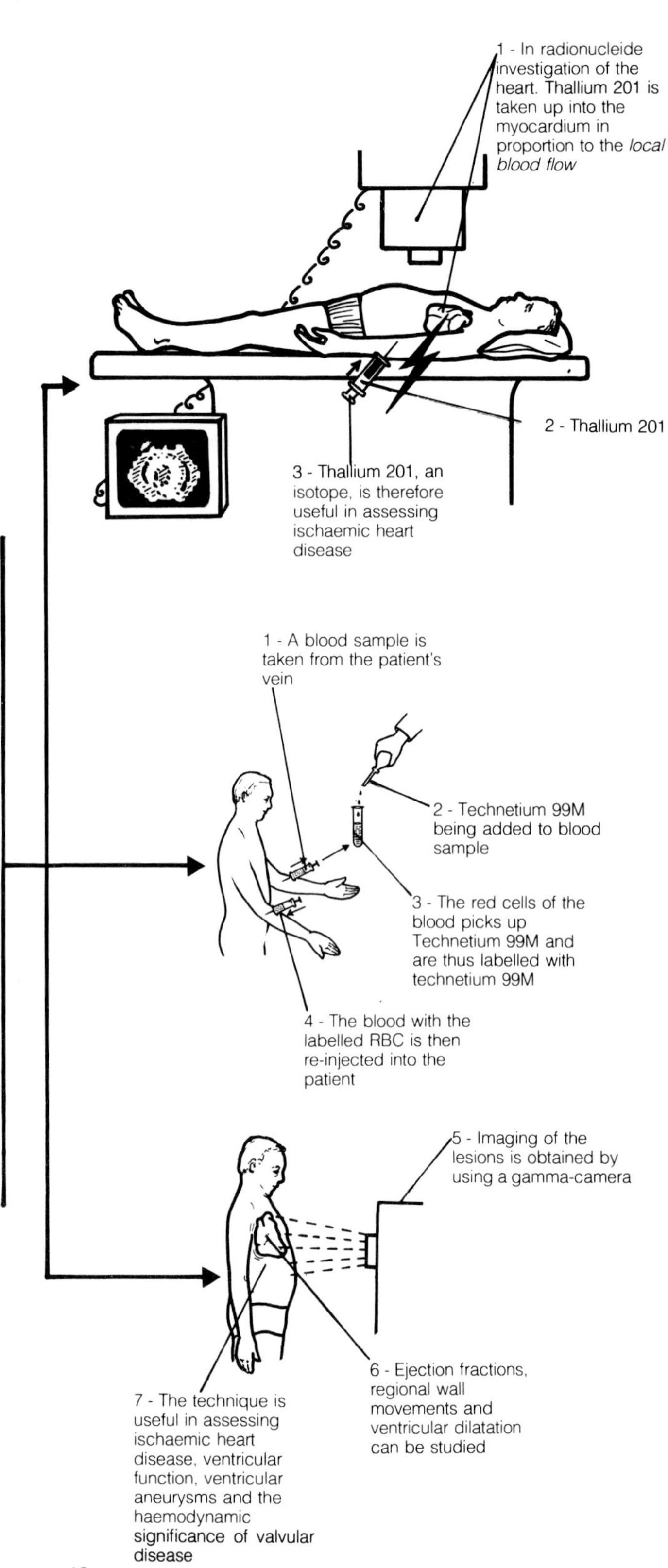
1 - In radionucleide investigation of the heart. Thallium 201 is taken up into the myocardium in proportion to the *local blood flow*
2 - Thallium 201
3 - Thallium 201, an isotope, is therefore useful in assessing ischaemic heart disease
1 - A blood sample is taken from the patient's vein
2 - Technetium 99M being added to blood sample
3 - The red cells of the blood picks up Technetium 99M and are thus labelled with technetium 99M
4 - The blood with the labelled RBC is then re-injected into the patient
5 - Imaging of the lesions is obtained by using a gamma-camera
6 - Ejection fractions, regional wall movements and ventricular dilatation can be studied
7 - The technique is useful in assessing ischaemic heart disease, ventricular function, ventricular aneurysms and the haemodynamic significance of valvular disease

Cardiac catheterization and selective angiography

The right and left sides of the heart can be investigated by inserting a catheter into a vein or artery in the arm or leg, and manipulating it under X-ray control, into the right atrium, right ventricle and pulmonary artery, or through the aortic valve into the left ventricle. This technique can be used to measure cardiac output, pulmonary vascular resistance and the magnitude of left to right shunts, as well as pressure gradients across heart valves. If contrast medium is introduced through the catheter, a direct view of left ventricular function can be obtained as well as the degree of regurgitation through heart valves. By placing the catheter tip in the origin of the coronary arteries and introducing X-ray contrast medium, the coronary arterial tree can be outlined and recorded on film. These techniques have a very low complication rate when performed by experienced staff, and are invaluable in assessing patients, especially prior to cardiac surgery.

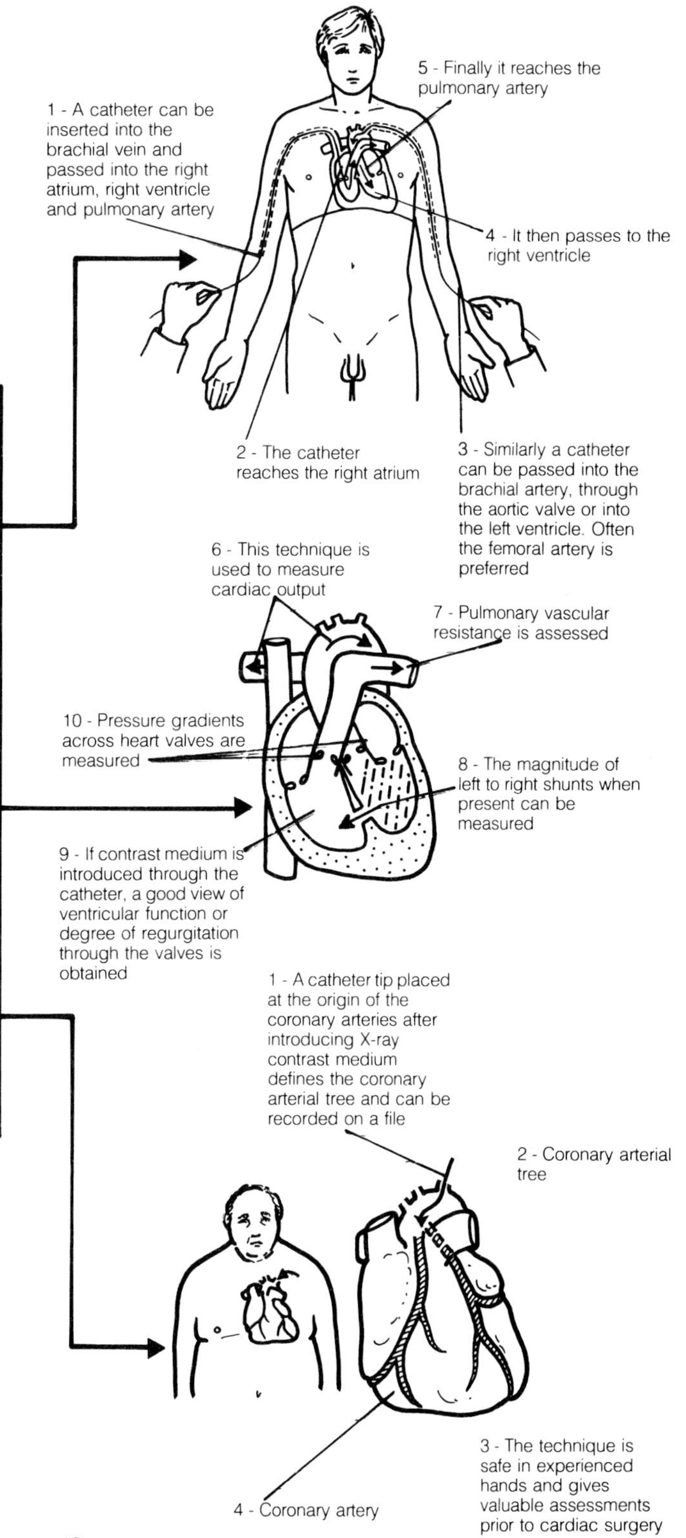
1 - A catheter can be inserted into the brachial vein and passed into the right atrium, right ventricle and pulmonary artery
2 - The catheter reaches the right atrium
3 - Similarly a catheter can be passed into the brachial artery, through the aortic valve or into the left ventricle. Often the femoral artery is preferred
4 - It then passes to the right ventricle
5 - Finally it reaches the pulmonary artery
6 - This technique is used to measure cardiac output
7 - Pulmonary vascular resistance is assessed
8 - The magnitude of left to right shunts when present can be measured
9 - If contrast medium is introduced through the catheter, a good view of ventricular function or degree of regurgitation through the valves is obtained
10 - Pressure gradients across heart valves are measured
1 - A catheter tip placed at the origin of the coronary arteries after introducing X-ray contrast medium defines the coronary arterial tree and can be recorded on a file
2 - Coronary arterial tree
3 - The technique is safe in experienced hands and gives valuable assessments prior to cardiac surgery
4 - Coronary artery

CORONARY ARTERY DISEASE

Coronary artery disease is one of the commonest causes of death in middle-aged men; in 1977, the mortality rate from coronary artery disease was 854/100,000 per annum in the United Kingdom. The main cause of coronary artery disease is atherosclerosis. As part of the ageing process, the arterial vessels develop fatty streaks and fibrous plaques in the intima, which can ulcerate, thrombose or become necrotic. When the patient exercises, blood flow through the narrow artery becomes insufficient to meet myocardial demand and **anginal pain** occurs. Complete occlusion of a coronary artery causes death of cardiac muscle (**myocardial infarction**) unless there is a good collateral blood flow.

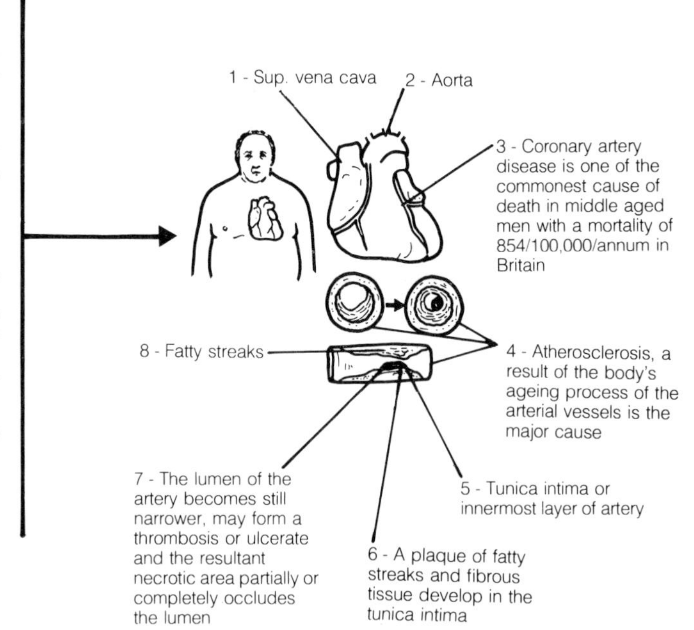
1 - Sup. vena cava
2 - Aorta
3 - Coronary artery disease is one of the commonest cause of death in middle aged men with a mortality of 854/100,000/annum in Britain
4 - Atherosclerosis, a result of the body's ageing process of the arterial vessels is the major cause
5 - Tunica intima or innermost layer of artery
6 - A plaque of fatty streaks and fibrous tissue develop in the tunica intima
7 - The lumen of the artery becomes still narrower, may form a thrombosis or ulcerate and the resultant necrotic area partially or completely occludes the lumen
8 - Fatty streaks

Risk factors for coronary artery disease

Although atherosclerosis is part of the ageing process, there are several major risk factors.

Sex Males have a much higher risk of coronary artery disease than females.

Cigarette smoking The risk of death from coronary artery disease is three times higher in cigarette smokers than in non-smokers.

Hypertension There is an increased risk of coronary artery disease in patients who are hypertensive and the risk increases progressively with increasing blood pressure. The risk is reduced when blood pressure is controlled.

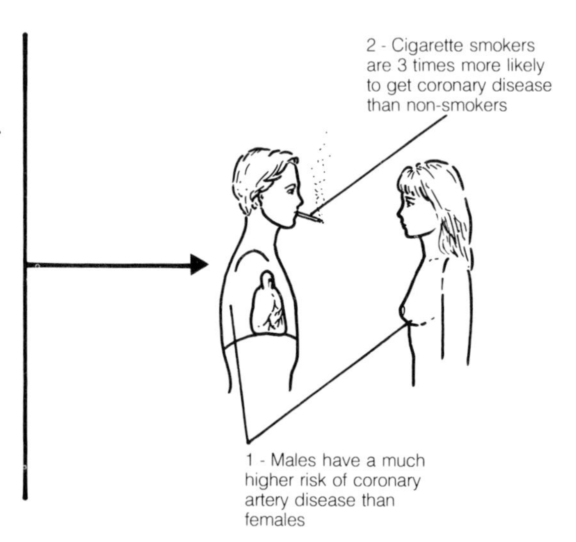
2 - Cigarette smokers are 3 times more likely to get coronary disease than non-smokers
1 - Males have a much higher risk of coronary artery disease than females

Serum cholesterol Hypercholesterolaemia is an important risk factor in atherosclerosis, and is often associated with raised low-density lipoprotein (LDL) concentration. It is important to screen for hyperlipidaemia in young patients with a strong family history of ischaemic heart disease and these patients should be given appropriate dietary advice.

Secondary risk factors

Other risk factors which are thought to be important, though less so than those listed above, include obesity, diabetes mellitus, family history and diet.

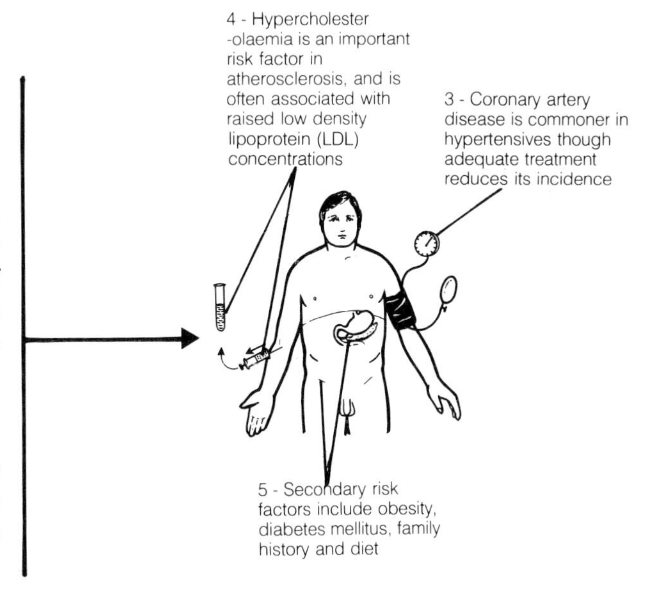
4 - Hypercholester -olaemia is an important risk factor in atherosclerosis, and is often associated with raised low density lipoprotein (LDL) concentrations
3 - Coronary artery disease is commoner in hypertensives though adequate treatment reduces its incidence
5 - Secondary risk factors include obesity, diabetes mellitus, family history and diet

Angina pectoris

The patient with angina pectoris complains of retrosternal chest pain, which is cramplike or constricting in character, radiating to the jaw, left arm, back or epigastrium. It is often precipitated by exercise and relieved within a few minutes by rest. Heavy meals, cold weather or emotion may also precipitate angina. Anginal attacks may be associated with dyspnoea.

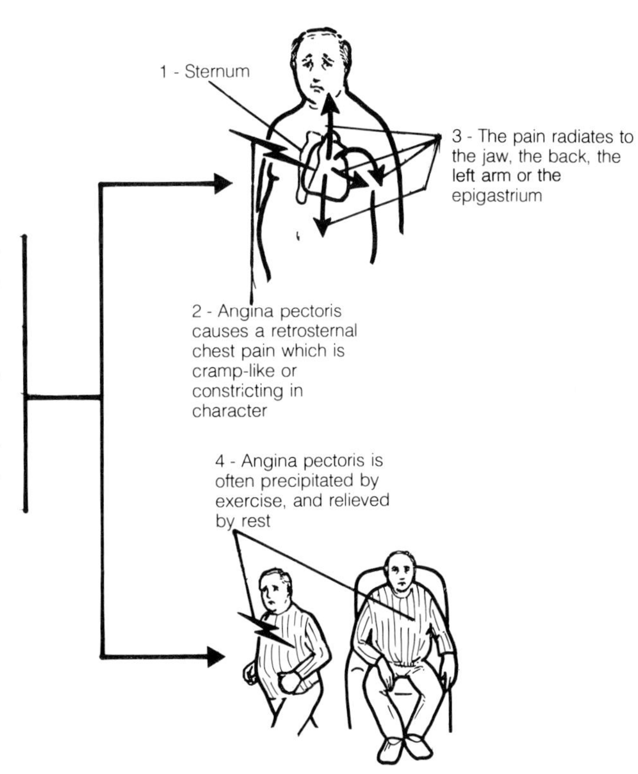
1 - Sternum
3 - The pain radiates to the jaw, the back, the left arm or the epigastrium
2 - Angina pectoris causes a retrosternal chest pain which is cramp-like or constricting in character
4 - Angina pectoris is often precipitated by exercise, and relieved by rest

Examination of patients with angina may reveal no abnormal physical signs. There may be signs of poor left ventricular function, with a third or fourth heart sound or left ventricular dilatation. If there is papillary muscle dysfunction there may be a mitral regurgitant murmur. Risk factors such as hypertension, hyperlipidaemia and diabetes mellitus should be looked for.

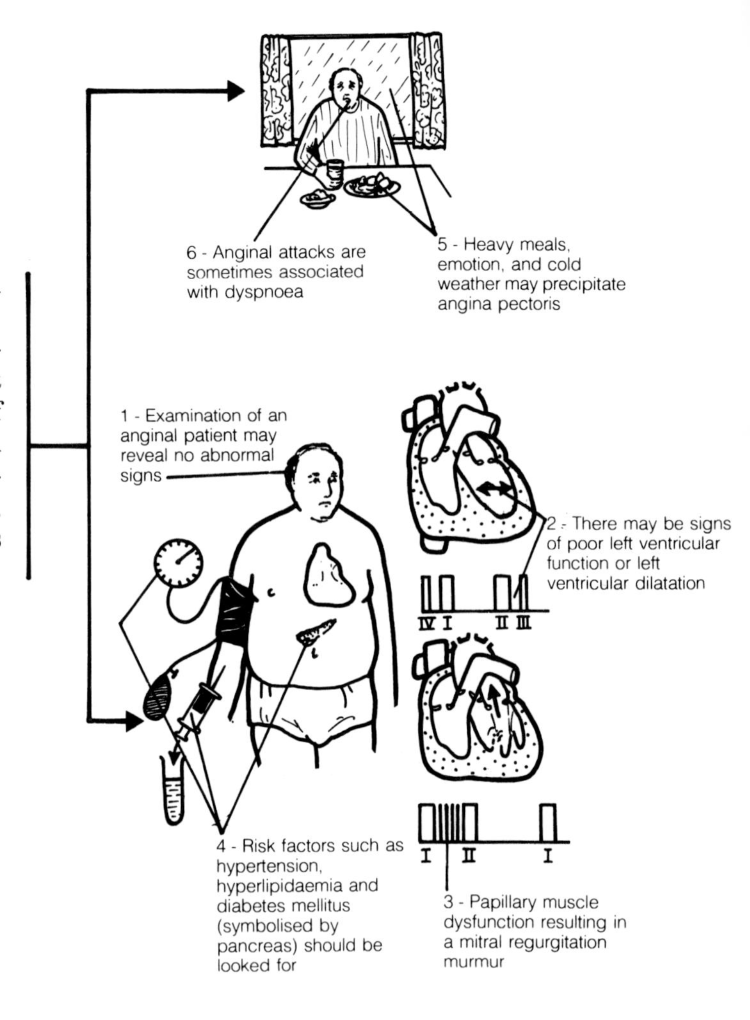
6 - Anginal attacks are sometimes associated with dyspnoea
5 - Heavy meals, emotion, and cold weather may precipitate angina pectoris
1 - Examination of an anginal patient may reveal no abnormal signs
2 - There may be signs of poor left ventricular function or left ventricular dilatation
IV I II III
I II I
4 - Risk factors such as hypertension, hyperlipidaemia and diabetes mellitus (symbolised by pancreas) should be looked for
3 - Papillary muscle dysfunction resulting in a mitral regurgitation murmur

Investigations

The resting ECG in patients with angina pectoris is often normal. There may be non-specific ECG changes such as flattening of the ST segment and T-wave inversion. An ECG taken while the patient is having pain may show ST segment depression.

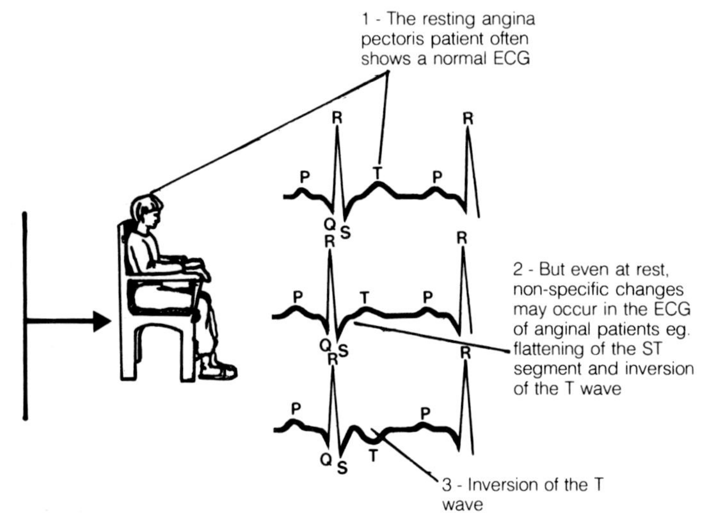
1 - The resting angina pectoris patient often shows a normal ECG
R
P
T
Q
S
2 - But even at rest, non-specific changes may occur in the ECG of anginal patients eg. flattening of the ST segment and inversion of the T wave
3 - Inversion of the T wave

The exercise ECG

A patient with a normal resting ECG may develop ST segment depression on exercise, which is very suggestive of myocardial ischaemia. However a normal exercise ECG does not exclude angina. Hypotension occuring during exercise usually indicates extensive coronary artery disease.

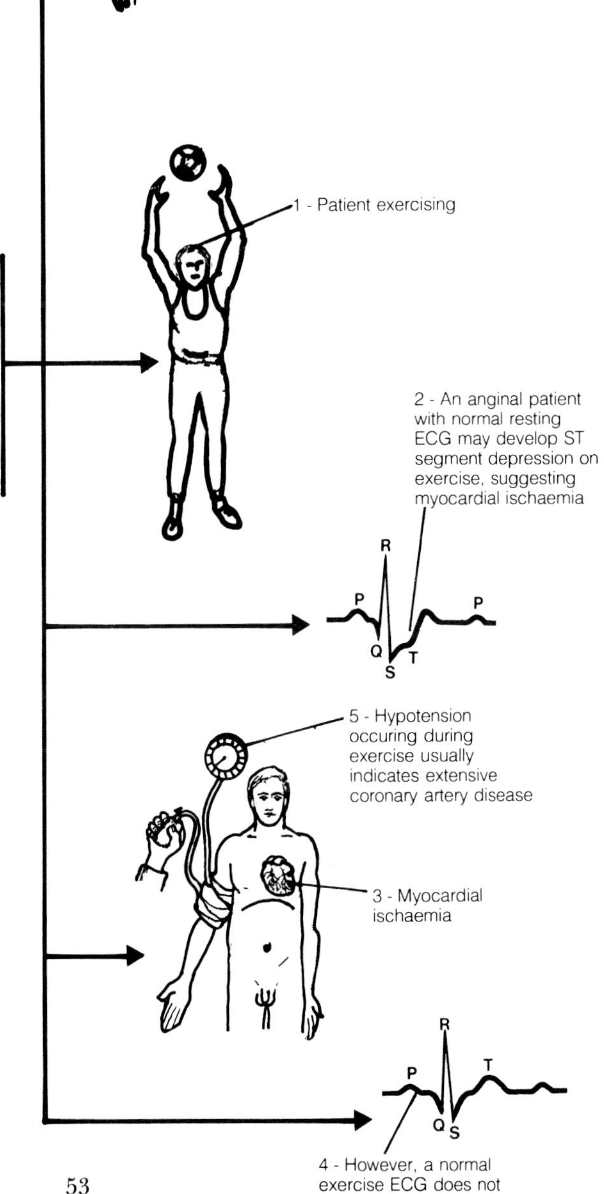
R
R
P
T
P
Q
S
Q
S
5 - ST depression during anginal attack
4 - During an anginal attack, the ECG may show ST depression
1 - Patient exercising
2 - An anginal patient with normal resting ECG may develop ST segment depression on exercise, suggesting myocardial ischaemia
R
P
P
Q
T
S
5 - Hypotension occuring during exercise usually indicates extensive coronary artery disease
3 - Myocardial ischaemia
R
P
T
Q
S
4 - However, a normal exercise ECG does not exclude angina pectoris

Coronary arteriography is very valuable in the diagnosis and management of patients with coronary artery disease.

Isotope scanning is another useful technique in assessing patients with angina.

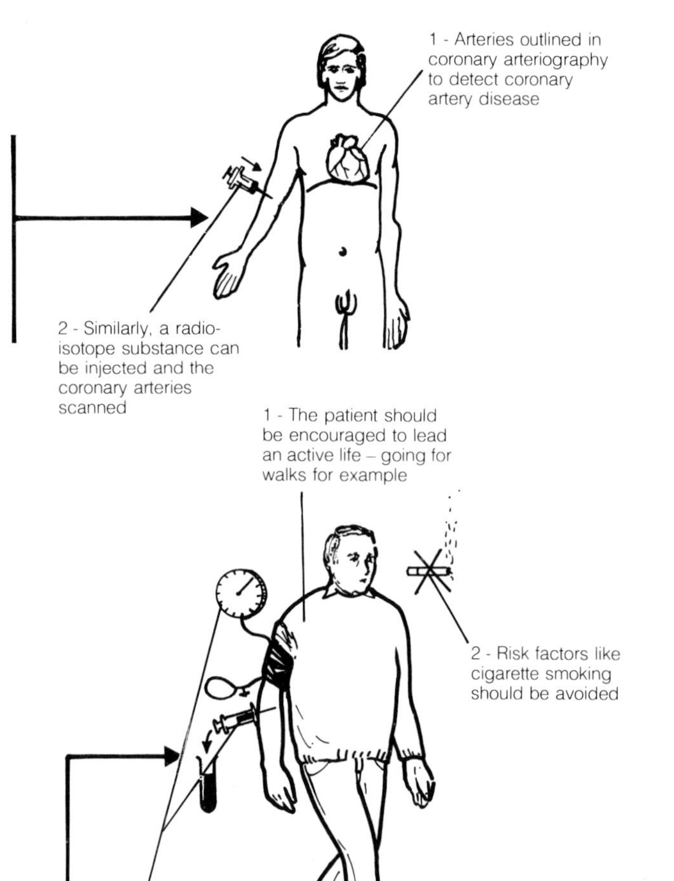
1 - Arteries outlined in coronary arteriography to detect coronary artery disease
2 - Similarly, a radio-isotope substance can be injected and the coronary arteries scanned

Treatment of angina pectoris

The patient should be encouraged to lead as active a life as possible. Risk factors such as cigarette smoking should be avoided and hypertension or hyperlipidaemias if present should be treated.

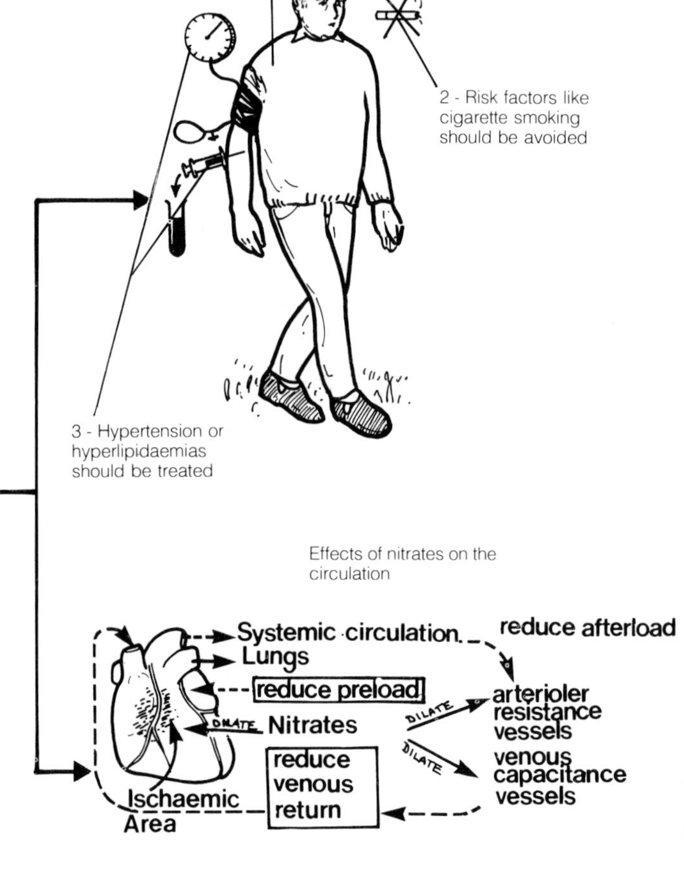
1 - The patient should be encouraged to lead an active life – going for walks for example
2 - Risk factors like cigarette smoking should be avoided
3 - Hypertension or hyperlipidaemias should be treated
Effects of nitrates on the circulation
Systemic circulation
reduce afterload
Lungs
reduce preload
Nitrates
DILATE
arterioler resistance vessels
venous capacitance vessels
reduce venous return
Ischaemic Area

Nitrates

Nitrates have been used for many years in treatment of angina pectoris. They cause vasodilation in ischaemic areas, as well as relieving coronary artery spasm. They also cause peripheral vasodilation, relieving both **preload** and **afterload** of the heart.

Sublingual glyceryl trinitrate (GTN) is very useful in relieving anginal attacks. It should be taken sublingually (0.3 – 0.5mg) every 3 minutes until the pain subsides. It is also effective if taken prophylactically before exertion. Side-effects include headaches, flushing, and occasionally hypotension. Isosorbide dinitrate is a long acting nitrate which has similar effects to GTN. The slow release preparation, given in a dosage of 20-40mg two-three times daily, is useful in angina prophylaxis.

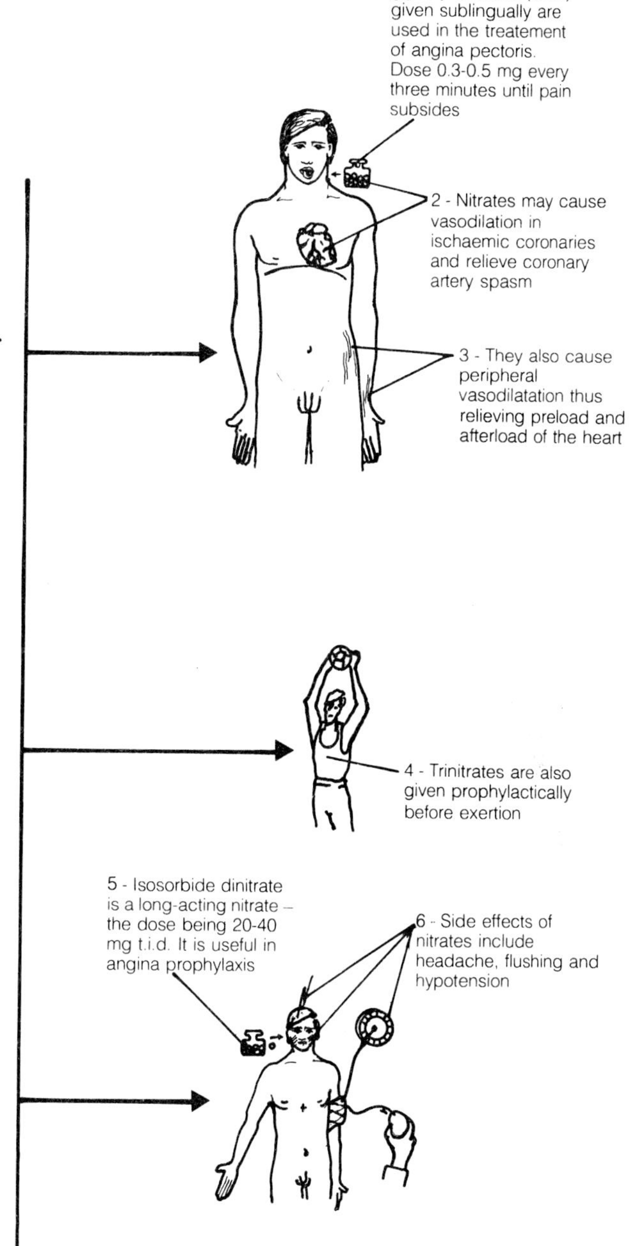
1 - Nitrates such as glyceryl trinitrate (GTN) given sublingually are used in the treatement of angina pectoris. Dose 0.3-0.5 mg every three minutes until pain subsides
2 - Nitrates may cause vasodilation in ischaemic coronaries and relieve coronary artery spasm
3 - They also cause peripheral vasodilatation thus relieving preload and afterload of the heart
4 - Trinitrates are also given prophylactically before exertion
5 - Isosorbide dinitrate is a long-acting nitrate – the dose being 20-40 mg t.i.d. It is useful in angina prophylaxis
6 - Side effects of nitrates include headache, flushing and hypotension

Betablockers

Betablockers are very effective in reducing anginal symptoms, and in combination with nitrates are standard therapy for angina pectoris. They reduce myocardial oxygen consumption by reducing the heart rate at any given exercise level and depressing contractility. Most betablockers seem to be equally effective against angina. The patient is usually started on a low dose (for example propranolol 40mg three times a day). The dosage is gradually increased until symptoms are controlled or until there is a bradycardia of 55-60 beats/minute.

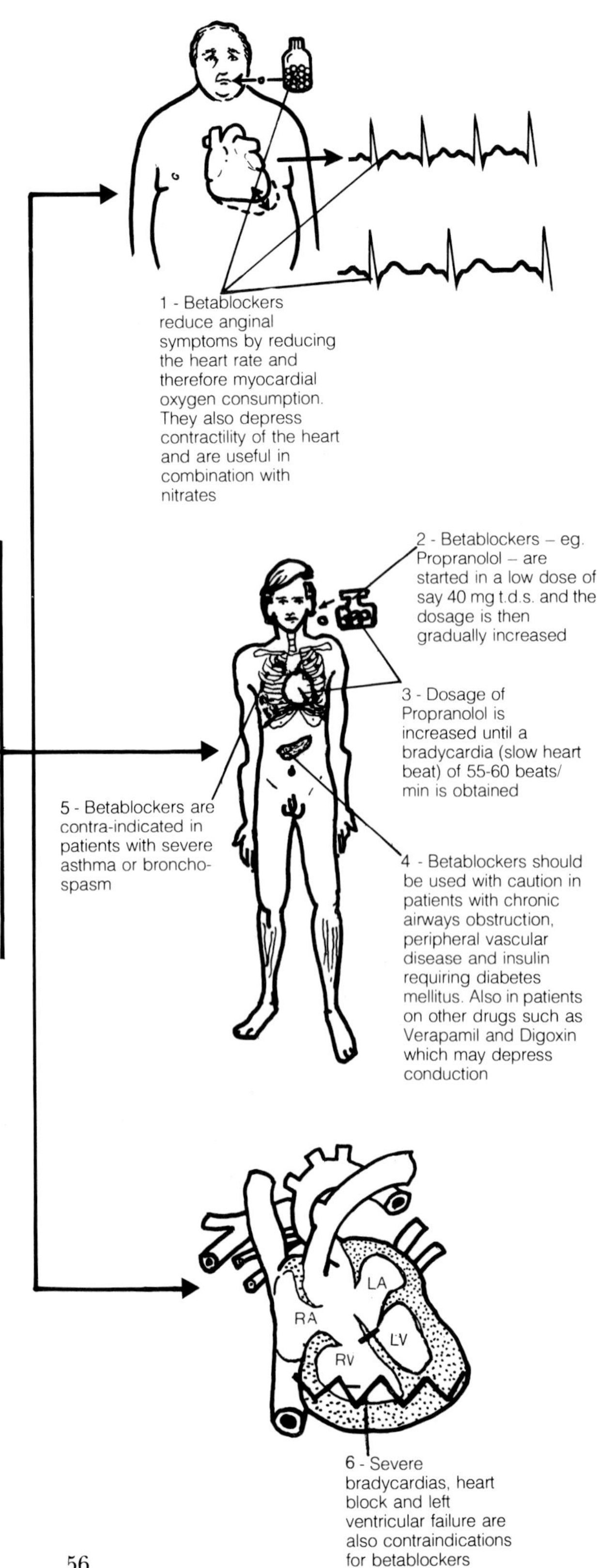
1 - Betablockers reduce anginal symptoms by reducing the heart rate and therefore myocardial oxygen consumption. They also depress contractility of the heart and are useful in combination with nitrates
2 - Betablockers – eg. Propranolol – are started in a low dose of say 40 mg t.d.s. and the dosage is then gradually increased
3 - Dosage of Propranolol is increased until a bradycardia (slow heart beat) of 55-60 beats/ min is obtained
4 - Betablockers should be used with caution in patients with chronic airways obstruction, peripheral vascular disease and insulin requiring diabetes mellitus. Also in patients on other drugs such as Verapamil and Digoxin which may depress conduction
5 - Betablockers are contra-indicated in patients with severe asthma or broncho-spasm
LA
RA
LV
RV
6 - Severe bradycardias, heart block and left ventricular failure are also contraindications for betablockers

Betablockers have many side-effects and are absolutely contra-indicated in patients with severe asthma or bronchospasm, severe bradycardias, left ventricular failure and heart block. They should be used with caution in patients with chronic airways obstruction, peripheral vascular disease, insulin-requiring diabetes mellitus, and in patients on other drugs which may depress conduction such as verapamil and Digoxin.

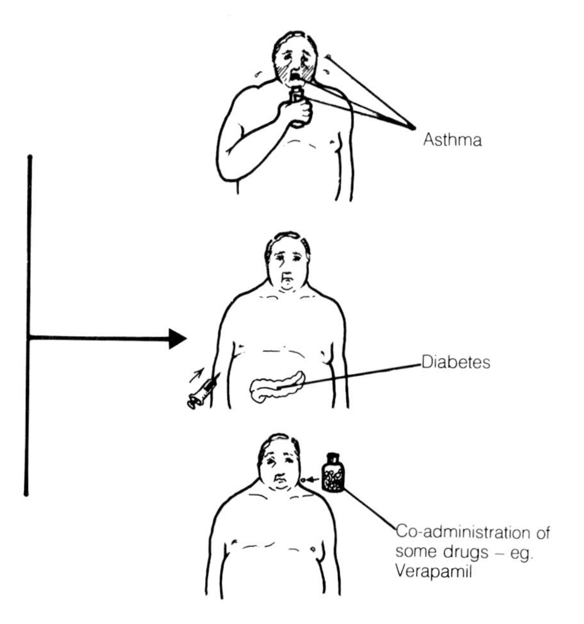
Asthma
Diabetes
Co-administration of some drugs – eg. Verapamil

Calcium antagonists

These drugs cause vasodilatation of the coronary arteries and peripheral vessels. They are generally used as third line agents in the treatment of angina. Nifedipine is used in a dosage of 10-20mg three times daily and is more powerful than verapamil in its action on coronary smooth muscle.

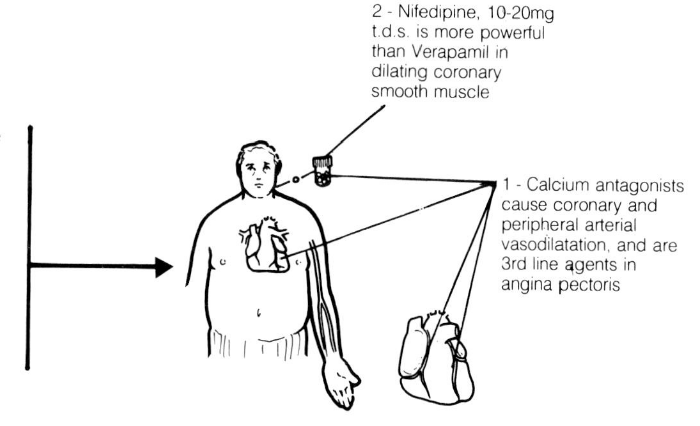
2 - Nifedipine, 10-20mg t.d.s. is more powerful than Verapamil in dilating coronary smooth muscle
1 - Calcium antagonists cause coronary and peripheral arterial vasodilatation, and are 3rd line agents in angina pectoris

Coronary artery bypass grafting is performed if medical treatment for angina fails or in patients demonstrated on angiography to have lesions with a serious prognosis. These include triple vessel disease, patients with over 50% stenosis of the left anterior descending artery as part of either two or three vessel disease, or patients with left main stem disease. Mortality from coronary artery bypass surgery is now low (1.3-2.3%), sixty per cent of patients achieve complete relief of angina after surgery and 90% are symptomatically improved. In the European vein graft trial, patients with extensive disease who had been surgically treated had a better eight year survival than the medically treated group.

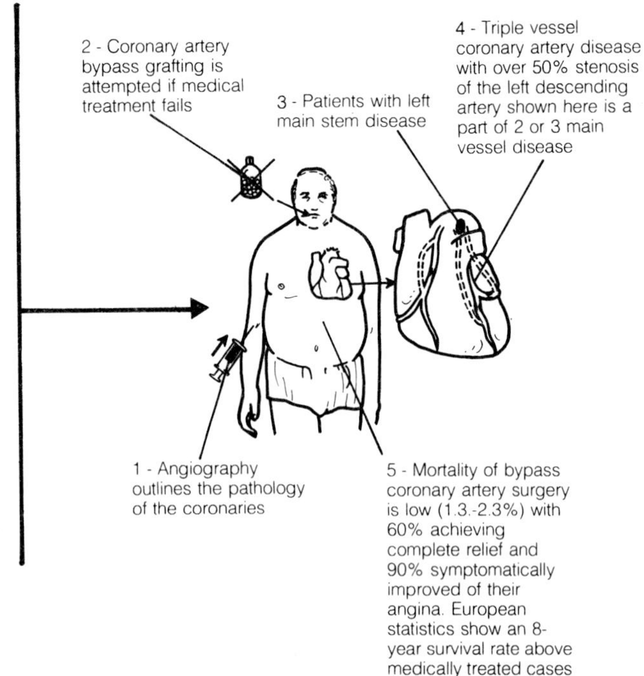
2 - Coronary artery bypass grafting is attempted if medical treatment fails
3 - Patients with left main stem disease
4 - Triple vessel coronary artery disease with over 50% stenosis of the left descending artery shown here is a part of 2 or 3 main vessel disease
1 - Angiography outlines the pathology of the coronaries
5 - Mortality of bypass coronary artery surgery is low (1.3.-2.3%) with 60% achieving complete relief and 90% symptomatically improved of their angina. European statistics show an 8-year survival rate above medically treated cases

Unstable angina pectoris

Patients with unstable angina are those who are at high risk of going on to develop myocardial infarction. They include;

(1) Angina of recent onset (within the preceding month).

(2) Changing pattern of angina of effort, with increasing frequency and severity of chest pains.

(3) Angina at rest, or lasting more than 15 minutes.

(4) Recurrent angina post-myocardial infarction.

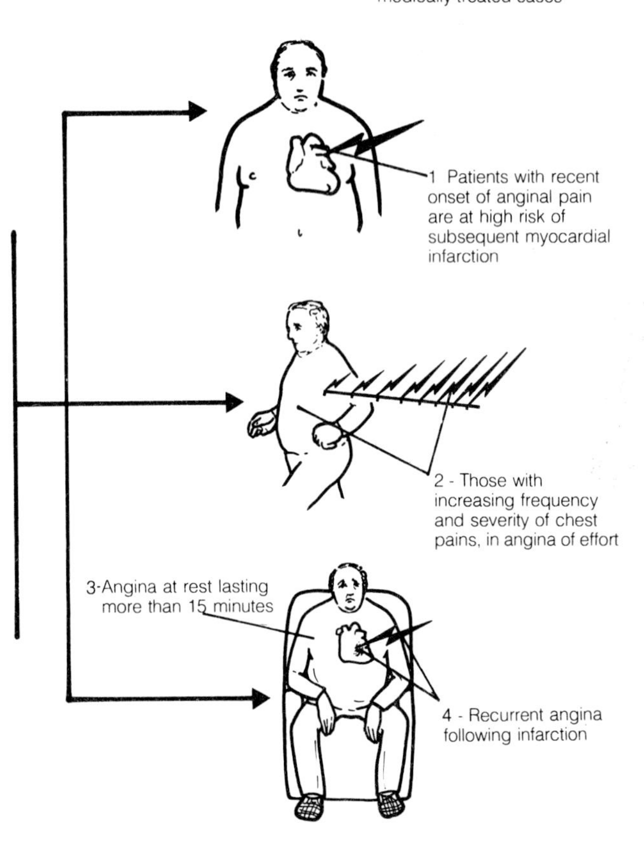
1 Patients with recent onset of anginal pain are at high risk of subsequent myocardial infarction
2 - Those with increasing frequency and severity of chest pains, in angina of effort
3-Angina at rest lasting more than 15 minutes
4 - Recurrent angina following infarction

These patients should be put on strict bed rest in a Coronary Care Unit and monitored. Serial ECGs and cardiac enzymes should be measured and therapy should include regular nitrates, betablockers or calcium antagonists if coronary spasm is suspected. If oral therapy fails to control the angina, intravenous nitrates can be infused. Most patients respond to this management but a small group require surgery.

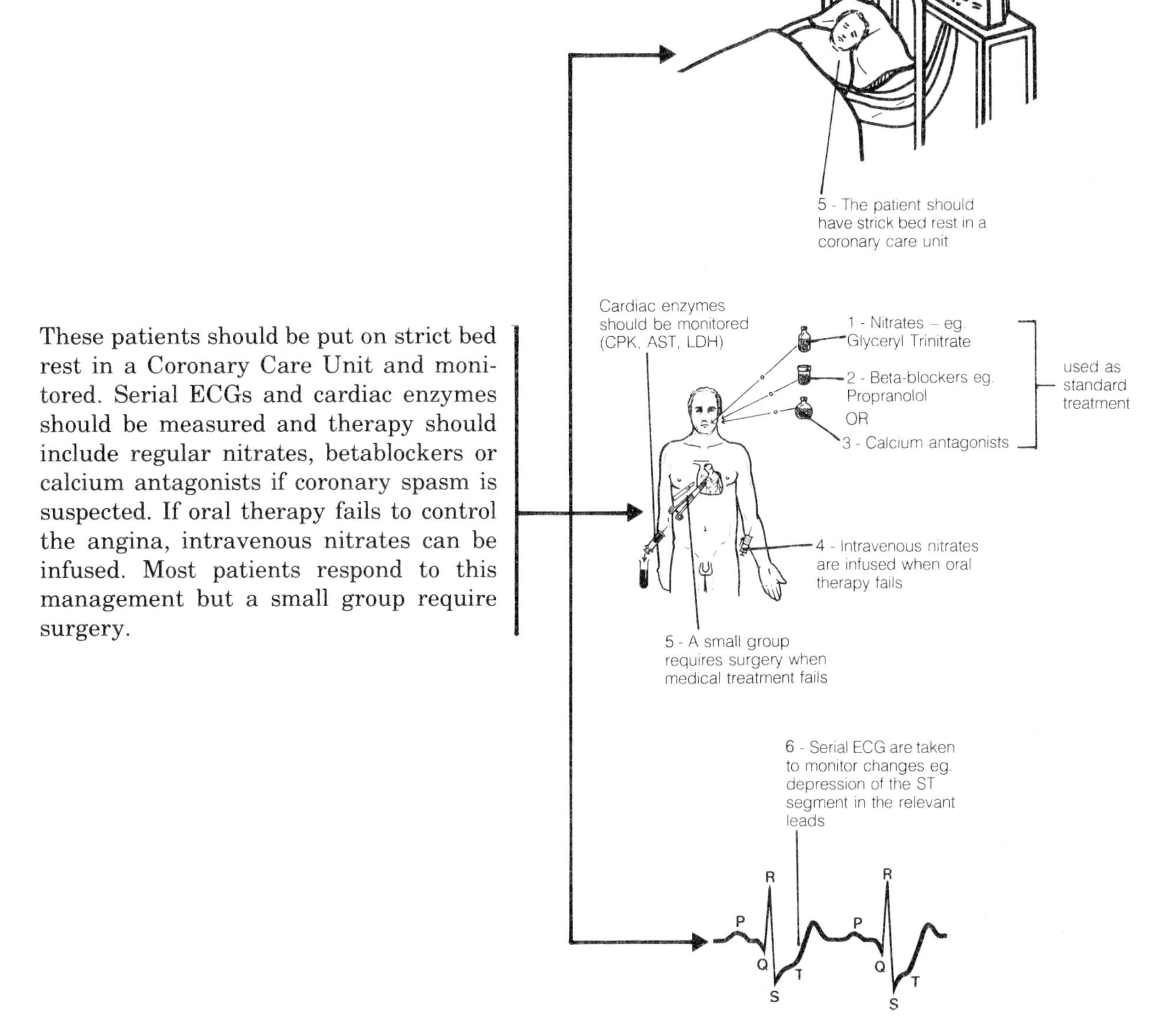
5 - The patient should have strick bed rest in a coronary care unit
Cardiac enzymes should be monitored (CPK, AST, LDH)
1 - Nitrates – eg. Glyceryl Trinitrate
2 - Beta-blockers eg. Propranolol
OR
3 - Calcium antagonists
used as standard treatment
4 - Intravenous nitrates are infused when oral therapy fails
5 - A small group requires surgery when medical treatment fails
6 - Serial ECG are taken to monitor changes eg. depression of the ST segment in the relevant leads
R
P
Q
S
T
R
P
Q
S
T

Prinzmetal's angina

These patients develop anginal chest pain associated with ST elevation on the ECG and no enzyme changes. Coronary angiography may show spasm, but no evidence of obstruction. The condition responds well to nitrates and calcium antagonists, but symptoms may worsen on betablockers.

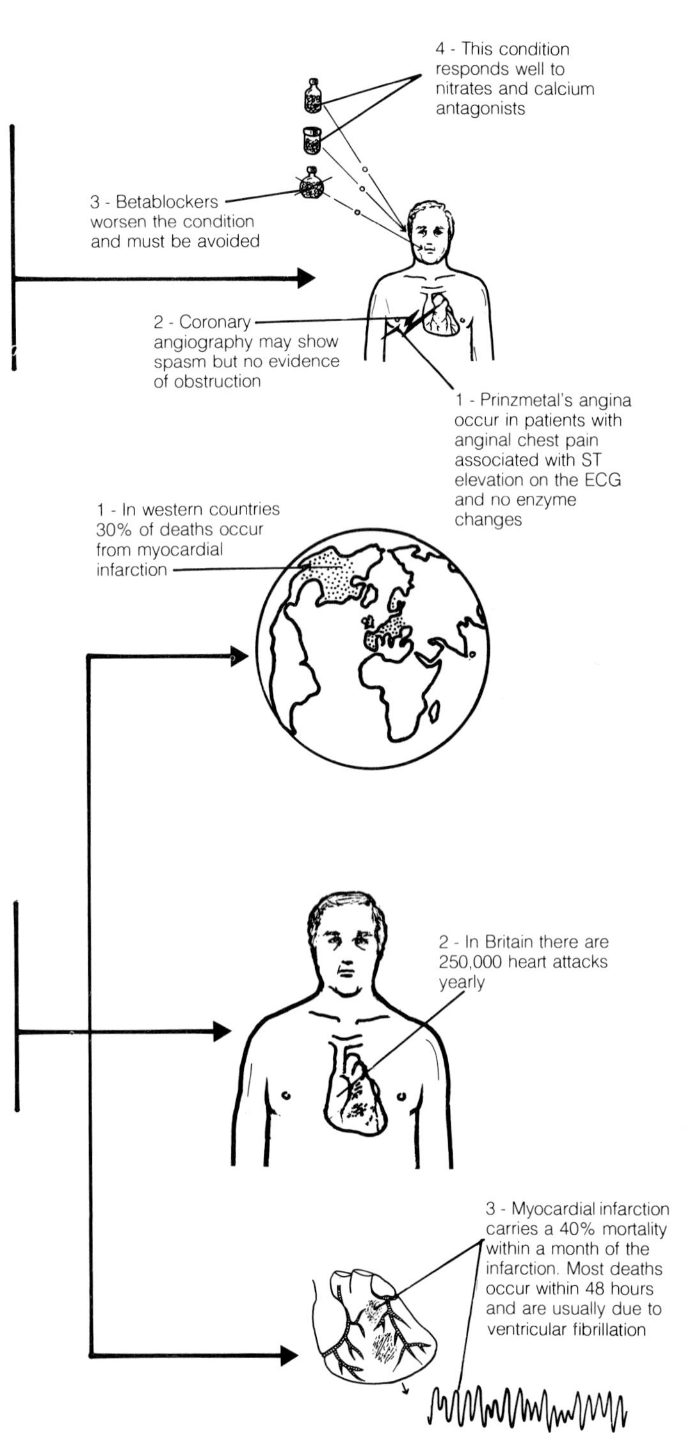
4 - This condition responds well to nitrates and calcium antagonists
3 - Betablockers worsen the condition and must be avoided
2 - Coronary angiography may show spasm but no evidence of obstruction
1 - Prinzmetal's angina occur in patients with anginal chest pain associated with ST elevation on the ECG and no enzyme changes
1 - In western countries 30% of deaths occur from myocardial infarction
2 - In Britain there are 250,000 heart attacks yearly
3 - Myocardial infarction carries a 40% mortality within a month of the infarction. Most deaths occur within 48 hours and are usually due to ventricular fibrillation

MYOCARDIAL INFARCTION

In western countries 30% of all deaths are due to myocardial infarction; in Britain there are 250,000 heart attacks each year. There is a 40% mortality within the first month after infarction; most deaths occur within the first 48 hours and are usually due to ventricular fibrillation.

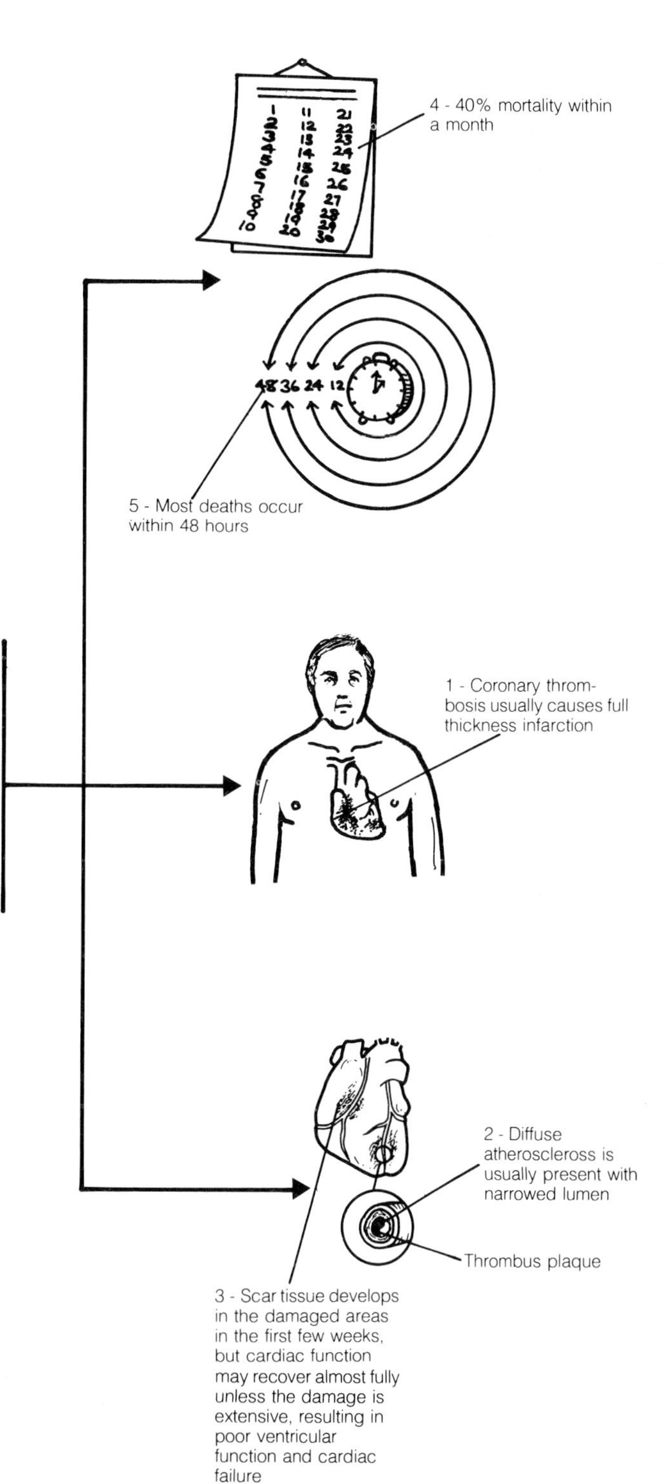
4 - 40% mortality within a month
5 - Most deaths occur within 48 hours
1 - Coronary thrombosis usually causes full thickness infarction
2 - Diffuse atheroscleross is usually present with narrowed lumen
Thrombus plaque
3 - Scar tissue develops in the damaged areas in the first few weeks, but cardiac function may recover almost fully unless the damage is extensive, resulting in poor ventricular function and cardiac failure

Pathology

Coronary thrombosis is usually the cause of full thickness infarction and these patients usually have diffuse atherosclerosis. After the first few weeks, scar tissue develops in the damaged areas of the heart. Cardiac function may recover almost fully, but if there has been extensive damage, ventricular function may remain poor, resulting in cardiac failure.

Clinical features of myocardial infarction

The patient usually develops severe central chest pain which he describes as crushing or choking in nature. The pain may radiate to the jaw and left arm or be accompanied by heaviness or tingling in both arms. The pain usually lasts for over half an hour and may be accompanied by nausea and profuse sweating. The patient may also complain of giddiness, faintness or dyspnoea.

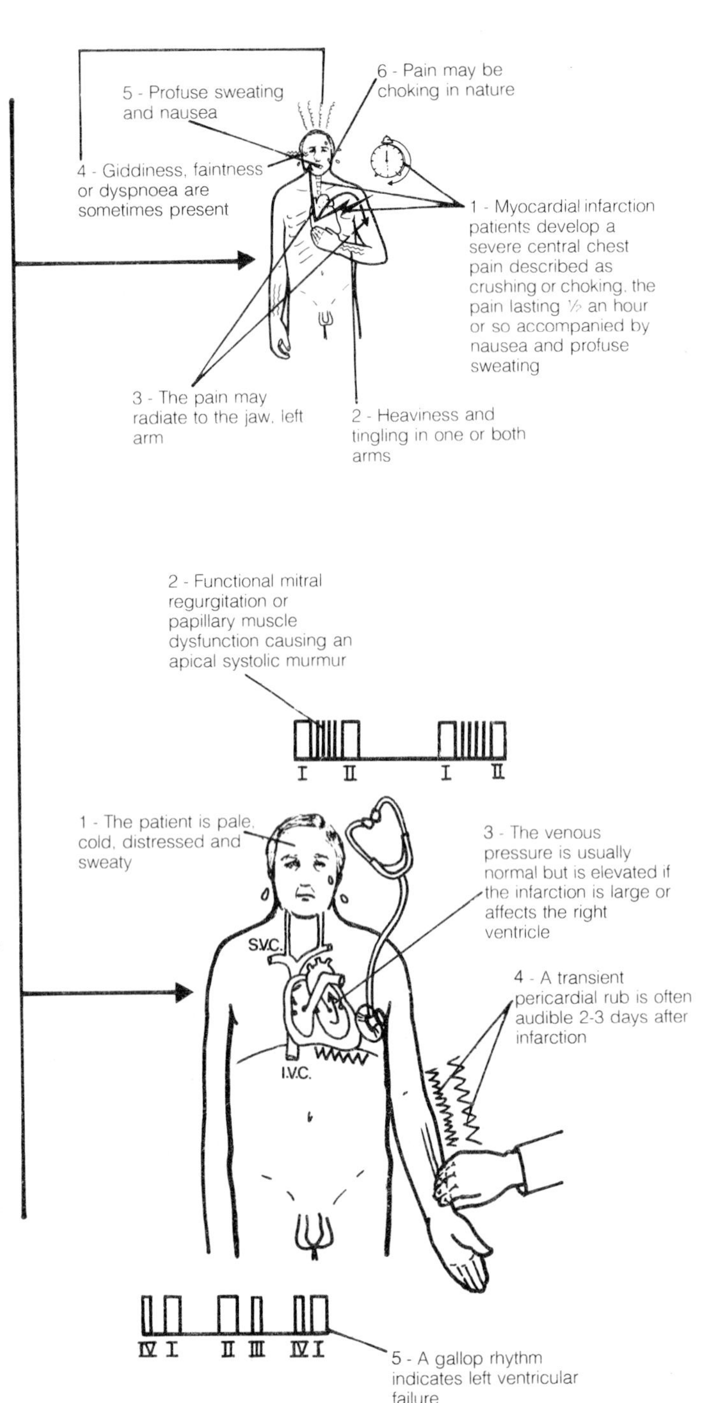
5 - Profuse sweating and nausea
6 - Pain may be choking in nature
4 - Giddiness, faintness or dyspnoea are sometimes present
1 - Myocardial infarction patients develop a severe central chest pain described as crushing or choking, the pain lasting ½ an hour or so accompanied by nausea and profuse sweating
3 - The pain may radiate to the jaw, left arm
2 - Heaviness and tingling in one or both arms
2 - Functional mitral regurgitation or papillary muscle dysfunction causing an apical systolic murmur
I II I II
1 - The patient is pale, cold, distressed and sweaty
3 - The venous pressure is usually normal but is elevated if the infarction is large or affects the right ventricle
S.V.C.
I.V.C.
4 - A transient pericardial rub is often audible 2-3 days after infarction
IV I II III IV I
5 - A gallop rhythm indicates left ventricular failure

Physical signs

The patient may be distressed, pale, sweaty and cold. There may be a tachycardia or bradycardia. Venous pressure is usually normal, but may be elevated if the infarction is large or affects the right ventricle. A gallop rhythm may be present, indicative of left ventricular failure and a transient apical systolic murmur sometimes occurs due to functional mitral regurgitation or papillary muscle dysfunction. Rarely, a loud pansystolic murmur can be heard due to rupture of the ventricular septum. A transient pericardial rub is sometimes audible two to three days after infarction. Crackles may be audible at the lung bases and if widespread and not cleared by coughing indicate left ventricular failure. Fever often occurs in patients within 24 hours of infarction and settles within a week.

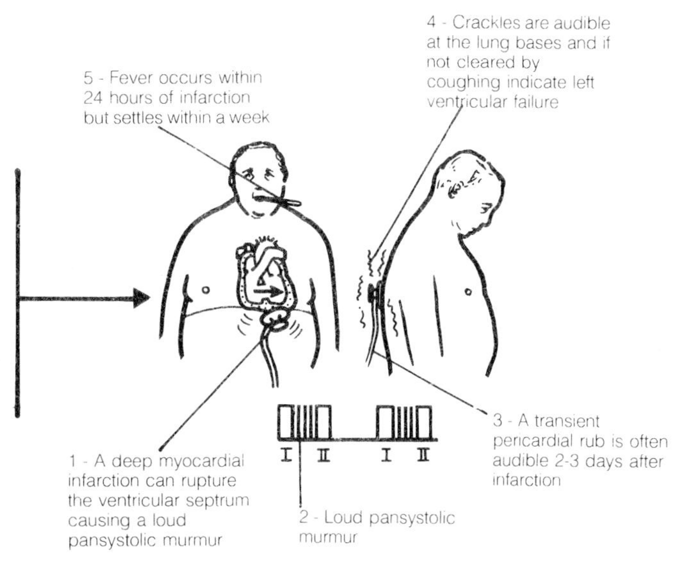
5 - Fever occurs within 24 hours of infarction but settles within a week
4 - Crackles are audible at the lung bases and if not cleared by coughing indicate left ventricular failure
I II I II
1 - A deep myocardial infarction can rupture the ventricular septrum causing a loud pansystolic murmur
2 - Loud pansystolic murmur
3 - A transient pericardial rub is often audible 2-3 days after infarction

Investigations

The ECG may be normal in the early stages, but within hours, ST elevation is seen in the leads over the affected surface of the heart. Deep, wide, Q waves usually develop within the first 24 hours, and indicate transmural infarction. Over the next few days, T wave inversion occurs, and ST elevation subsides.

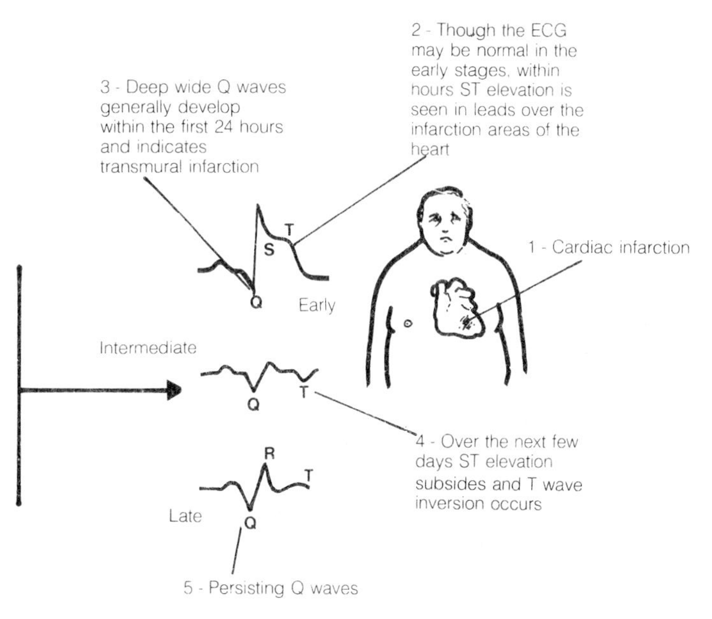
3 - Deep wide Q waves generally develop within the first 24 hours and indicates transmural infarction
2 - Though the ECG may be normal in the early stages, within hours ST elevation is seen in leads over the infarction areas of the heart
S
T
Q
Early
1 - Cardiac infarction
Intermediate
Q
T
4 - Over the next few days ST elevation subsides and T wave inversion occurs
R
T
Late
Q
5 - Persisting Q waves

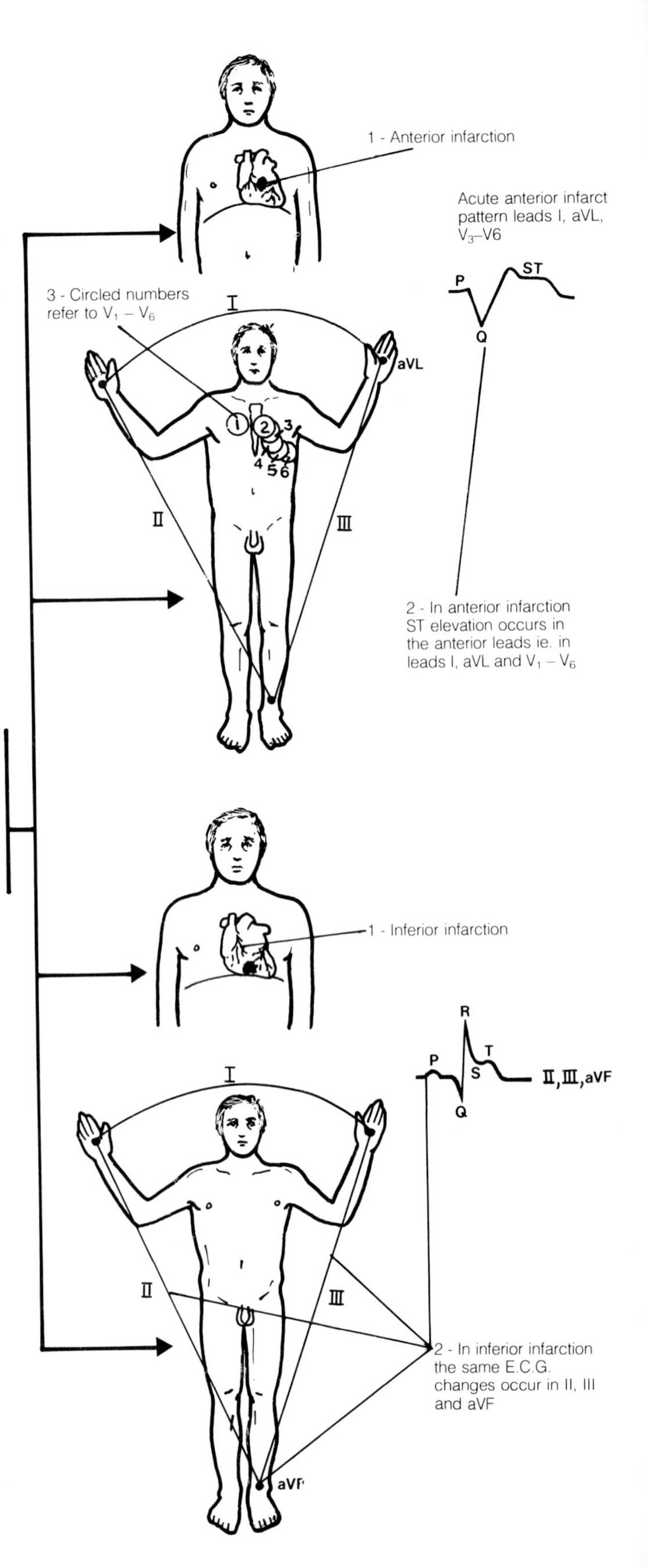
1 - Anterior infarction
Acute anterior infarct pattern leads I, aVL, V3–V6
P
ST
Q
3 - Circled numbers refer to V1 – V6
I
aVL
1
2
3
4
5
6
II
III
2 - In anterior infarction ST elevation occurs in the anterior leads ie. in leads I, aVL and V1 – V6
1 - Inferior infarction
R
P
T
S
Q
II, III, aVF
I
II
III
2 - In inferior infarction the same E.C.G. changes occur in II, III and aVF
aVF

Both usually return to normal within a few weeks, although the Q waves persist. In **anterior infarction,** ST elevation occurs in the anterior leads, ie. leads I, aVL and VI-V6. In **inferior infarction** the changes occur in leads II, II and aVF.

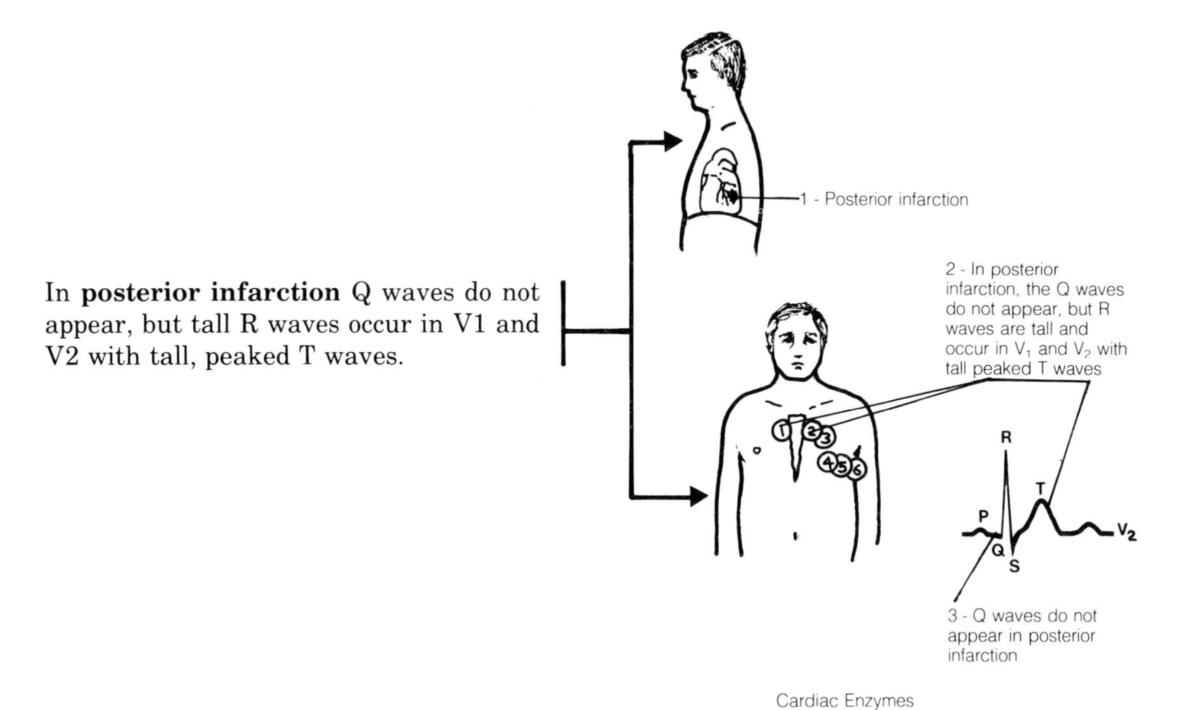
1 - Posterior infarction
2 - In posterior infarction, the Q waves do not appear, but R waves are tall and occur in V_1 and V_2 with tall peaked T waves
R
T
P
Q
S
V_2
3 - Q waves do not appear in posterior infarction

In **posterior infarction** Q waves do not appear, but tall R waves occur in V1 and V2 with tall, peaked T waves.

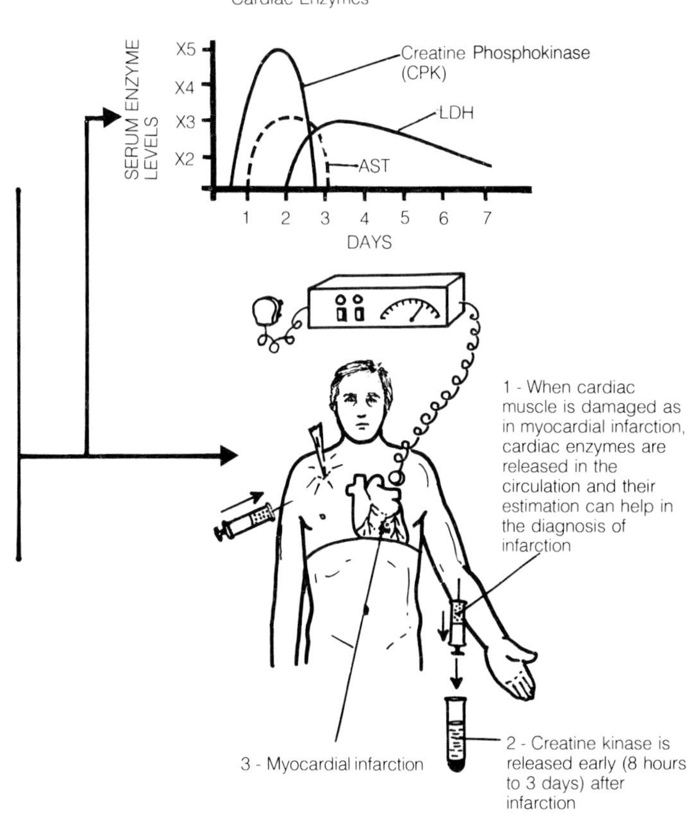
Cardiac Enzymes
SERUM ENZYME LEVELS
X5
X4
X3
X2
Creatine Phosphokinase (CPK)
LDH
AST
1 2 3 4 5 6 7
DAYS
1 - When cardiac muscle is damaged as in myocardial infarction, cardiac enzymes are released in the circulation and their estimation can help in the diagnosis of infarction
2 - Creatine kinase is released early (8 hours to 3 days) after infarction
3 - Myocardial infarction

Cardiac enzymes

When cardiac muscle damage occurs, enzymes are released into the circulation, and their estimation can help in making the diagnosis of myocardial infarction. Creatine kinase is released early (8 hours – 3 days) after infarction, but it is also present in skeletal muscle and levels will be raised after an intramuscular injection, muscle trauma or electrical cardioversion. In these cases estimation of the CK-NB isoenzyme fraction is useful, as this is specific to cardiac muscle.

Aspartate aminotransferase (AST) is elevated 1-5 days after myocardial infarction. It is found in the liver as well as in the heart and high levels can occur in hepatic congestion.

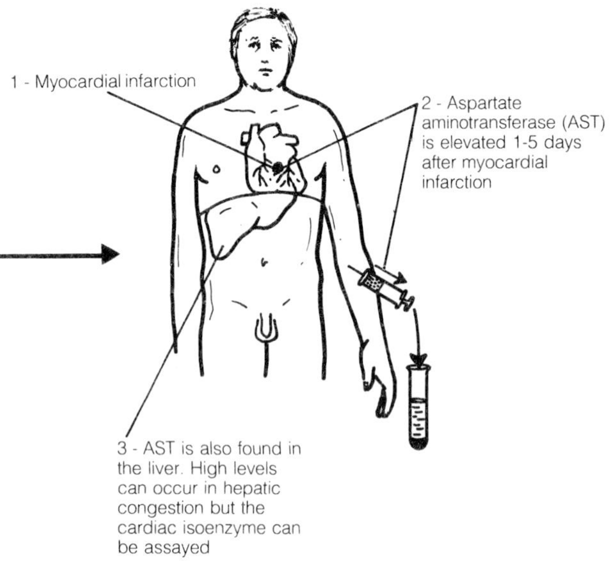

1 - Myocardial infarction

2 - Aspartate aminotransferase (AST) is elevated 1-5 days after myocardial infarction

3 - AST is also found in the liver. High levels can occur in hepatic congestion but the cardiac isoenzyme can be assayed

Lactate dehydrogenase (LDH) is released 2-10 days after infarction, but false positives can occur. A cardiac isoenzyme is available which can be assayed if necessary.

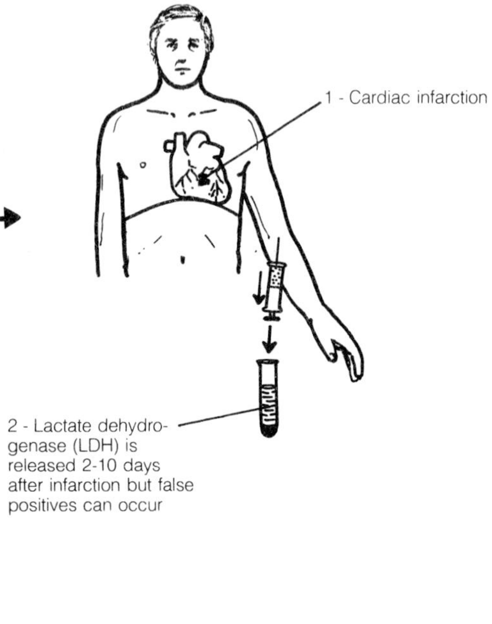

1 - Cardiac infarction

2 - Lactate dehydrogenase (LDH) is released 2-10 days after infarction but false positives can occur

Electrolytes

It has been shown that patients are more likely to develop arrhythmias after infarction if they are hypokalaemic. It is therefore important to estimate the serum potassium and correct it if necessary.

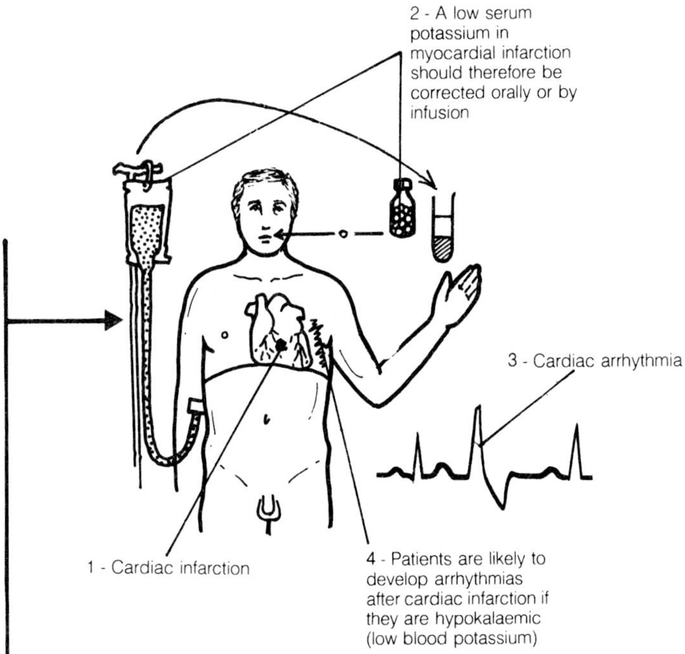

2 - A low serum potassium in myocardial infarction should therefore be corrected orally or by infusion
3 - Cardiac arrhythmia
1 - Cardiac infarction
4 - Patients are likely to develop arrhythmias after cardiac infarction if they are hypokalaemic (low blood potassium)

Chest X-ray may show evidence of pulmonary oedema. There may be evidence of aortic dissection which is an important differential diagnosis of chest pain.

3 - Aortic dissection is an important differential diagnosis of chest pain
1 - Chest X-ray showing evidence of pulmonary oedema
2 - "Bat-wing" shadows radiating form the hilar area

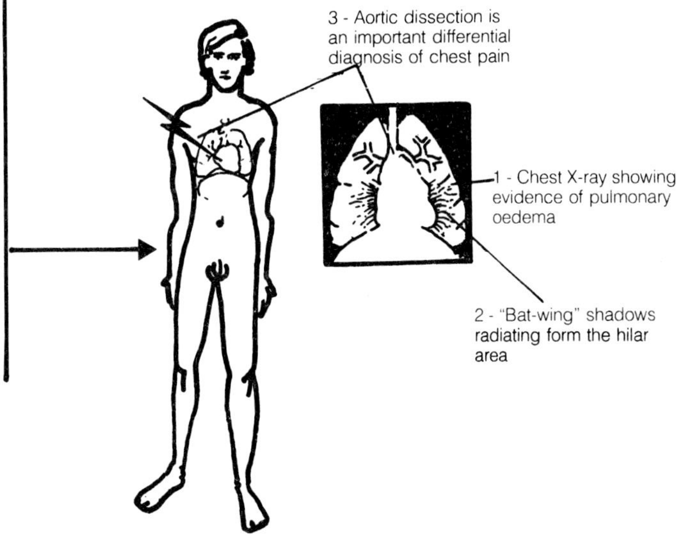

Complete heart block in inferior infarction is usually benign, and pacing is only required if there is hypotension. In anterior infarction, the outlook is less good, and pacing is usually required.

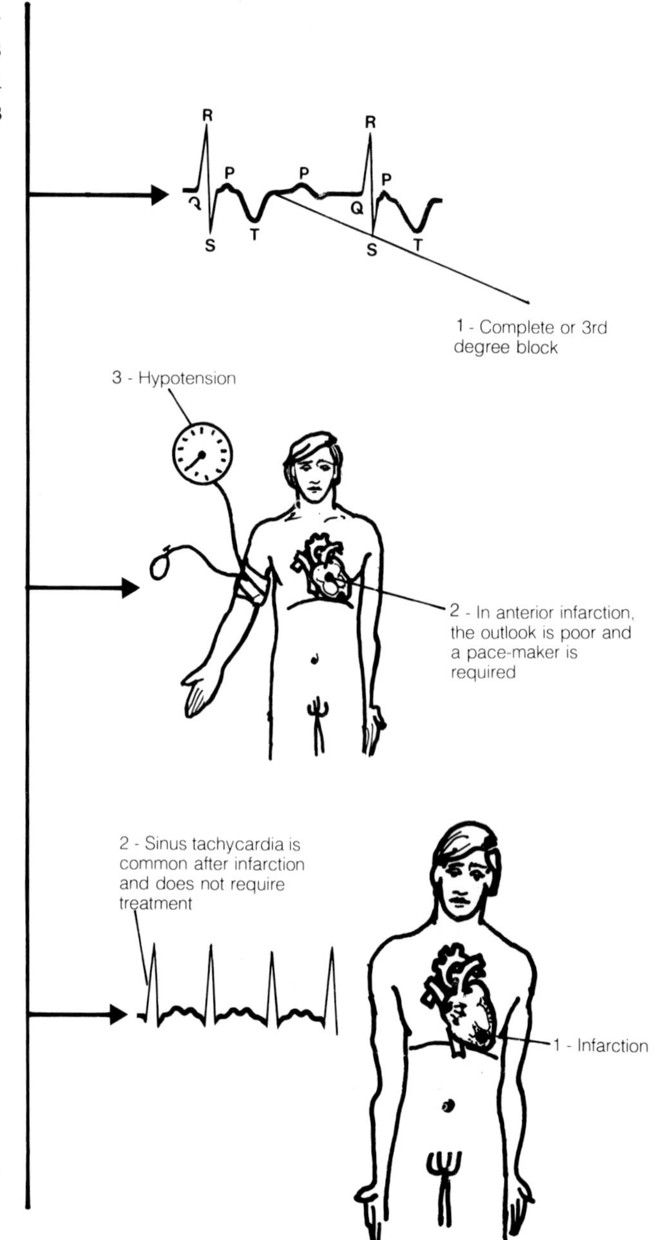
R
P
Q
S
T
1 - Complete or 3rd degree block
3 - Hypotension
2 - In anterior infarction, the outlook is poor and a pace-maker is required
2 - Sinus tachycardia is common after infarction and does not require treatment
1 - Infarction

Sinus tachycardia is common after infarction and does not require treatment.

Treatment

Strict bed rest is needed for the first 24 hours, when the patient should be monitored on a Coronary Care Unit. Pain relief is important and diamorphine should be given intravenously in doses of 2.5 to 5mg as often as necessary. An antiemetic such as prochlorperazine 12.5mg iv or im is usually required. Heparin is often given subcutaneously (5000 units twice daily) while the patient is on bed rest. Provided there are no complications, the patient is allowed to sit out of bed after 24 hours and use the commode at the bedside. Gentle mobilization should begin at 3 days and the patient should be ready for discharge by the eighth or tenth day.

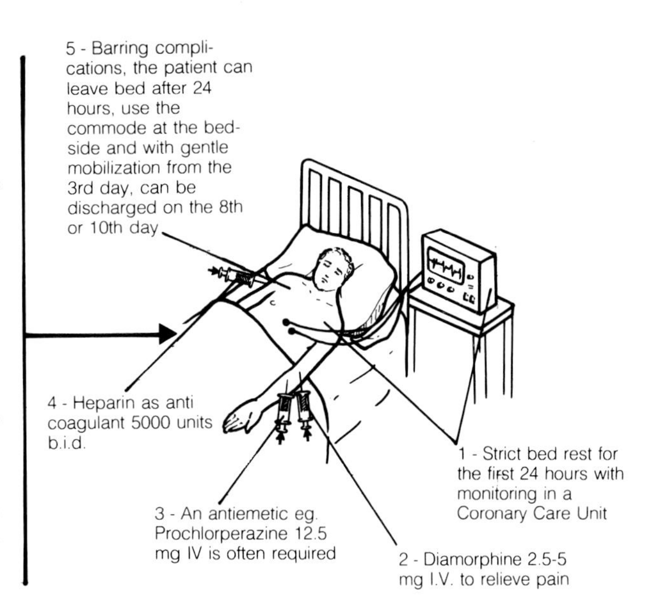
5 - Barring complications, the patient can leave bed after 24 hours, use the commode at the bedside and with gentle mobilization from the 3rd day, can be discharged on the 8th or 10th day
4 - Heparin as anti coagulant 5000 units b.i.d.
3 - An antiemetic eg. Prochlorperazine 12.5 mg IV is often required
1 - Strict bed rest for the first 24 hours with monitoring in a Coronary Care Unit
2 - Diamorphine 2.5-5 mg I.V. to relieve pain

2 - Hypotension occurs at times with sinus bradycardia

3 - Atropine 0.6-1.2 mg I.V. is effective in treating sinus bradycardia

1 - Coronary infarction

4 - Sinus bradycardia

Complications of infarction

Arrhythmias are common after infarction, and should be treated if the cardiac output is compromised. **Sinus bradycardia** is common and requires treatment only if causing hypotension. Atropine 0.6-1.2mg iv is usually effective.

First degree heart block is common with inferior infarction and treatment with intravenous atropine is only necessary if hypotension occurs.

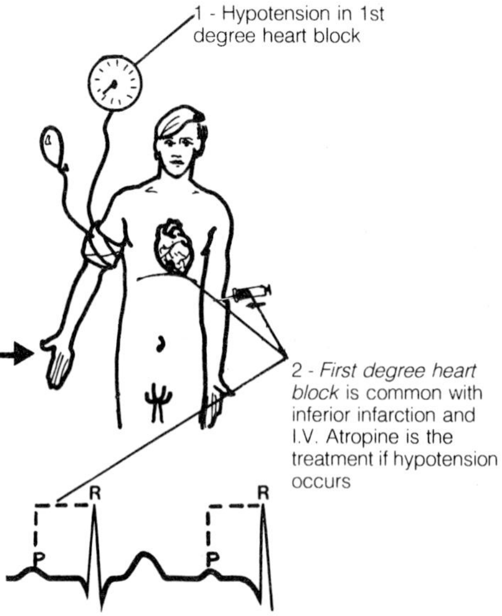

1 - Hypotension in 1st degree heart block

2 - *First degree heart block* is common with inferior infarction and I.V. Atropine is the treatment if hypotension occurs

R
R
P
P

Second degree AV block is usually associated with inferior infarction and has a good prognosis. If the patient has had an anterior infarction, the prognosis is poor and pacing may be required.

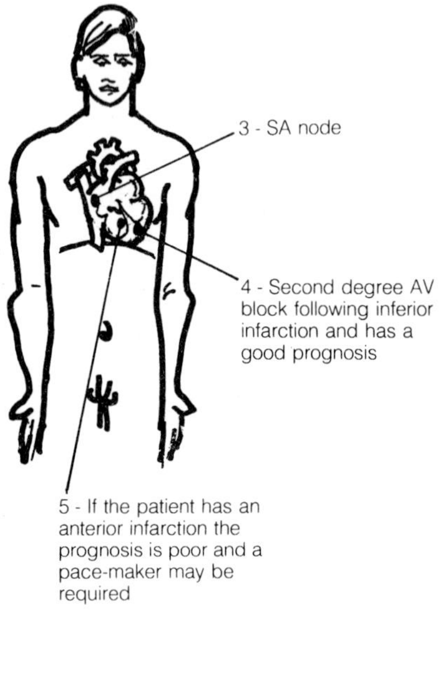

3 - SA node

4 - Second degree AV block following inferior infarction and has a good prognosis

5 - If the patient has an anterior infarction the prognosis is poor and a pace-maker may be required

Supraventricular tachycardias

These are uncommon following myocardial infarction. The initial treatment consists of carotid sinus massage. This increases vagal tone and the patient may revert to sinus rhythm. Failing this, verapamil can be given rapidly intravenously 5mg at a time, up to a total of 15mg. In patients on digoxin or betablockers however, intravenous verapamil requires great caution as the drugs all inhibit the AV node and severe bradycardias and hypotension can occur. An intravenous betablocker (such as practolol 5-10mg iv) or oral digoxin are alternative treatments. If the patient's cardiac output is compromised, then DC cardioversion is the treatment of choice. This is carried out under a short acting general anaesthetic, supervised by an anaesthetist. An initial shock of 50 joules is given, gradually increasing to 400 joules until sinus rhythm is obtained.

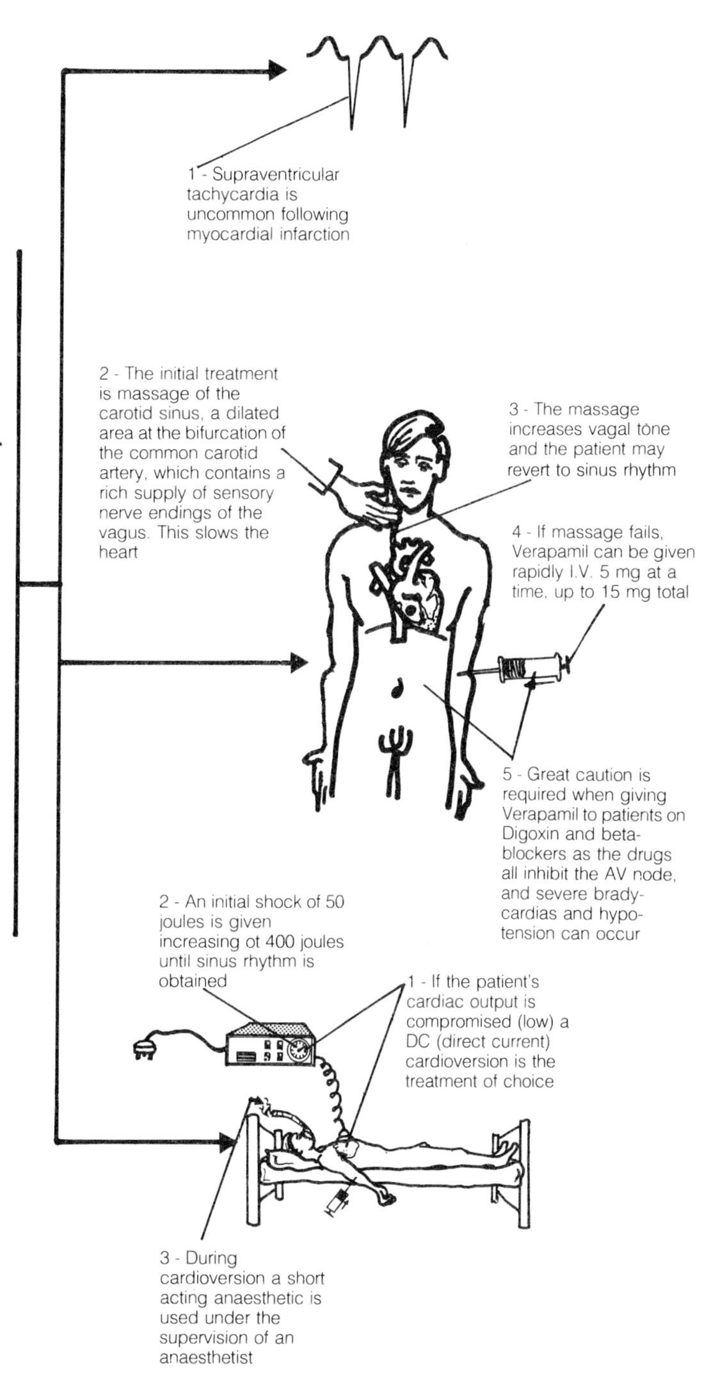
1 - Supraventricular tachycardia is uncommon following myocardial infarction
2 - The initial treatment is massage of the carotid sinus, a dilated area at the bifurcation of the common carotid artery, which contains a rich supply of sensory nerve endings of the vagus. This slows the heart
3 - The massage increases vagal tône and the patient may revert to sinus rhythm
4 - If massage fails, Verapamil can be given rapidly I.V. 5 mg at a time, up to 15 mg total
5 - Great caution is required when giving Verapamil to patients on Digoxin and beta-blockers as the drugs all inhibit the AV node, and severe brady-cardias and hypo-tension can occur
2 - An initial shock of 50 joules is given increasing ot 400 joules until sinus rhythm is obtained
1 - If the patient's cardiac output is compromised (low) a DC (direct current) cardioversion is the treatment of choice
3 - During cardioversion a short acting anaesthetic is used under the supervision of an anaesthetist

3 - If oral Digoxin fails Verapamil or Practolol may be required

Atrial fibrillation

This occurs in 10% of patients and can usually be controlled with oral digoxin. If the ventricular rate is very rapid, intravenous verapamil or practolol may be required.

2 - If oral Digoxin fails, IV Verapamil or Practolol may be required

1 - Atrial fibrillation

1 - Ventricular tachycardia occurs when a ventricular focus depolarises at high frequency. The QRS complexes are wide and abnormal and no P waves are visible

Ventricular tachycardia

This occurs when a ventricular focus depolarises at high frequency. The QRS complexes are wide and abnormal, and no P waves are visible. Treatment is by electrical cardioversion. Alternatively, a 100mg intravenous bolus of Lignocaine is given, followed by a lignocaine infusion at 4mg/minute. To prevent recurrence disopyramide (100mg qds), quinidine 1.2-2g daily, amiodarone (300-600mg daily), or tocainide (a lignocaine analogue), (400mg 8 hourly) can be given orally.

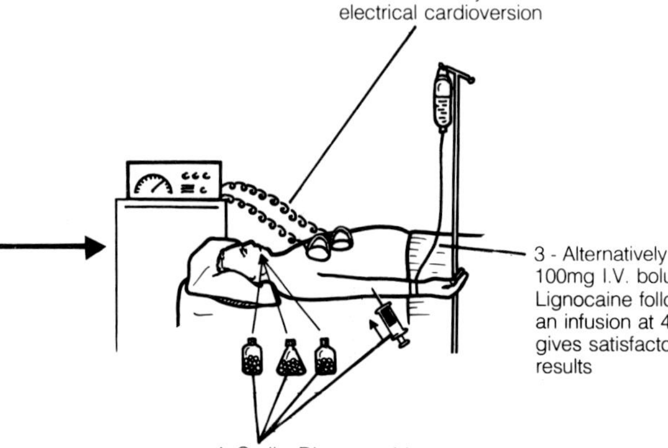
2 - Treatment by electrical cardioversion
3 - Alternatively a 100mg I.V. bolus of Lignocaine followed by an infusion at 4mg/min gives satisfactory results
4 Orally, Disopyramide 100mg qds, Quinidine 1.2-2 g daily, Amiodarone 300-600 mg daily or Tocainide 400mg 8 hourly may individually be used to prevent recurrence

Ventricular fibrillation

May follow ventricular tachycardia or be initiated by an R on T ventricular ectopic beat. Treatment is by electrical cardioversion.

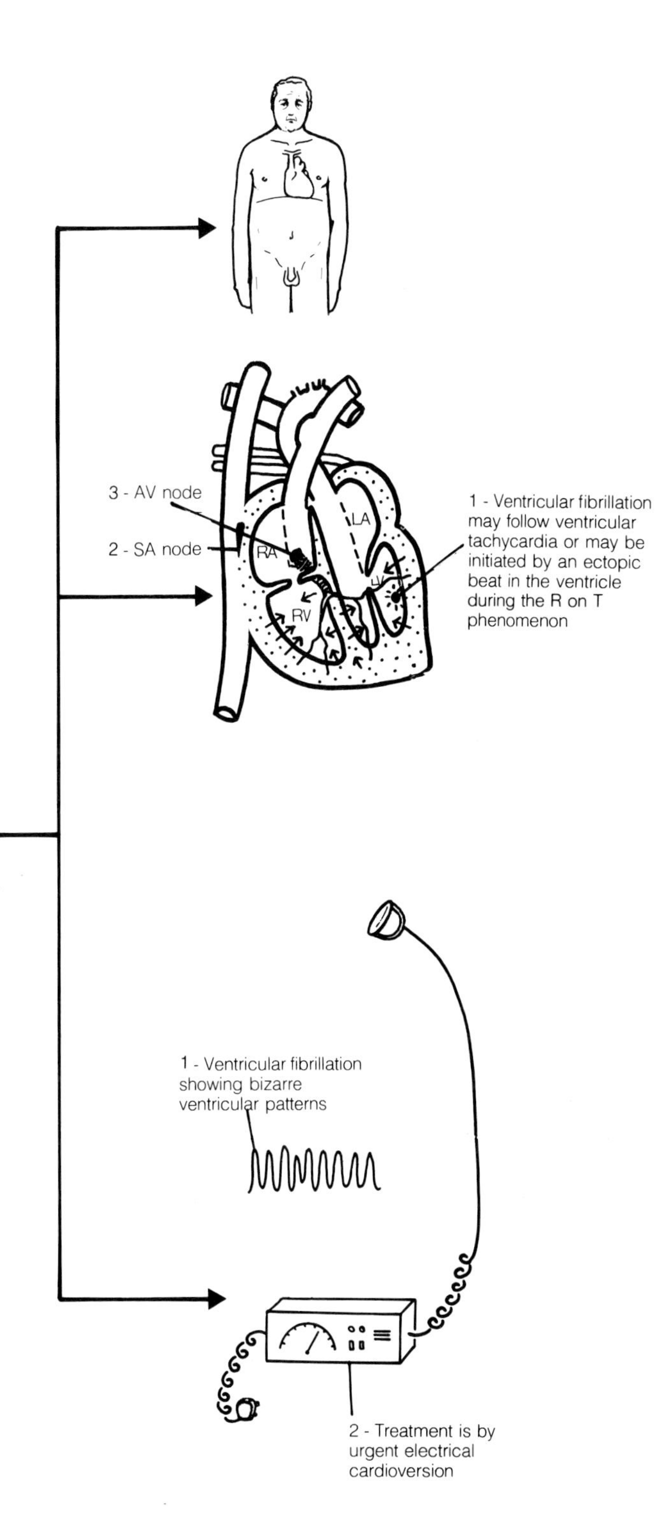
3 - AV node
2 - SA node
LA
RA
RV
1 - Ventricular fibrillation may follow ventricular tachycardia or may be initiated by an ectopic beat in the ventricle during the R on T phenomenon
1 - Ventricular fibrillation showing bizarre ventricular patterns
2 - Treatment is by urgent electrical cardioversion

Cardiac arrest

May be due to ventricular asystole or ventricular fibrillation. The patient loses consciousness, pulses are impalpable, and respiration ceases. Resuscitation must be commenced immediately. The patient should be put on the floor or fracture boards, and cardiac massage commenced at 60 compressions/minute. Effective massage at the lower end of the sternum should produce a palpable femoral pulse. The airway should be cleared, and the lungs inflated by mouth to mouth resuscitation or using a Brooke's airway or an Ambubag. A cuffed endotracheal tube should be inserted when the patient is pink. 100mls of 8.4% bicarbonate solution should be given intravenously as soon as possible, then 60mls every 15 minutes.

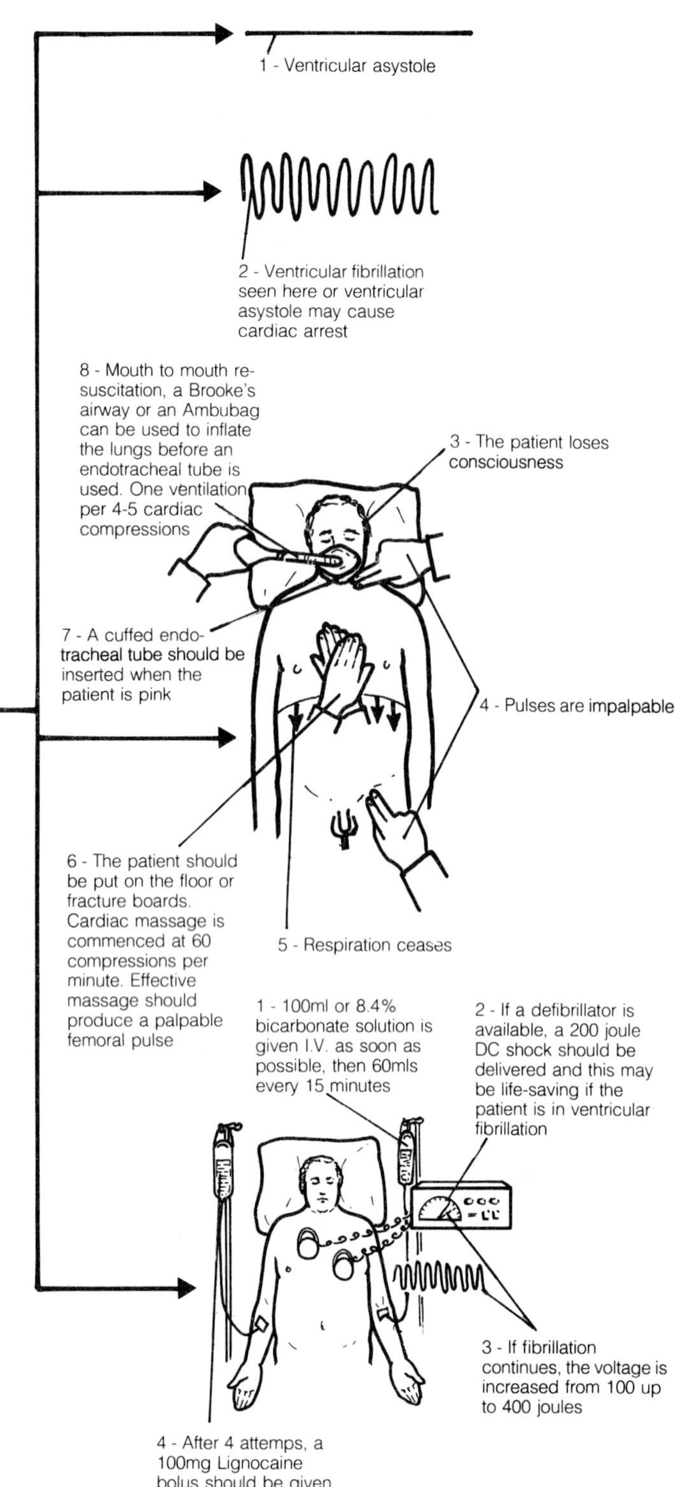
1 - Ventricular asystole
2 - Ventricular fibrillation seen here or ventricular asystole may cause cardiac arrest
8 - Mouth to mouth resuscitation, a Brooke's airway or an Ambubag can be used to inflate the lungs before an endotracheal tube is used. One ventilation per 4-5 cardiac compressions
3 - The patient loses consciousness
7 - A cuffed endotracheal tube should be inserted when the patient is pink
4 - Pulses are impalpable
6 - The patient should be put on the floor or fracture boards. Cardiac massage is commenced at 60 compressions per minute. Effective massage should produce a palpable femoral pulse
5 - Respiration ceases
1 - 100ml or 8.4% bicarbonate solution is given I.V. as soon as possible, then 60mls every 15 minutes
2 - If a defibrillator is available, a 200 joule DC shock should be delivered and this may be life-saving if the patient is in ventricular fibrillation
3 - If fibrillation continues, the voltage is increased from 100 up to 400 joules
4 - After 4 attemps, a 100mg Lignocaine bolus should be given I.V. and defibrillation repeated. Once fibrillation is under control, the Lignocaine infusion should be maintained

When the defibrillator is available, a 200 joule DC shock should be delivered; this may be life-saving if the patient is in ventricular fibrillation. If fibrillation continues, the voltage should be increased by 100 joules at a time to 400 joules. After four attempts a 100mg lignocaine bolus should be given intravenously and defibrillation repeated. Once the fibrillation is under control, the lignocaine infusion should be maintained.

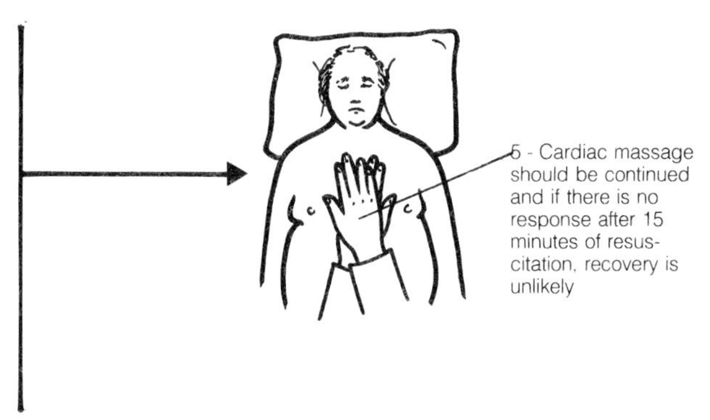
5 - Cardiac massage should be continued and if there is no response after 15 minutes of resuscitation, recovery is unlikely

If the heart is in asystole, ventricular fibrillation can be provoked by giving 10mls of adrenaline (1:10,000) and 10mls of 10% calcium gluconate. Defibrillation is then performed to restore sinus rhythm. If asystole persists, further boluses of adrenaline and calcium gluconate should be given intravenously. If this is unsuccessful the drugs may be given directly into the heart by inserting a needle into the left fourth intercostal space to the left of the midline. Cardiac massage should be continued. If there is no cardiac output after 15 minutes of resuscitation, recovery is unlikely.

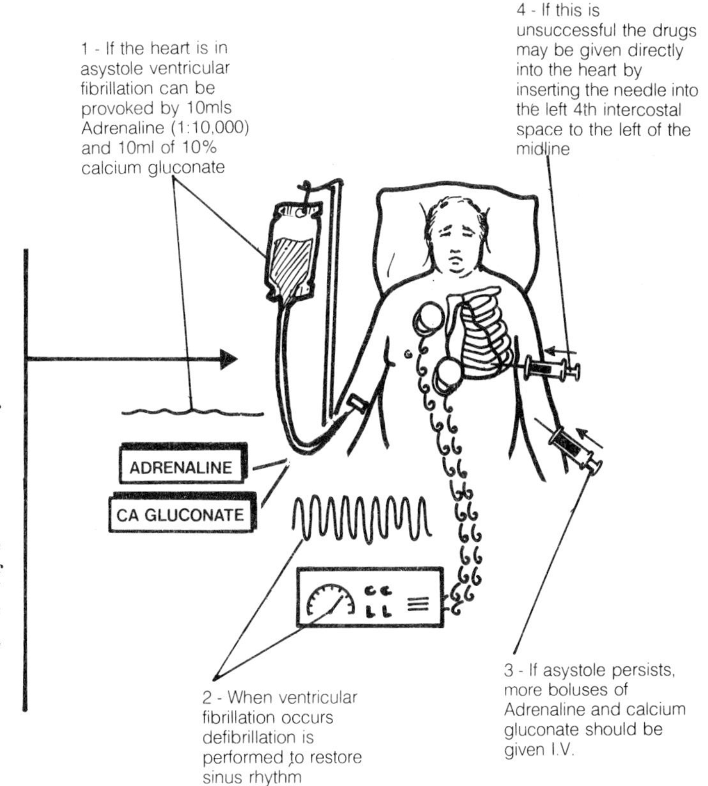
1 - If the heart is in asystole ventricular fibrillation can be provoked by 10mls Adrenaline (1:10,000) and 10ml of 10% calcium gluconate
4 - If this is unsuccessful the drugs may be given directly into the heart by inserting the needle into the left 4th intercostal space to the left of the midline
ADRENALINE
CA GLUCONATE
2 - When ventricular fibrillation occurs defibrillation is performed to restore sinus rhythm
3 - If asystole persists, more boluses of Adrenaline and calcium gluconate should be given I.V.

Heart failure

Heart failure is common following myocardial infarction and the patient may develop tachypnoea, a gallop rhythm, and bilateral basal inspiratory crackles. The patient should be nursed sitting upright in bed and given 40-60% oxygen via a mask. Oral thiazide diuretics are useful in mild cases; in more severe cases, intravenous frusemide may be necessary (20mg initially, but up to 120mg may be required). Nitrates are useful in reducing preload, and diamorphine is often helpful in treating failure as well as relieving pain.

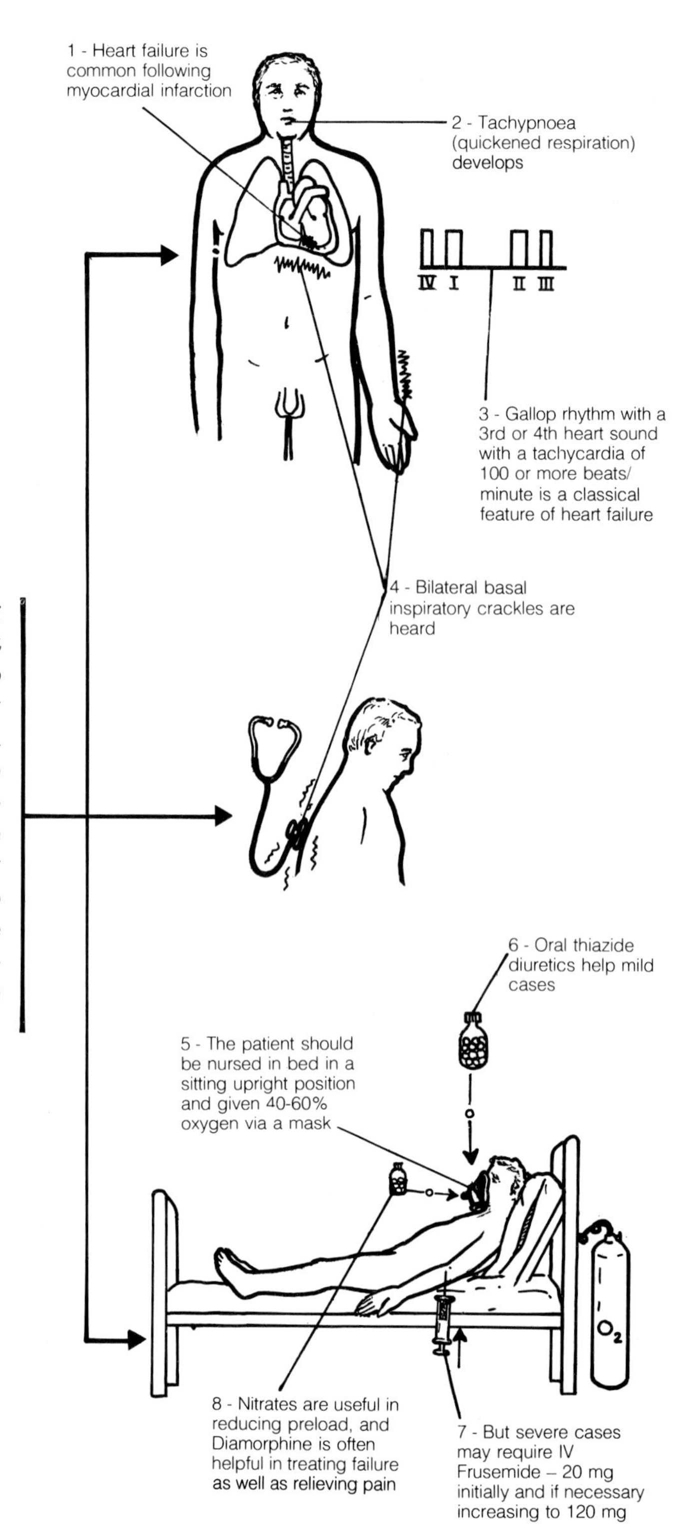
1 - Heart failure is common following myocardial infarction
2 - Tachypnoea (quickened respiration) develops
IV I II III
3 - Gallop rhythm with a 3rd or 4th heart sound with a tachycardia of 100 or more beats/ minute is a classical feature of heart failure
4 - Bilateral basal inspiratory crackles are heard
6 - Oral thiazide diuretics help mild cases
5 - The patient should be nursed in bed in a sitting upright position and given 40-60% oxygen via a mask
O2
8 - Nitrates are useful in reducing preload, and Diamorphine is often helpful in treating failure as well as relieving pain
7 - But severe cases may require IV Frusemide – 20 mg initially and if necessary increasing to 120 mg

Cardiogenic shock occurs in 12% of patients following myocardial infarction and has a very poor prognosis. Hypotension (systolic blood pressure of less than 90 mmHg) occurs and is associated with oliguria, poor peripheral perfusion and acidosis. Dopamine, a precursor of noradrenaline, increases blood flow in the renal, mesenteric, coronary and cerebral beds. When given intravenously in low doses (5-15μg/Kg/minute), it has an inotropic affect and increases renal blood flow. At higher doses however, peripheral vasoconstriction can occur. If severe congestive cardiac failure is also present, Dobutamine, a synthetic analogue of dopamine, at doses of 2.5 - 10μg/Kg/minute intravenously, is useful. In certain cases where there is no aortic regurgitation, the intra-aortic balloon pump may be helpful. A balloon catheter is introduced into the descending thoracic aorta via the femoral artery. It is inflated during diastole and deflates during systole using an ECG triggered control. This improves coronary perfusion and reduces afterload. Patients occasionally respond well but 90% of patients with cardiogenic shock following infarction will die.

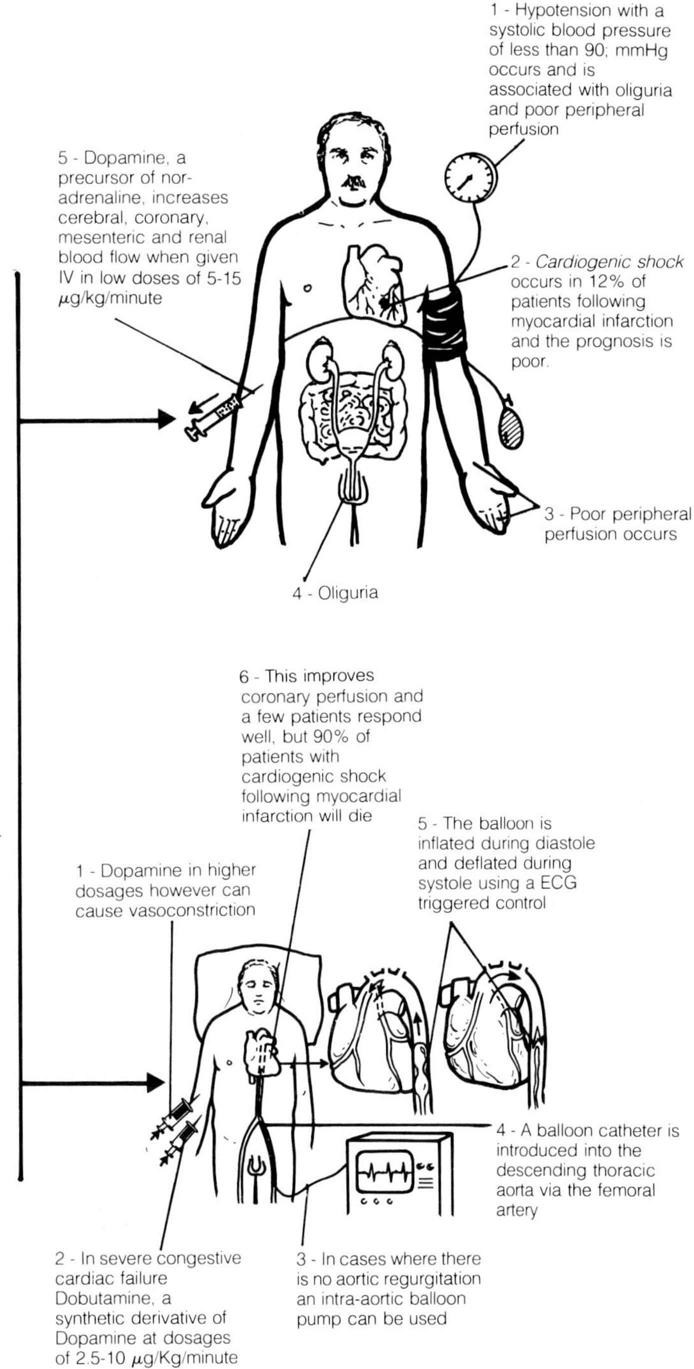
1 - Hypotension with a systolic blood pressure of less than 90; mmHg occurs and is associated with oliguria and poor peripheral perfusion
2 - *Cardiogenic shock* occurs in 12% of patients following myocardial infarction and the prognosis is poor.
3 - Poor peripheral perfusion occurs
4 - Oliguria
5 - Dopamine, a precursor of nor-adrenaline, increases cerebral, coronary, mesenteric and renal blood flow when given IV in low doses of 5-15 μg/kg/minute
1 - Dopamine in higher dosages however can cause vasoconstriction
2 - In severe congestive cardiac failure Dobutamine, a synthetic derivative of Dopamine at dosages of 2.5-10 μg/Kg/minute I.V. is useful
3 - In cases where there is no aortic regurgitation an intra-aortic balloon pump can be used
4 - A balloon catheter is introduced into the descending thoracic aorta via the femoral artery
5 - The balloon is inflated during diastole and deflated during systole using a ECG triggered control
6 - This improves coronary perfusion and a few patients respond well, but 90% of patients with cardiogenic shock following myocardial infarction will die

Recurrent angina after myocardial infarction carries a poor prognosis. It should be treated with betablockers, nitrates and calcium antagonists. Coronary angiography should be considered in younger patients with a view to coronary artery bypass surgery. However, little data is available on the results of operation in this group.

Pericarditis

This commonly occurs after myocardial infarction and presents with chest pain which is positional in nature and worse on inspiration. Treatment is with aspirin or non-steroidal anti-inflammatory drugs. In severe cases, intravenous hydrocortisone may be required. (100mg qds).

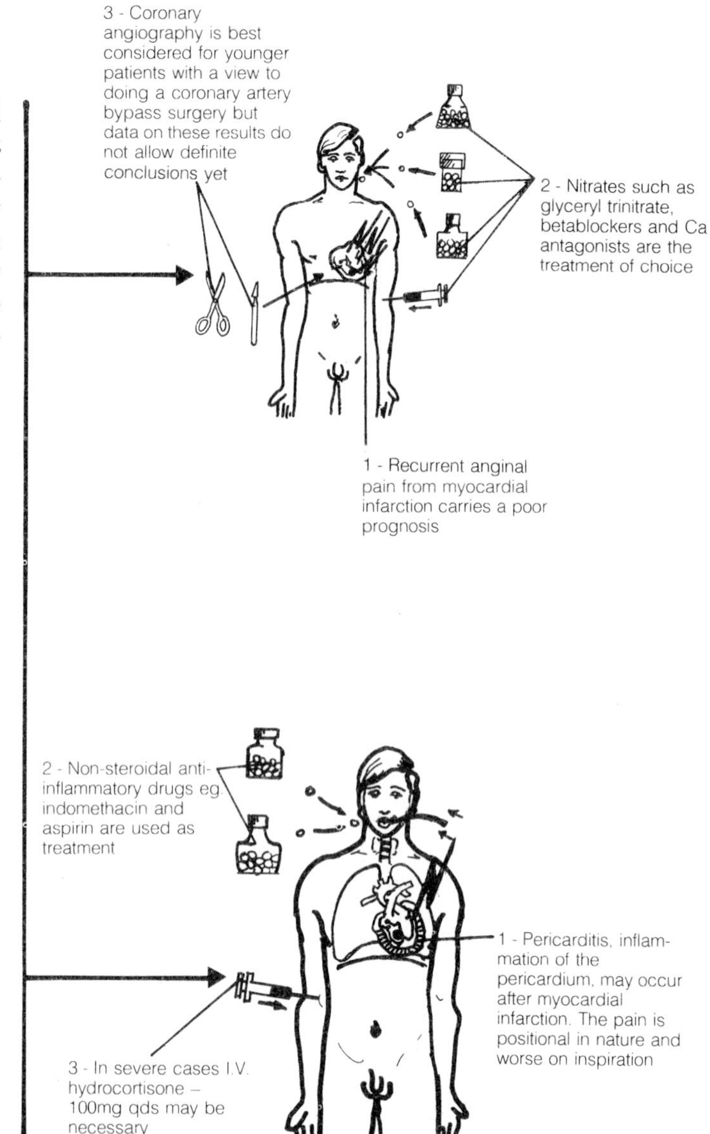
3 - Coronary angiography is best considered for younger patients with a view to doing a coronary artery bypass surgery but data on these results do not allow definite conclusions yet
2 - Nitrates such as glyceryl trinitrate, betablockers and Ca antagonists are the treatment of choice
1 - Recurrent anginal pain from myocardial infarction carries a poor prognosis
2 - Non-steroidal anti-inflammatory drugs eg indomethacin and aspirin are used as treatment
1 - Pericarditis, inflammation of the pericardium, may occur after myocardial infarction. The pain is positional in nature and worse on inspiration
3 - In severe cases I.V. hydrocortisone – 100mg qds may be necessary

Dressler's syndrome is pericarditis occuring approximately 10-12 days after infarction, associated with fever, pleural or pericardial pain and a raised ESR. Treatment is symptomatic; occasionally a short course of steroids may be required. Dressler's syndrome has an anti-immune basis.

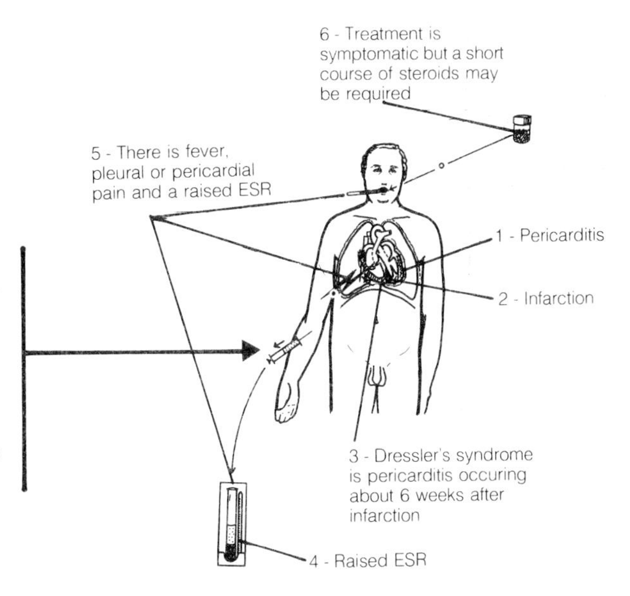
6 - Treatment is symptomatic but a short course of steroids may be required
5 - There is fever, pleural or pericardial pain and a raised ESR
1 - Pericarditis
2 - Infarction
3 - Dressler's syndrome is pericarditis occuring about 6 weeks after infarction
4 - Raised ESR

Acute mitral regurgitation can occur secondary to papillary muscle dysfunction, when a transient apical systolic murmur can be heard. Rupture of the papillary muscle can cause severe mitral regurgitation with a loud murmur and pulmonary oedema. The patient should be stabilized with medical treatment before cardiac surgery is attempted. An acute ventricular septal defect can occur and should also be treated medically before surgery.

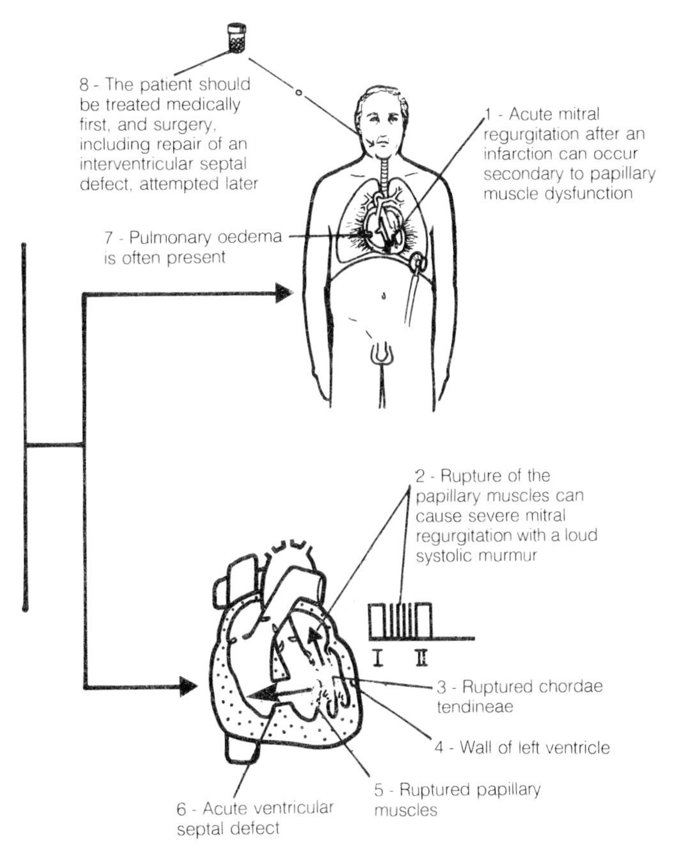
8 - The patient should be treated medically first, and surgery, including repair of an interventricular septal defect, attempted later
1 - Acute mitral regurgitation after an infarction can occur secondary to papillary muscle dysfunction
7 - Pulmonary oedema is often present
2 - Rupture of the papillary muscles can cause severe mitral regurgitation with a loud systolic murmur
I
II
3 - Ruptured chordae tendineae
4 - Wall of left ventricle
5 - Ruptured papillary muscles
6 - Acute ventricular septal defect

Ventricular aneurysm may develop, commonly after anterior infarction. The patient may develop persistent cardiac failure or angina. Persistent ST elevation is usually present on the ECG, and there may be a bulging left cardiac border on the chest X-ray. The aneurysm can be confirmed by 2D echo-cardiography or radioisotope ventriculography. In patients with severe symptoms, surgery may be required, with resection of the aneurysm and coronary artery bypass grafting.

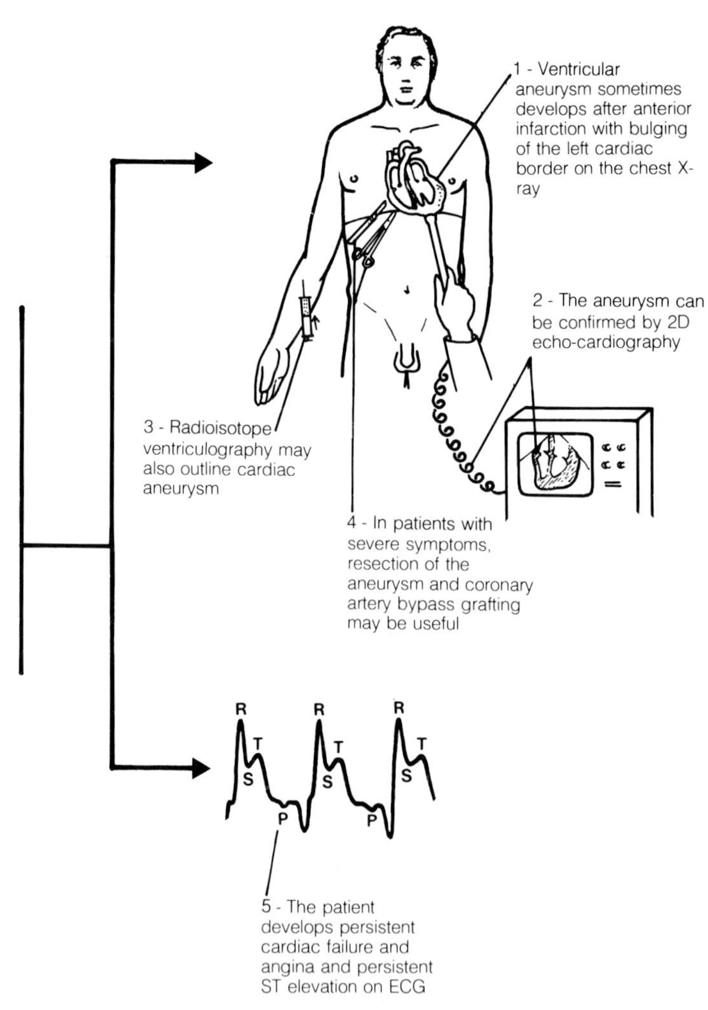
1 - Ventricular aneurysm sometimes develops after anterior infarction with bulging of the left cardiac border on the chest X-ray
2 - The aneurysm can be confirmed by 2D echo-cardiography
3 - Radioisotope ventriculography may also outline cardiac aneurysm
4 - In patients with severe symptoms, resection of the aneurysm and coronary artery bypass grafting may be useful
R
T
S
P
5 - The patient develops persistent cardiac failure and angina and persistent ST elevation on ECG

Rehabilitation after myocardial infarction

Provided there are no complications, the patient can be walking on the flat within the first week after a heart attack. Thereafter the patient should be encouraged to take as much exercise as he finds comfortable. Driving should be avoided

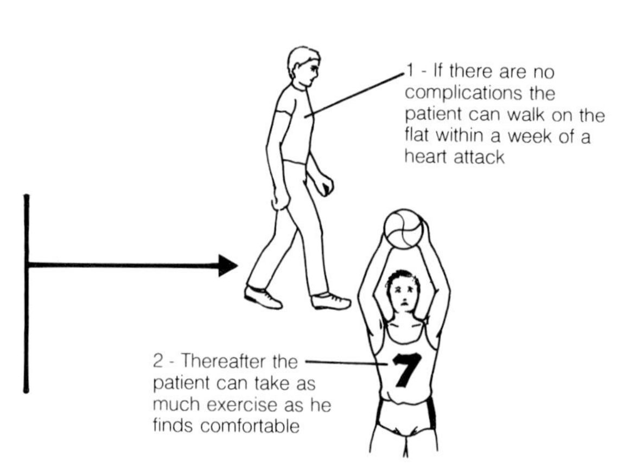
1 - If there are no complications the patient can walk on the flat within a week of a heart attack
2 - Thereafter the patient can take as much exercise as he finds comfortable
7

for one month, and patients are usually able to return to work. HGV and public service vehicle (PSV) licence holders are required to surrender them. If exercise tolerance is satisfactory, return to sexual activity is permissible after a few weeks. Smoking must be avoided and weight lost where necessary. Depending on the patient's occupation, he should be able to return to work within 8-12 weeks.

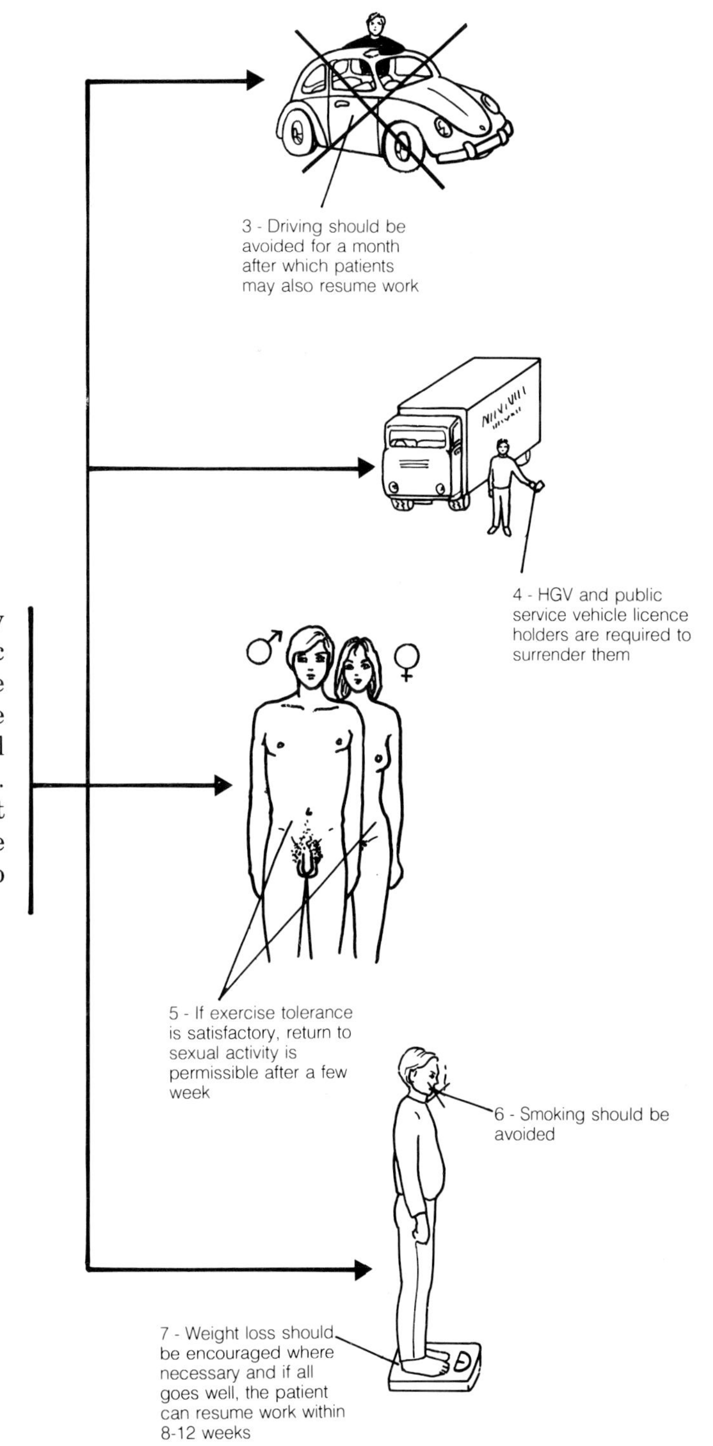
3 - Driving should be avoided for a month after which patients may also resume work
4 - HGV and public service vehicle licence holders are required to surrender them
5 - If exercise tolerance is satisfactory, return to sexual activity is permissible after a few week
6 - Smoking should be avoided
7 - Weight loss should be encouraged where necessary and if all goes well, the patient can resume work within 8-12 weeks

Secondary prevention

Recent studies with timolol and propranolol have shown a significant reduction in reinfarction and mortality rate in patients given these betablockers after infarction. The role of betablockade after infarction still has to be fully evaluated, but it would seem wise to put patients on betablockers if they are hypertensive or develop angina after infarction. There are no conclusive studies to recommend long term anticoagulant treatment after infarction. Studies with aspirin and other anti-platelet agents are in progress at the moment.

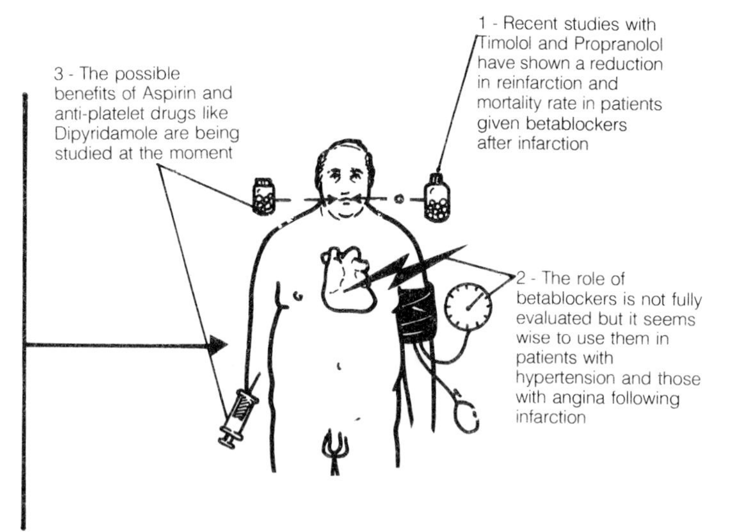

1 - Recent studies with Timolol and Propranolol have shown a reduction in reinfarction and mortality rate in patients given betablockers after infarction
2 - The role of betablockers is not fully evaluated but it seems wise to use them in patients with hypertension and those with angina following infarction
3 - The possible benefits of Aspirin and anti-platelet drugs like Dipyridamole are being studied at the moment

HYPERTENSION

Blood pressure is influenced by age, sex and genetic factors. The "average" blood pressure is 120/80 at 20 years and 160/90 at 60 years. For insurance examination, 150/90 is taken as the upper limit of normal. Middle aged patients with a diastolic blood pressure of 120 mmHg have a 70% five year survival and those with diastolic blood pressures of 100 mmHg or more can expect a 16 year reduction in life expectancy. There is definite evidence that treatment of hypertension reduces morbidity and mortality from strokes and coronary artery disease. It is therefore important to detect and treat hypertension.

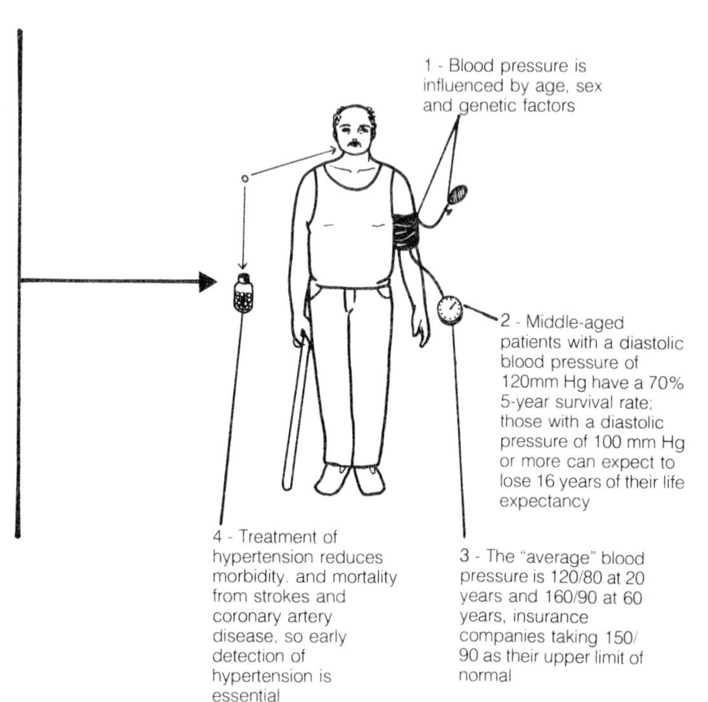

1 - Blood pressure is influenced by age, sex and genetic factors
2 - Middle-aged patients with a diastolic blood pressure of 120mm Hg have a 70% 5-year survival rate; those with a diastolic pressure of 100 mm Hg or more can expect to lose 16 years of their life expectancy
3 - The "average" blood pressure is 120/80 at 20 years and 160/90 at 60 years, insurance companies taking 150/90 as their upper limit of normal
4 - Treatment of hypertension reduces morbidity, and mortality from strokes and coronary artery disease, so early detection of hypertension is essential

Causes of Hypertension

In about 95% of cases of hypertension, no underlying cause can be found.

Essential hypertension tends to run in families and is more common in negroes than in caucasians. Obese people tend to have higher blood pressure, and alcoholics are often hypertensive. Excessive dietary salt is also thought to be important.

Secondary hypertension In about 5% of cases of hypertension an underlying cause can be found.

Renal hypertension Hypertension is associated with many renal diseases, including renal artery stenosis, chronic pyelonephritis, glomerulonephritis, following renal transplantation and analgesic nephropathy.

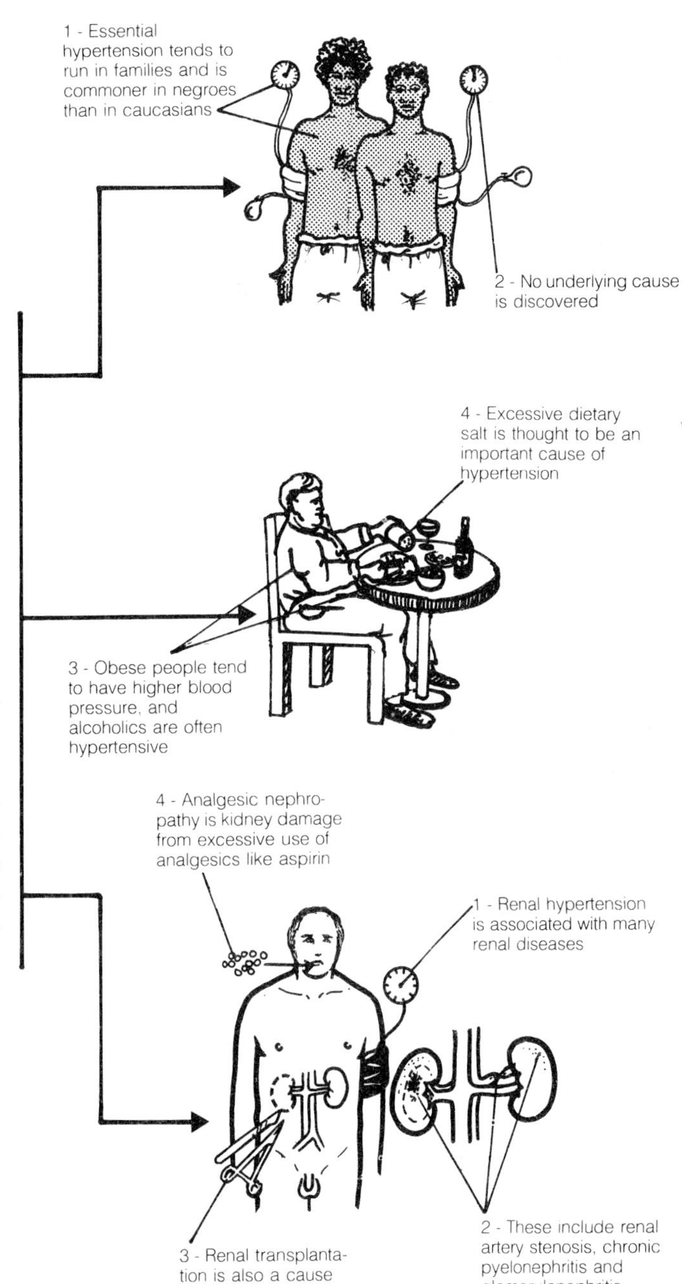
1 - Essential hypertension tends to run in families and is commoner in negroes than in caucasians
2 - No underlying cause is discovered
4 - Excessive dietary salt is thought to be an important cause of hypertension
3 - Obese people tend to have higher blood pressure, and alcoholics are often hypertensive
4 - Analgesic nephropathy is kidney damage from excessive use of analgesics like aspirin
1 - Renal hypertension is associated with many renal diseases
2 - These include renal artery stenosis, chronic pyelonephritis and glomerulonephritis
3 - Renal transplantation is also a cause

Adrenal hypertension Cushing's syndrome is associated with hypertension, which may be due to raised cortisol levels, or increased activity of the renin angiotensin system. Conn's syndrome (primary hyperaldosteronism) is caused by an adrenal tumour which produces aldosterone autonomously. Phaeochromocytoma is an adrenal tumour associated with raised plasma levels of adrenaline and noradrenaline; it is associated with very labile blood pressure.

1 - Cushing's syndrome is associated with hypertension, which may be due to raised cortisol levels or increased activity of the renin angiotensin system

2 - Patients with Cushing's syndrome present with striae on the abdomen and thighs, a buffalo hump, osteoporotic bones which fracture easily, diabetes (symbolised by pancreas) and obesity of the trunk and thiness of the limbs

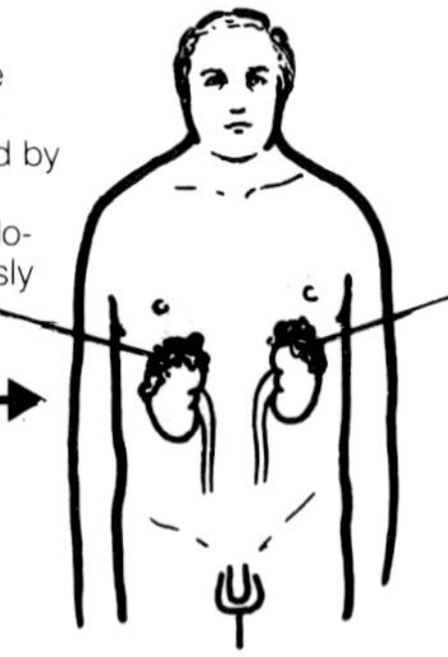

1 Conn's syndrome (primary hyperaldosteronism) is caused by an adrenal tumour which produces aldosterone autonomously

2 - Phaeochromocytoma is an adrenal tumour associated with raised plasma levels of adrenaline and noradrenaline. It is associated with very labile blood pressure

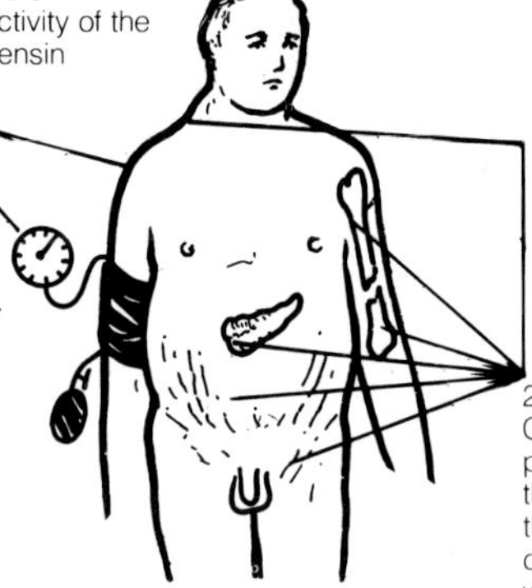

Coarctation of the aorta is a rare but curable cause of hypertension. Clinically there is marked delay between the radial and femoral pulses, and the chest X-ray shows notching of the inferior surfaces of the ribs.

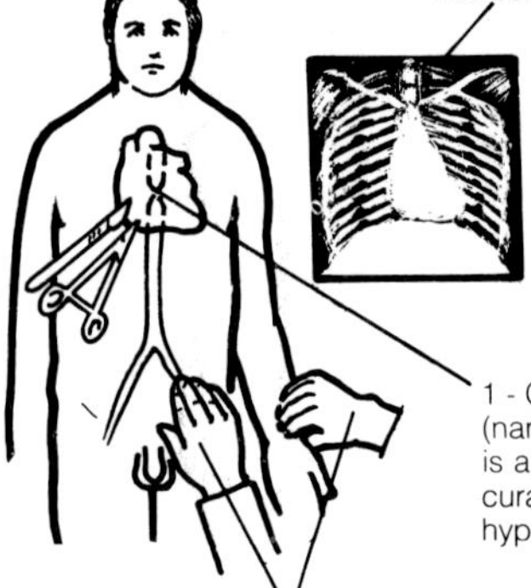

3 - X-ray shows notching of the inferior surface of the ribs due to the enlarged intercostal arteries.

1 - Coarctation (narrowing) of the aorta is a rare but surgically curable cause of hypertension

2 - There is a marked delay between radial and femoral pulses

Toxaemia of pregnancy Hypertension may be caused or worsened by pregnancy.

Drugs

Drugs such as oral contraceptives, steroids and tricyclic antidepressants can cause reversible hypertension.

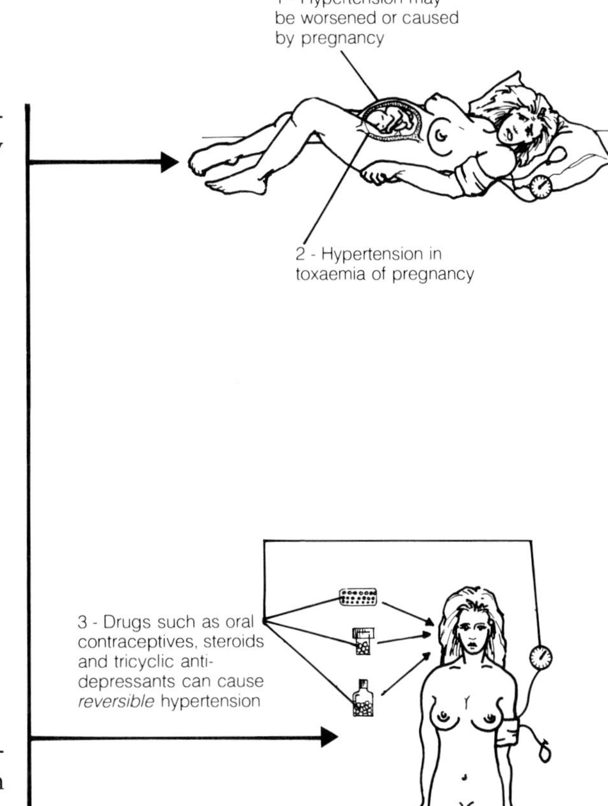
1 - Hypertension may be worsened or caused by pregnancy
2 - Hypertension in toxaemia of pregnancy
3 - Drugs such as oral contraceptives, steroids and tricyclic anti-depressants can cause *reversible* hypertension

Clinical features of hypertension

Many hypertensive patients are asymptomatic and there may be no abnormalities on examination apart from the raised blood pressure. There may be symptoms of end organ damage such as angina, exertional and paroxysmal nocturnal dyspnoea, transient ischaemic attacks or a history of stroke. Nocturia may indicate renal disease. Patients with phaeochromocytoma may complain of palpitations and sweating, while muscle weakness may be a feature of Conn's syndrome. The drug history is important.

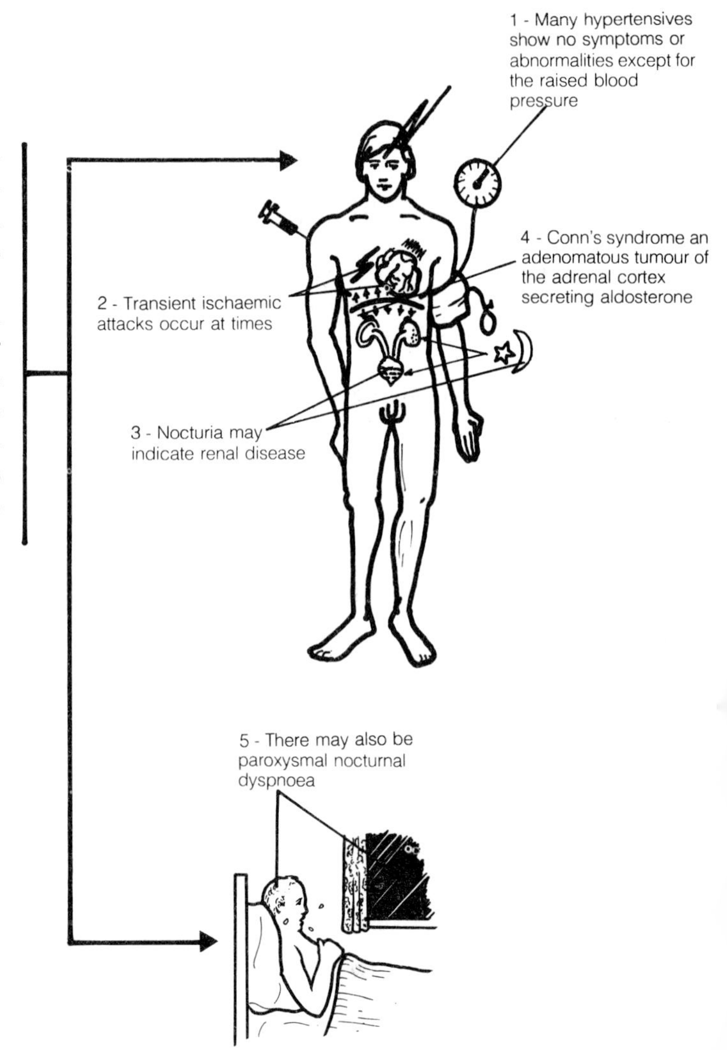
1 - Many hypertensives show no symptoms or abnormalities except for the raised blood pressure
2 - Transient ischaemic attacks occur at times
3 - Nocturia may indicate renal disease
4 - Conn's syndrome an adenomatous tumour of the adrenal cortex secreting aldosterone
5 - There may also be paroxysmal nocturnal dyspnoea

Examination may reveal features of left ventricular hypertrophy with a heaving apex beat which may be displaced, and a triple rhythm. There may be features of left ventricular failure and the chest should be examined and a careful history taken to exclude asthma (this will preclude the use of betablockers). Radial and femoral pulses should be palpated simultaneously to exclude coarctation. The abdomen should be examined for renal enlargement, and the bruits of renal artery stenosis should be listened for; (these are best heard 5cms lateral to the umbilicus). The optic fundi should be examined. Hypertensive retinopathy is classified as follows:

GRADES OF HYPERTENSIVE RETINOPATHY

Grade I – narrowed tortuous retinal vessels, "silver wiring"

Grade II – Arteriovenous nipping

Grade III is Grades I and II with haemorrhages and exudates

Grade IV is Grades I to III with papilloedema

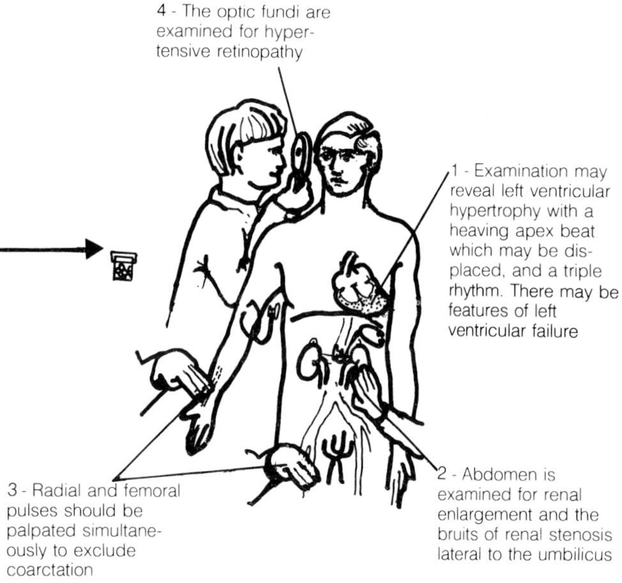

4 - The optic fundi are examined for hypertensive retinopathy

1 - Examination may reveal left ventricular hypertrophy with a heaving apex beat which may be displaced, and a triple rhythm. There may be features of left ventricular failure

3 - Radial and femoral pulses should be palpated simultaneously to exclude coarctation

2 - Abdomen is examined for renal enlargement and the bruits of renal stenosis lateral to the umbilicus

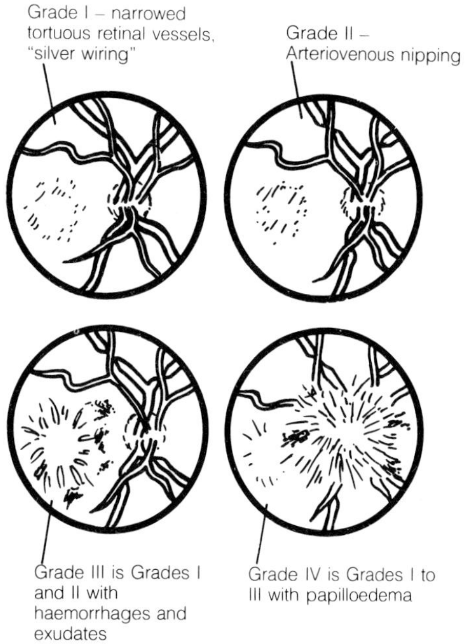

Grade I; narrowed tortuous retinal vessels.
Grade II; arteriovenous nipping.
Grade III; Grades I and II with haemorrhages and exudates.
Grade IV; Grades I, II and III with papilloedema.
However, Grades I and II can occur in older patients without hypertension, and it is probably better to describe any features seen in each eye (for example retinal haemorrhage at 6 o'clock in the right eye). Routine urinalysis for microscopy, glucose and protein should always be performed.

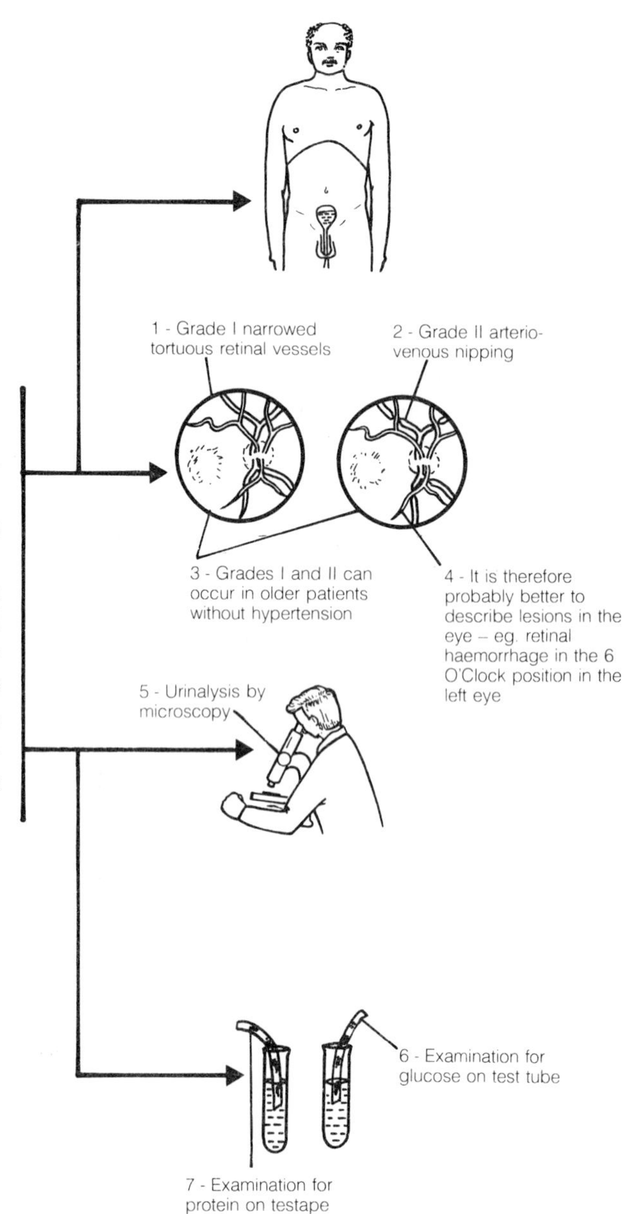
1 - Grade I narrowed tortuous retinal vessels
2 - Grade II arterio-venous nipping
3 - Grades I and II can occur in older patients without hypertension
4 - It is therefore probably better to describe lesions in the eye – eg. retinal haemorrhage in the 6 O'Clock position in the left eye
5 - Urinalysis by microscopy
6 - Examination for glucose on test tube
7 - Examination for protein on testape

Investigations

Serum electrolytes should be measured as the patient may have primary or secondary hyperaldosteronism with hypokalaemia. **Serum creatinine and urea levels** will be elevated in renal disease. **A chest X-ray** should be performed with measurement of the cardiothoracic ratio. **The ECG** may show changes of left ventricular hypertrophy. Urinalysis should be performed for cells, casts and protein in young patients or those in whom there is reason to suspect secondary hypertension. Further investigations include:

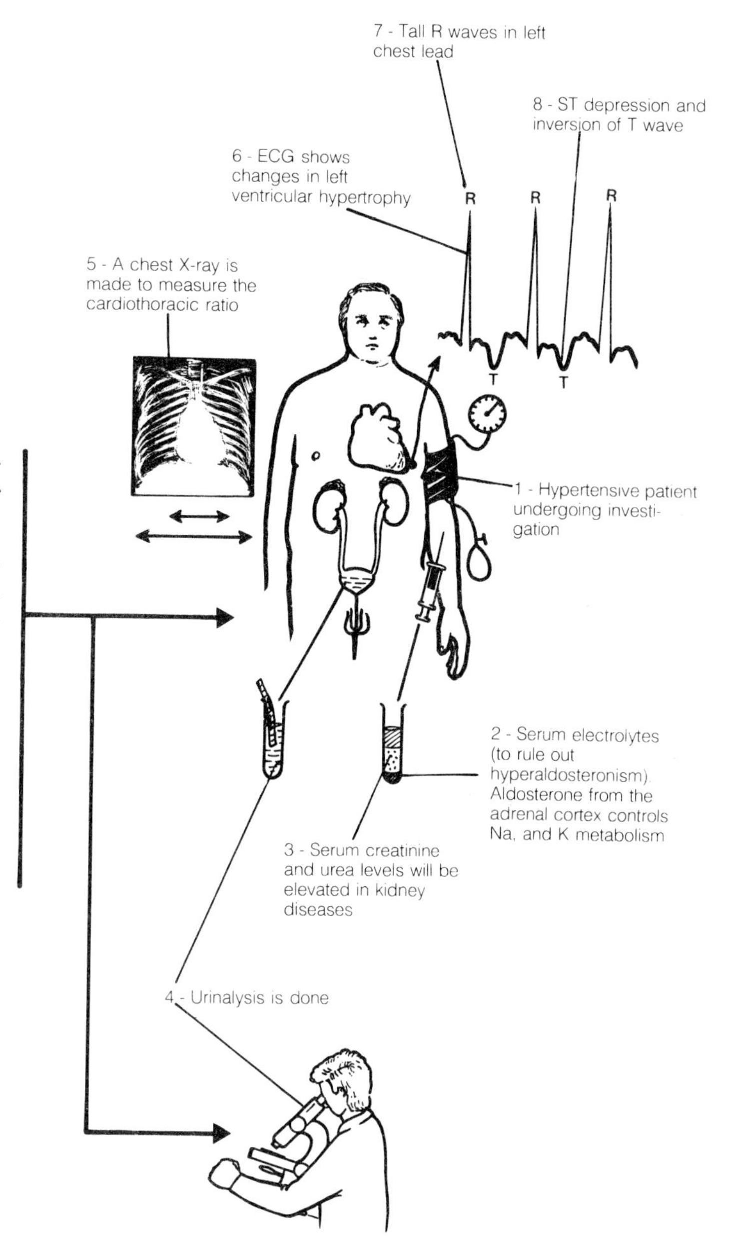
7 - Tall R waves in left chest lead
8 - ST depression and inversion of T wave
6 - ECG shows changes in left ventricular hypertrophy
R
R
R
T
T
5 - A chest X-ray is made to measure the cardiothoracic ratio
1 - Hypertensive patient undergoing investigation
2 - Serum electrolytes (to rule out hyperaldosteronism) Aldosterone from the adrenal cortex controls Na, and K metabolism
3 - Serum creatinine and urea levels will be elevated in kidney diseases
4 - Urinalysis is done

1. **Intravenous urography**
   This may be indicated to provide evidence of renal disease. In renal artery stenosis there is classically a delayed, dense nephrogram on the IVU. Renin concentrations can be measured in the vein draining the affected kidney and in the IVC. A renal vein to IVC renin ratio of 1.5:1 or more is indicative of significant stenosis, and these patients respond well to surgery.

2. **Urinary VMA levels**
   Are raised in patients with phaeochromocytoma and plasma levels of adrenaline and noradrenaline are also high. An abdominal CT scan is useful in localising tumours.

3. **Plasma aldosterone levels**
   These may be high with low plasma renin levels and hypokalaemia in patients with Conn's syndrome.

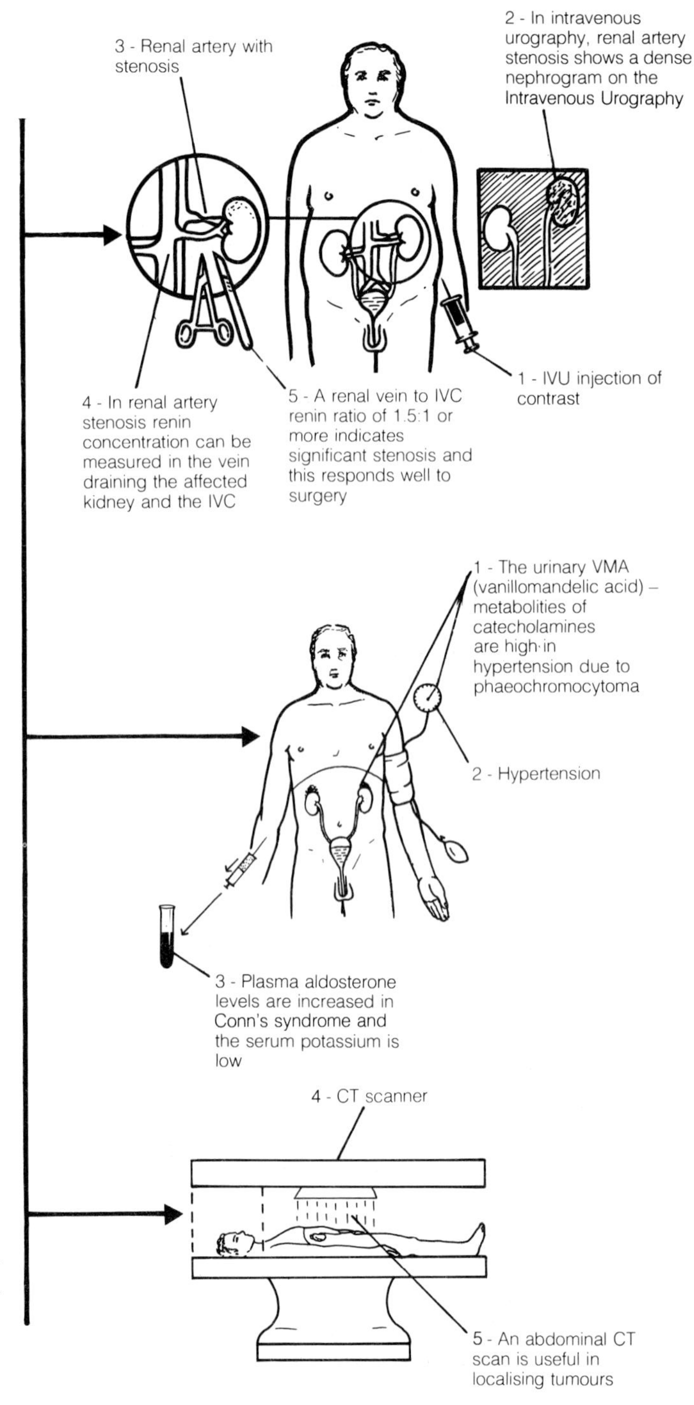
3 - Renal artery with stenosis
2 - In intravenous urography, renal artery stenosis shows a dense nephrogram on the Intravenous Urography
4 - In renal artery stenosis renin concentration can be measured in the vein draining the affected kidney and the IVC
5 - A renal vein to IVC renin ratio of 1.5:1 or more indicates significant stenosis and this responds well to surgery
1 - IVU injection of contrast
1 - The urinary VMA (vanillomandelic acid) – metabolities of catecholamines are high in hypertension due to phaeochromocytoma
2 - Hypertension
3 - Plasma aldosterone levels are increased in Conn's syndrome and the serum potassium is low
4 - CT scanner
5 - An abdominal CT scan is useful in localising tumours

Malignant hypertension This is an uncommon condition which is a medical emergency. The diastolic blood pressure is usually over 140 mmHg and papilloedema is present. The patient may complain of headaches, vomiting and fits, or impaired consciousness may occur.

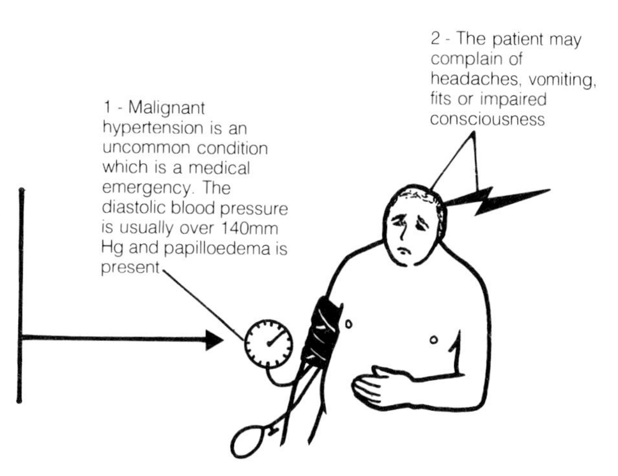
1 - Malignant hypertension is an uncommon condition which is a medical emergency. The diastolic blood pressure is usually over 140mm Hg and papilloedema is present
2 - The patient may complain of headaches, vomiting, fits or impaired consciousness

Treatment of hypertension

The obese hypertensive patient should be encouraged to lose weight. Any drugs which aggravate hypertension such as the contraceptive pill should be withdrawn.

Thiazide diuretics are useful as a first line agent in older patients and are often effective in negroes. Side-effects include hypokalaemia, hyperuricaemia and hyperglycaemia. A small dose is often effective such as bendrofluazide 2.5-5mg daily.

Betablockers are effective in reducing blood pressure, especially in the younger patient. The patient should be started on a small dosage (for example, propranolol 40mg three times daily) this is gradually increased until there is adequate contol. Long acting betablockers such as atenolol are taken only once daily and are useful where compliance is a problem. Betablockers should be avoided in patients with bronchospasm or heart failure.

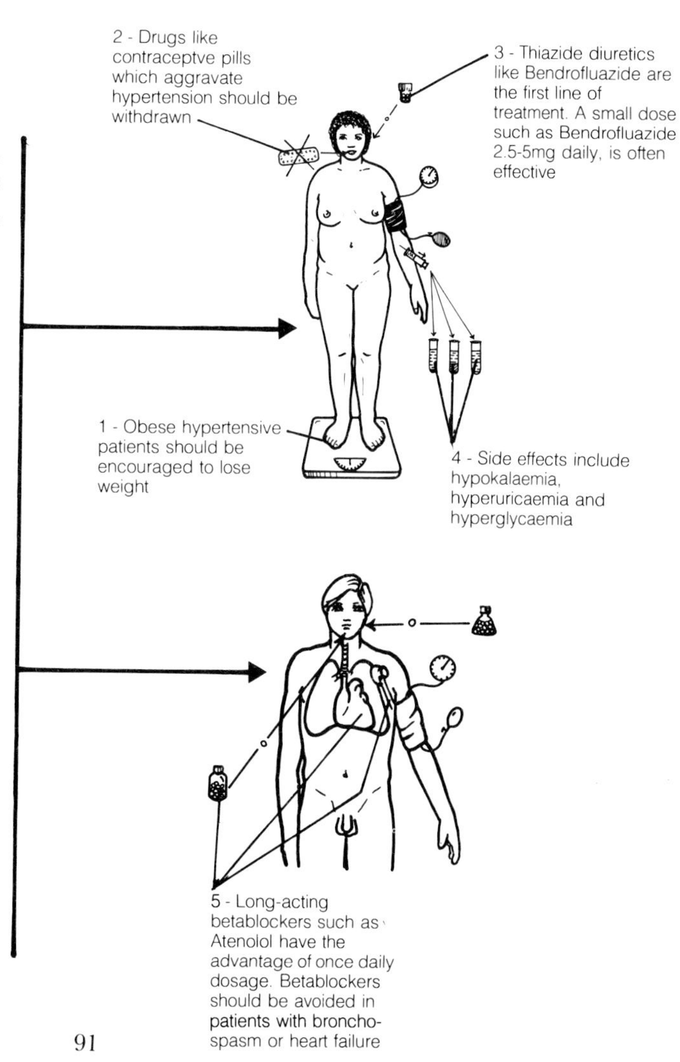
2 - Drugs like contraceptve pills which aggravate hypertension should be withdrawn
3 - Thiazide diuretics like Bendrofluazide are the first line of treatment. A small dose such as Bendrofluazide 2.5-5mg daily, is often effective
1 - Obese hypertensive patients should be encouraged to lose weight
4 - Side effects include hypokalaemia, hyperuricaemia and hyperglycaemia
5 - Long-acting betablockers such as Atenolol have the advantage of once daily dosage. Betablockers should be avoided in patients with broncho-spasm or heart failure

Thiazides and betablockers act synergistically and the combination is useful in the patient whose blood pressure is difficult to control on a single agent.

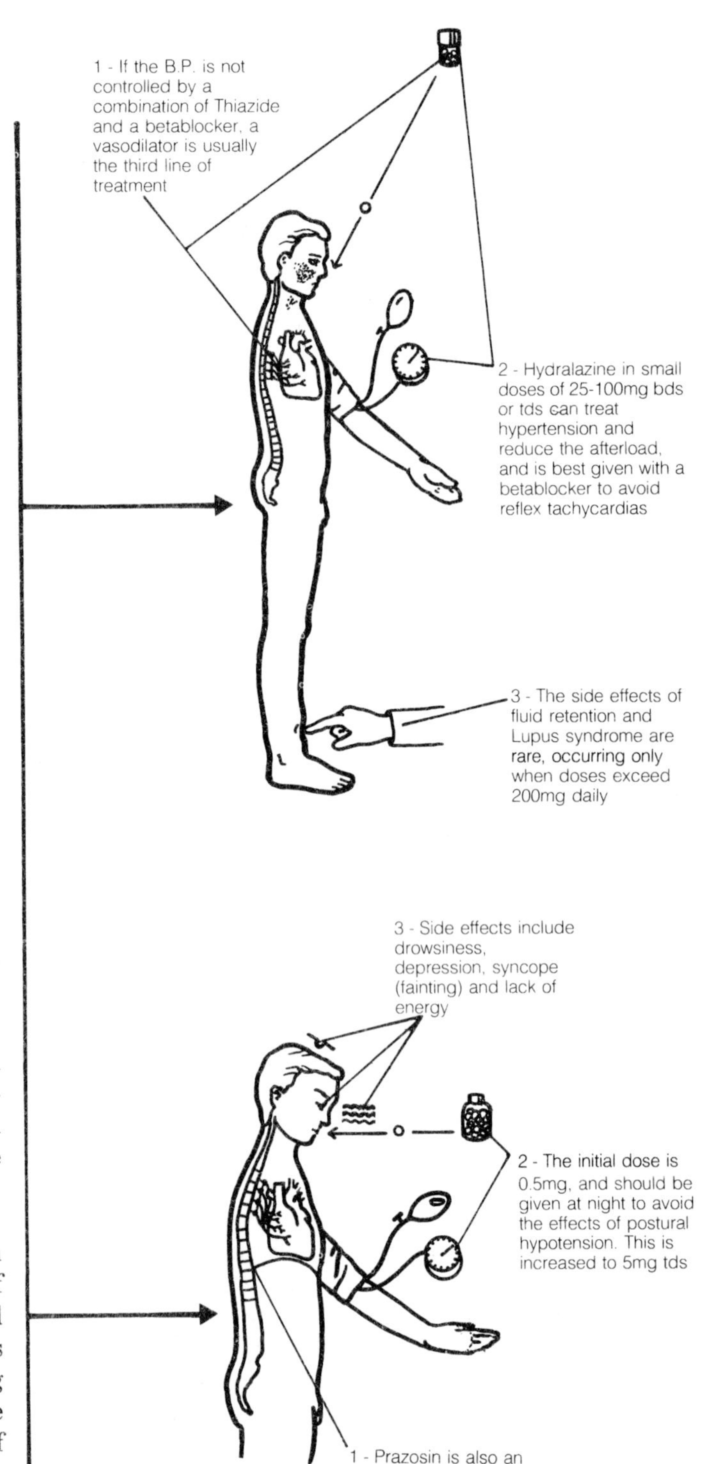
1 - If the B.P. is not controlled by a combination of Thiazide and a betablocker, a vasodilator is usually the third line of treatment
2 - Hydralazine in small doses of 25-100mg bds or tds can treat hypertension and reduce the afterload, and is best given with a betablocker to avoid reflex tachycardias
3 - The side effects of fluid retention and Lupus syndrome are rare, occurring only when doses exceed 200mg daily
3 - Side effects include drowsiness, depression, syncope (fainting) and lack of energy
2 - The initial dose is 0.5mg, and should be given at night to avoid the effects of postural hypotension. This is increased to 5mg tds
1 - Prazosin is also an alphablocking drug

Vasodilators If the blood pressure is not controlled on a combination of thiazide and betablocker, a vasodilator is usually the third line of treatment:—

Hydralazine in small doses (25-100mg twice to three times daily) is useful in treating hypertension and reducing afterload. It is best given in combination with a betablocker to avoid reflex tachycardias. Side-effects include fluid retention and the lupus syndrome but this is rare in doses below 200mg daily.

Prazosin is a vasodilator with partial alphablocking effects. The initial dose of 0.5mg should be given at night to avoid the effects of postural hypotension. This can gradually be increased to up to 5mg three times a day. Side-effects include first dose syncope, drowsiness, lack of energy and depression.

Calcium antagonists have a vasodilator property and can be useful in treating hypertension. A slow release preparation of nifedipine is now available, the dosage being 20mg twice daily. Side-effects include headaches, facial flushing and dizziness.

Centrally acting drugs These are not commonly used nowadays because of their severe side-effects.

Methyldopa is believed to act as a central alpha stimulator. The dosage is 0.5-3g daily in divided doses but it can cause drowsiness, depression, nightmares, impotence, haemolytic anaemia and jaundice.

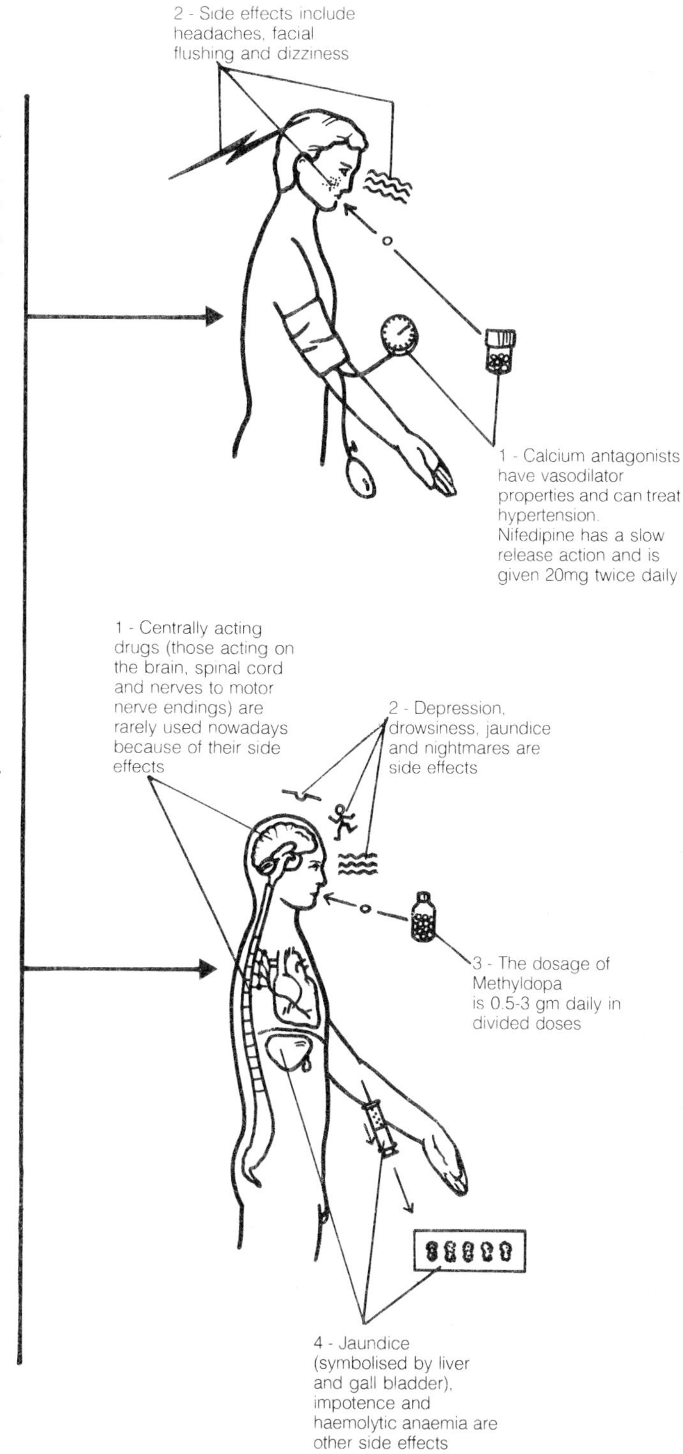
2 - Side effects include headaches, facial flushing and dizziness
1 - Calcium antagonists have vasodilator properties and can treat hypertension. Nifedipine has a slow release action and is given 20mg twice daily
1 - Centrally acting drugs (those acting on the brain, spinal cord and nerves to motor nerve endings) are rarely used nowadays because of their side effects
2 - Depression, drowsiness, jaundice and nightmares are side effects
3 - The dosage of Methyldopa is 0.5-3 gm daily in divided doses
4 - Jaundice (symbolised by liver and gall bladder), impotence and haemolytic anaemia are other side effects

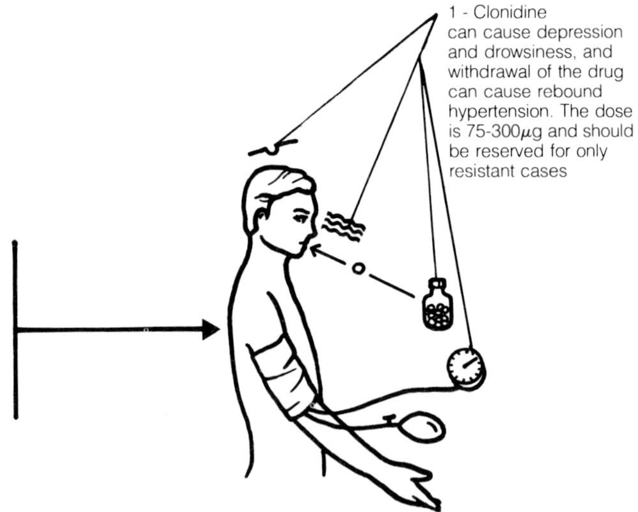
1 - Clonidine can cause depression and drowsiness, and withdrawal of the drug can cause rebound hypertension. The dose is 75-300μg and should be reserved for only resistant cases

Clonidine causes drowsiness and depression and sudden withdrawal of the drug can cause rebound hypertension. It is usually reserved for resistant cases of hypertension (75-300μg daily in divided doses).

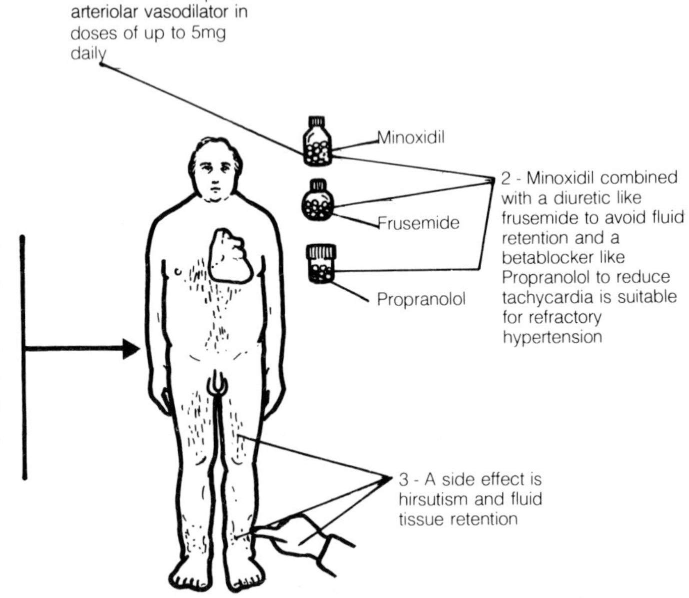
1 - Minoxidil is a potent arteriolar vasodilator in doses of up to 5mg daily
Minoxidil
Frusemide
Propranolol
2 - Minoxidil combined with a diuretic like frusemide to avoid fluid retention and a betablocker like Propranolol to reduce tachycardia is suitable for refractory hypertension
3 - A side effect is hirsutism and fluid tissue retention

Resistant cases of hypertension

Minoxidil is a potent arteriolar vasodilator which can be used in combination with a diuretic to avoid fluid retention and a betablocker to reduce tachycardia, in doses of up to 45mg daily. It is only suitable for refractory hypertension as hirsutism and fluid retention are side-effects.

Captopril acts by inhibiting angiotensin converting enzymes (ACE) as well as inhibiting angiotensin II at receptor sites. It acts as a vasodilator and can cause a dramatic drop in blood pressure. An initial test dose of 6.25mg should therefore be given before the dosage is gradually increased to a maximum of 150mg/day. This drug is effective when plasma renin levels are high as in renal artery stenosis but also reduces blood pressure in ordinary essential hypertension. Other ACE inhibitors will soon be available.

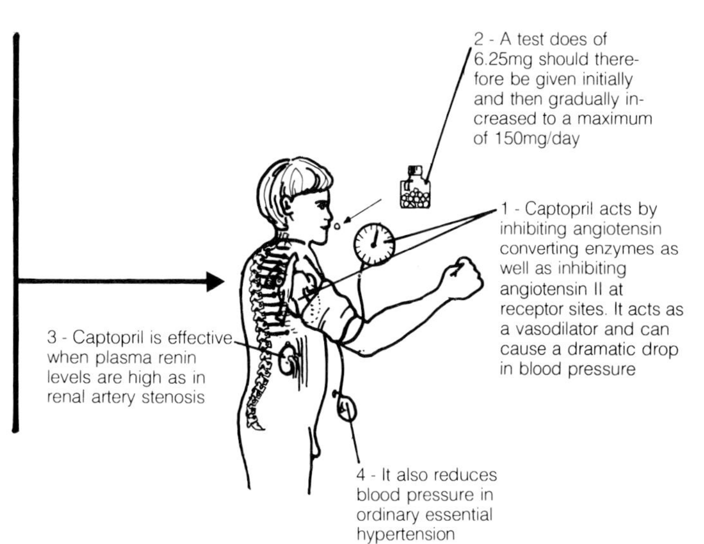
2 - A test does of 6.25mg should therefore be given initially and then gradually increased to a maximum of 150mg/day
1 - Captopril acts by inhibiting angiotensin converting enzymes as well as inhibiting angiotensin II at receptor sites. It acts as a vasodilator and can cause a dramatic drop in blood pressure
3 - Captopril is effective when plasma renin levels are high as in renal artery stenosis
4 - It also reduces blood pressure in ordinary essential hypertension

Malignant Hypertension

Intravenous drugs are rarely necessary in malignant hypertension as control of blood pressure can usually be achieved with bed rest and oral medications. However, if the patient is fitting, has fluctuating neurological signs or is in severe left ventricular failure, intravenous drugs may be necessary. The blood pressure should not be lowered too rapidly or cerebral ischaemia can result. Nitroprusside is a potent vasodilator and is therefore useful if severe heart failure is present. The dosage is initially 10 μg/minute increasing to 75 μg/minute as an intravenous infusion. The blood pressure should be continuously monitored preferably on an Intensive Care Unit. Hydralazine can be given cautiously as in intravenous infusion of 5-10mg over 20 minutes. There is less danger of serious hypotension than with nitroprusside, but the blood pressure should be carefully monitored. Labetalol can be infused at a rate of 2mg/minute to a total of 2mg/kg body weight. It may worsen heart failure but causes a smooth fall in blood pressure which is dose related.

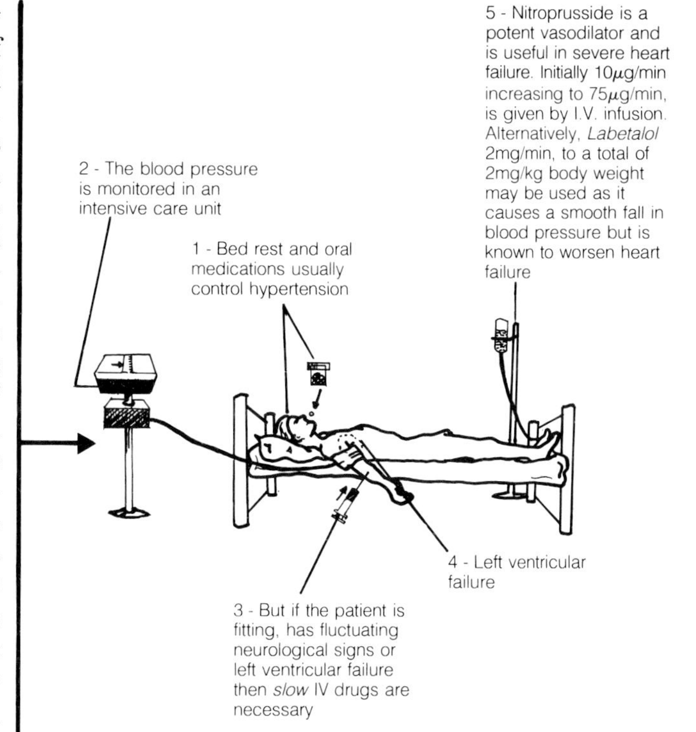
5 - Nitroprusside is a potent vasodilator and is useful in severe heart failure. Initially 10μg/min increasing to 75μg/min, is given by I.V. infusion. Alternatively, *Labetalol* 2mg/min, to a total of 2mg/kg body weight may be used as it causes a smooth fall in blood pressure but is known to worsen heart failure
2 - The blood pressure is monitored in an intensive care unit
1 - Bed rest and oral medications usually control hypertension
4 - Left ventricular failure
3 - But if the patient is fitting, has fluctuating neurological signs or left ventricular failure then *slow* IV drugs are necessary

The renin-angiotensin-aldosterone system

Renin is a proteolytic enzyme which is released from the juxtaglomerular apparatus of the kidney in response to a fall in the blood flow. Renin acts on a renin substrate (alpha II globulin) to release angiotensin I, which under the influence of converting enzyme becomes angiotensin II. This substance causes vasoconstriction, and also stimulates aldosterone release which causes sodium and water retention. Both these actions tend to raise the blood pressure. Plasma renin is low in about 20% of patients with essential hypertension, and high in about 10%. Drugs may alter renin levels (see diagram). In Conn's syndrome (primary hyperaldosteronism) aldosterone levels are high and renin levels low due to negative feedback. In renal disease with secondary hyperaldosteronism, renin and aldosterone levels are high.

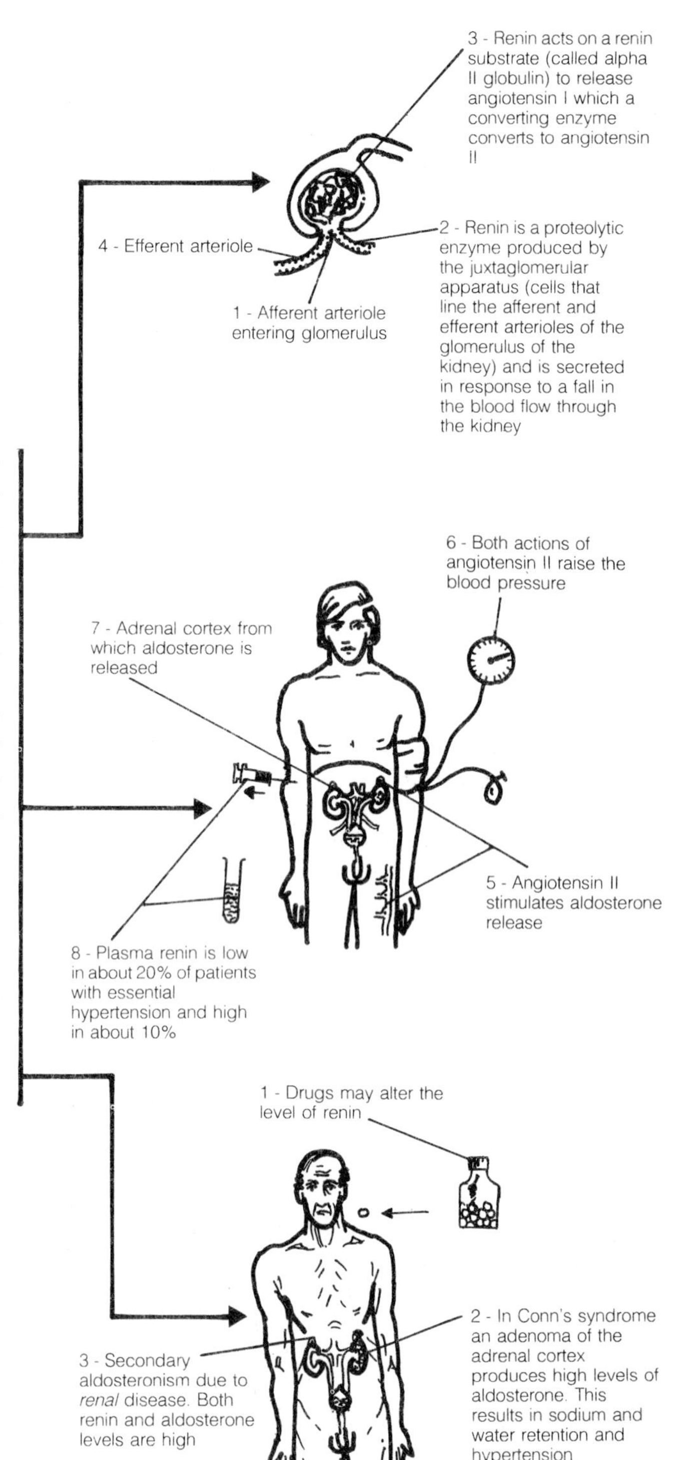
3 - Renin acts on a renin substrate (called alpha II globulin) to release angiotensin I which a converting enzyme converts to angiotensin II
4 - Efferent arteriole
2 - Renin is a proteolytic enzyme produced by the juxtaglomerular apparatus (cells that line the afferent and efferent arterioles of the glomerulus of the kidney) and is secreted in response to a fall in the blood flow through the kidney
1 - Afferent arteriole entering glomerulus
6 - Both actions of angiotensin II raise the blood pressure
7 - Adrenal cortex from which aldosterone is released
5 - Angiotensin II stimulates aldosterone release
8 - Plasma renin is low in about 20% of patients with essential hypertension and high in about 10%
1 - Drugs may alter the level of renin
3 - Secondary aldosteronism due to *renal* disease. Both renin and aldosterone levels are high
2 - In Conn's syndrome an adenoma of the adrenal cortex produces high levels of aldosterone. This results in sodium and water retention and hypertension

HEART FAILURE

Heart failure is the failure of cardiac output to meet the body's circulatory demands. Cardiac output depends on stroke volume, which in turn depends on the filling pressure (preload, afterload, and myocardial contractility).

Preload: In the normal heart, an increase in ventricular volume increases the force of contraction in proportion to the stretch of the fibres (the Frank-Starling effect). In the failing heart, this relationship fails, and left venticular filling pressure or preload rises.

Myocardial contractility: In heart failure, myocardial contractility is impaired. This is usually due to myocardial strain due to longstanding increase in mechanical load. The ventricle usually hypertrophies in response to load, and this results in increasing stiffness of its wall and ultimately impaired efficiency.

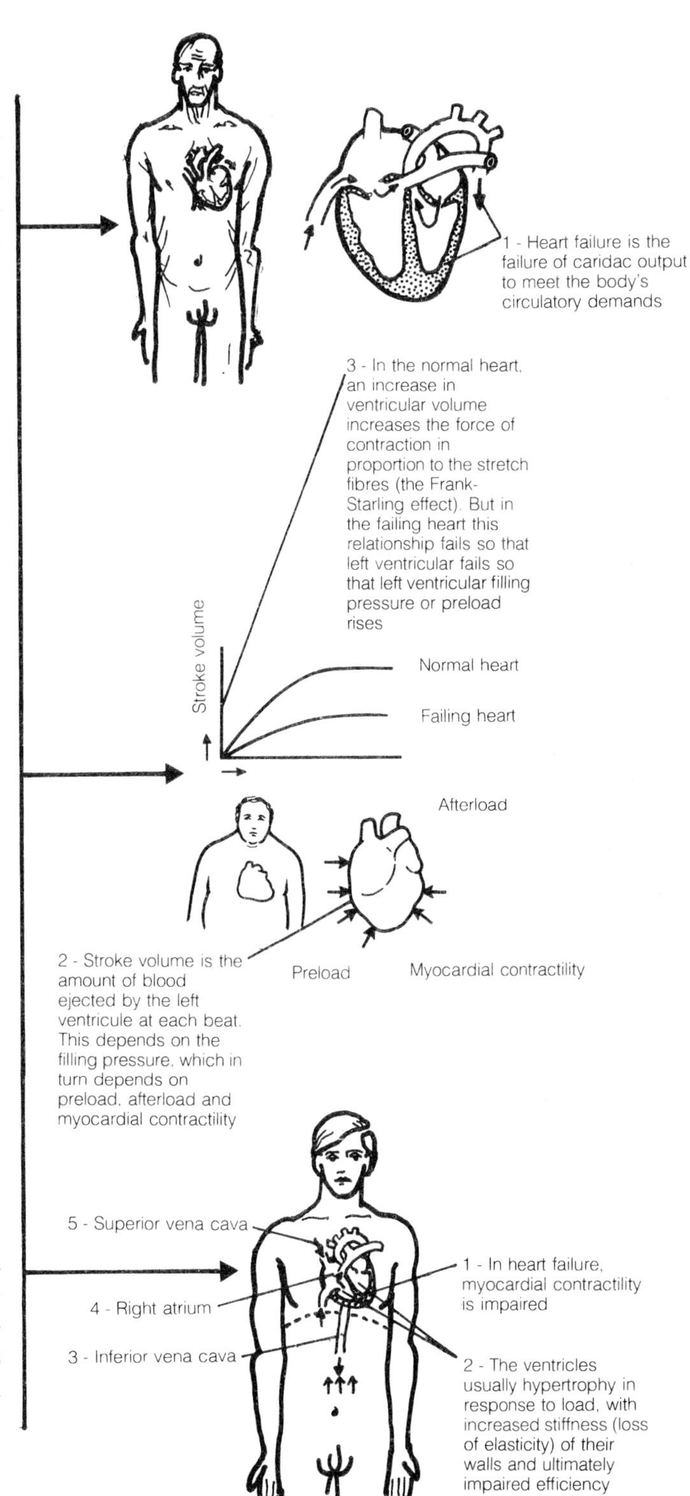
1 - Heart failure is the failure of caridac output to meet the body's circulatory demands
3 - In the normal heart, an increase in ventricular volume increases the force of contraction in proportion to the stretch fibres (the Frank-Starling effect). But in the failing heart this relationship fails so that left ventricular fails so that left ventricular filling pressure or preload rises
Stroke volume
Normal heart
Failing heart
Afterload
Preload
Myocardial contractility
2 - Stroke volume is the amount of blood ejected by the left ventricule at each beat. This depends on the filling pressure, which in turn depends on preload, afterload and myocardial contractility
5 - Superior vena cava
4 - Right atrium
3 - Inferior vena cava
1 - In heart failure, myocardial contractility is impaired
2 - The ventricles usually hypertrophy in response to load, with increased stiffness (loss of elasticity) of their walls and ultimately impaired efficiency

Afterload This is the resistance encountered by the ventricles; that is pulmonary and arterial resistance.

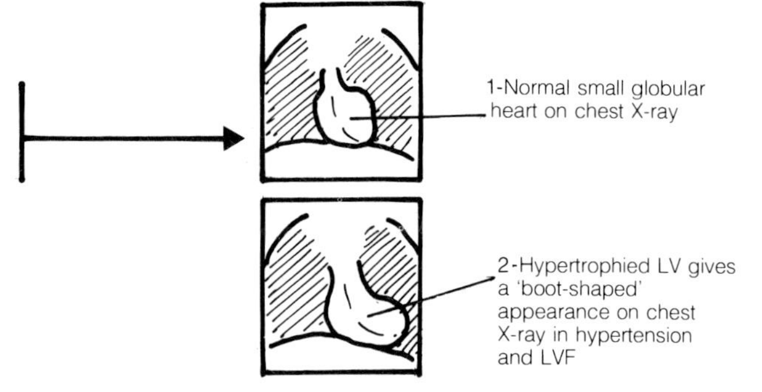
1-Normal small globular heart on chest X-ray
2-Hypertrophied LV gives a 'boot-shaped' appearance on chest X-ray in hypertension and LVF

Features of cardiac failure

Fluid retention In heart failure there is reduced renal blood flow. As a compensatory homeostatic mechanism sodium is retained, with fluid retention and oedema. The oedema puts on pressure, and collects at the ankles if the patient is ambulant and at the sacrum if he is bed bound.

Tachycardia Heart rate increases as a compensatory mechanism in heart failure and is a major factor in increasing cardiac output.

Raised venous pressure occurs and is best assessed from the internal jugular veins. The liver may be enlarged and tender due to congestion.

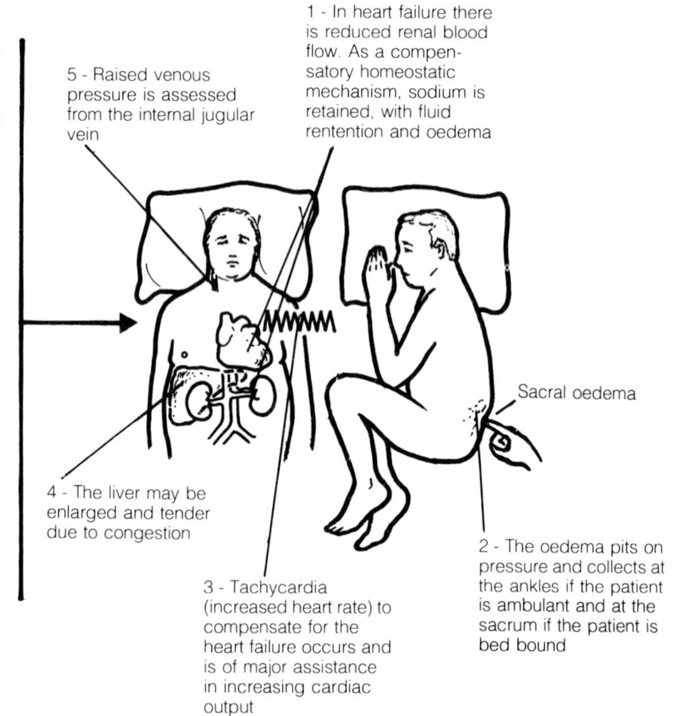
1 - In heart failure there is reduced renal blood flow. As a compensatory homeostatic mechanism, sodium is retained, with fluid rentention and oedema
5 - Raised venous pressure is assessed from the internal jugular vein
Sacral oedema
4 - The liver may be enlarged and tender due to congestion
3 - Tachycardia (increased heart rate) to compensate for the heart failure occurs and is of major assistance in increasing cardiac output
2 - The oedema pits on pressure and collects at the ankles if the patient is ambulant and at the sacrum if the patient is bed bound

Left ventricular hypertrophy and dilatation The heart adapts to increased afterload by hypertrophy (increase in cell size). This initially increases contractility, but severe hypertrophy is ultimately negatively inotropic. The heart adapts to increased diastolic volumes by chronic dilatation.

Dyspnoea Difficulty with breathing is a common feature of heart failure. It is commonly seen in patients with raised left atrial and pulmonary venous pressures. Dyspnoea initially develops on exertion, but as heart failure progresses, orthopnoea (breathlessness on lying flat) and paroxysmal nocturnal dyspnoea (severe attacks of shortness of breath at night) occur.

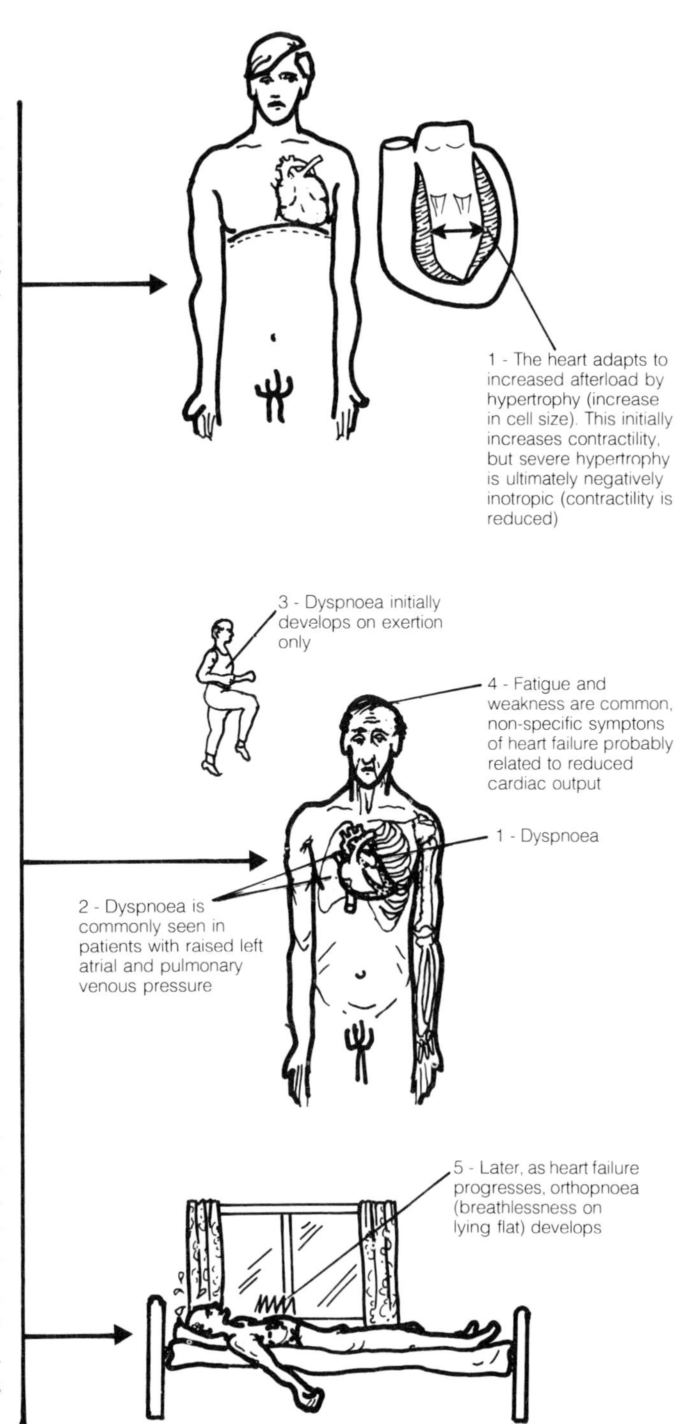
1 - The heart adapts to increased afterload by hypertrophy (increase in cell size). This initially increases contractility, but severe hypertrophy is ultimately negatively inotropic (contractility is reduced)
3 - Dyspnoea initially develops on exertion only
4 - Fatigue and weakness are common, non-specific symptons of heart failure probably related to reduced cardiac output
1 - Dyspnoea
2 - Dyspnoea is commonly seen in patients with raised left atrial and pulmonary venous pressure
5 - Later, as heart failure progresses, orthopnoea (breathlessness on lying flat) develops

Fatigue and weakness These are common, non-specific symptoms of heart failure, probably related to reduced cardiac output.

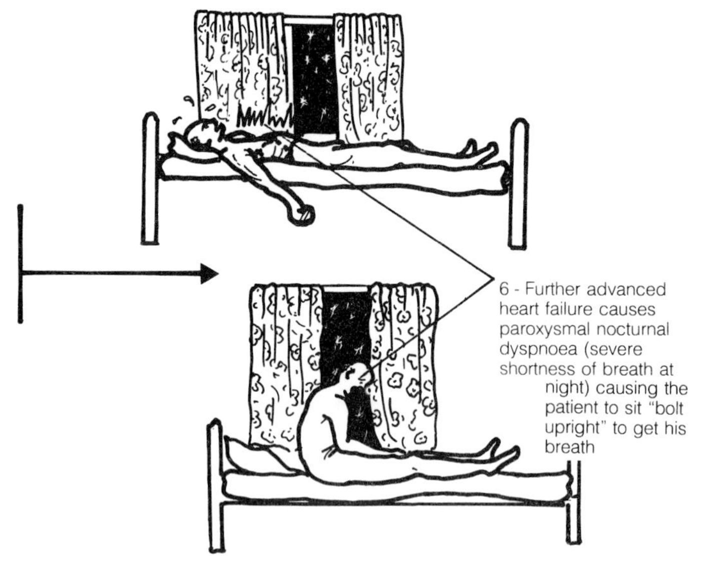
6 - Further advanced heart failure causes paroxysmal nocturnal dyspnoea (severe shortness of breath at night) causing the patient to sit "bolt upright" to get his breath

LEFT VENTRICULAR FAILURE

This may be produced by:
1. Pressure load as in hypertension or aortic stenosis.
2. Volume load, for example mitral or aortic regurgitation.
3. Myocardial disease predominantly affecting the left ventricle, for example ischaemic heart disease, myocardial infarction, cardiomyopathy..

Symptoms

Progressive breathlessness is the main feature of left ventricular failure. This may come on insidiously, with a gradual onset of exertional dyspnoea, orthopnoea and paroxysmal nocturnal dyspnoea. Acute left ventricular failure sometimes occurs after myocardial infarction and may present with acute breathlessness and wheezing.

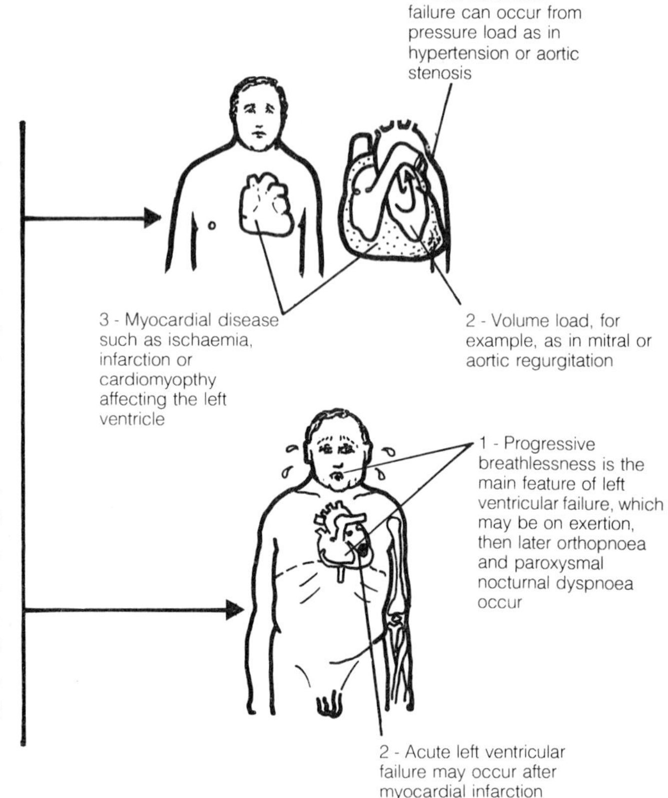
1 - Left ventricular failure can occur from pressure load as in hypertension or aortic stenosis
3 - Myocardial disease such as ischaemia, infarction or cardiomyopthy affecting the left ventricle
2 - Volume load, for example, as in mitral or aortic regurgitation
1 - Progressive breathlessness is the main feature of left ventricular failure, which may be on exertion, then later orthopnoea and paroxysmal nocturnal dyspnoea occur
2 - Acute left ventricular failure may occur after myocardial infarction

Signs

On examination there is a tachycardia and evidence of left ventricular hypertrophy. A triple rhythm is usually present and on auscultation of the chest, fine bilateral inspiratory crackles are present at the bases. In severe left ventricular failure with pulmonary oedema, there is severe breathlessness, cyanosis and cough productive of blood stained frothy sputum. Wheezing may be marked and fine inspiratory crackles can be heard all over the chest. The ECG may show features of left ventricular hypertrophy or T wave changes over the left ventricular lead. The chest X-ray may show features of pulmonary oedema.

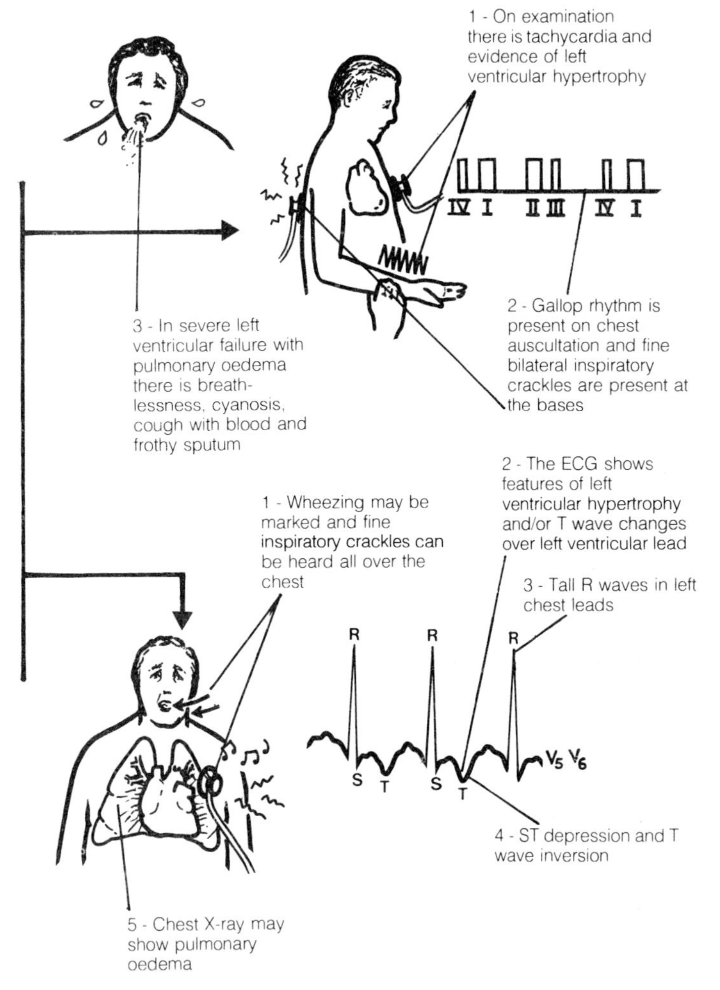
1 - On examination there is tachycardia and evidence of left ventricular hypertrophy
IV I
II III
IV I
2 - Gallop rhythm is present on chest auscultation and fine bilateral inspiratory crackles are present at the bases
3 - In severe left ventricular failure with pulmonary oedema there is breathlessness, cyanosis, cough with blood and frothy sputum
1 - Wheezing may be marked and fine inspiratory crackles can be heard all over the chest
2 - The ECG shows features of left ventricular hypertrophy and/or T wave changes over left ventricular lead
3 - Tall R waves in left chest leads
R
R
R
S
T
S
T
V5 V6
4 - ST depression and T wave inversion
5 - Chest X-ray may show pulmonary oedema

Treatment

The patient should be sat upright in bed and given oxygen at high concentration (or in controlled quantities where there is lung disease). Frusemide initially in doses of 40mg intravenously (but up to 120mg may be required) produces a prompt diuresis with removal of oedema fluid from the lungs. Diamorphine 5-20mg intravenously has analgesic and sedative actions as well causing venous and arteriolar dilation. Nitrates are effective in reducing preload and also have a mild arteriolar dilating action. They can be given orally but in acute pulmonary oedema intravenous nitrates are probably more effective. In critically ill patients sodium nitroprusside can be given as an intravenous infusion. This is so potent that intra-arterial and pulmonary capillary wedge pressures should ideally be monitored during its administraiton.

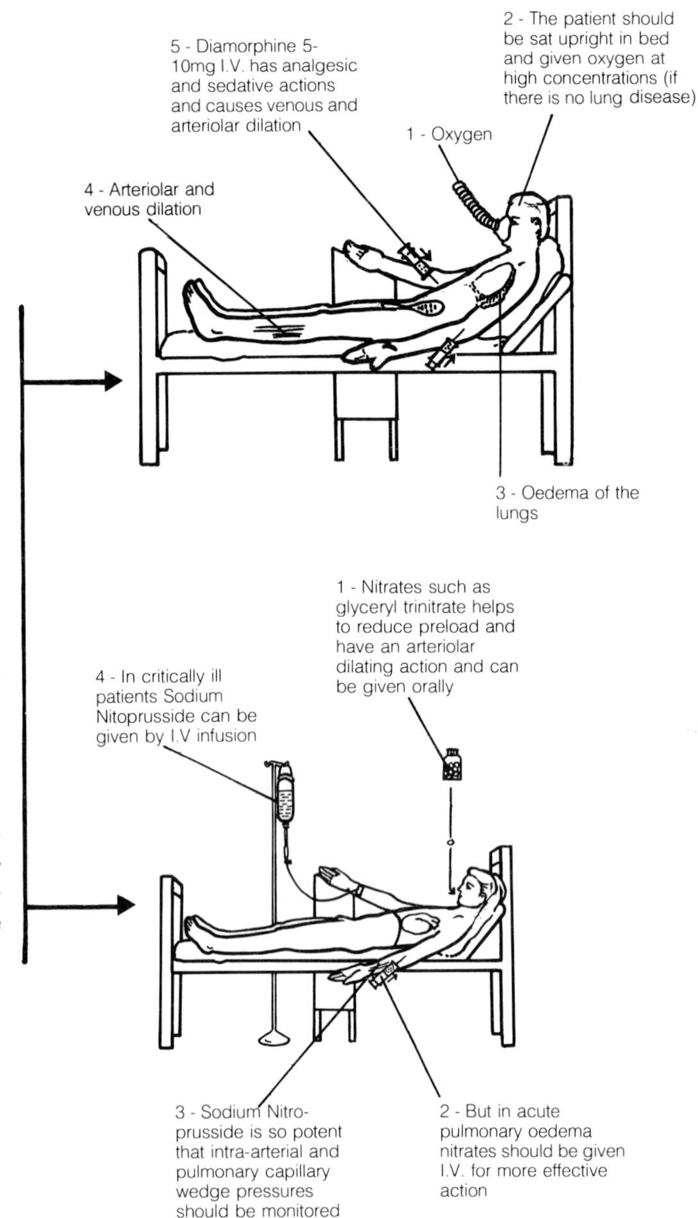
5 - Diamorphine 5-10mg I.V. has analgesic and sedative actions and causes venous and arteriolar dilation
2 - The patient should be sat upright in bed and given oxygen at high concentrations (if there is no lung disease)
1 - Oxygen
4 - Arteriolar and venous dilation
3 - Oedema of the lungs
1 - Nitrates such as glyceryl trinitrate helps to reduce preload and have an arteriolar dilating action and can be given orally
4 - In critically ill patients Sodium Nitoprusside can be given by I.V infusion
3 - Sodium Nitro-prusside is so potent that intra-arterial and pulmonary capillary wedge pressures should be monitored during its administrtion
2 - But in acute pulmonary oedema nitrates should be given I.V. for more effective action

RIGHT VENTRICULAR FAILURE

Right ventricular failure can occur secondary to left ventricular failure or longstanding mitral valve disease.. It can also develop secondary to diseases in the lung such as chronic airways obstruction, pulmonary fibrosis or pulmonary emboli. Valvular lesions of the right heart, such as pulmonary stenosis or atrial septal defect can also lead to right ventricular failure.

2 - It may follow lung diseases such as chronic airways obstructon, pulmonary fibrosis or emboli
1 - Right ventricular failure can occur secondary to left ventricular failure or after long standing mitral valve disease

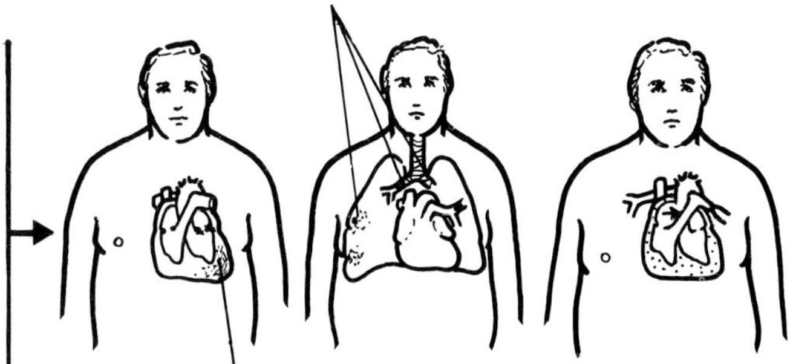

Clinical features

The clinical features of right ventricular failure are raised venous pressure, peripheral oedema and an enlarged tender liver. There may be mild jaundice due to the liver congestion, and stomach congestion can lead to nausea and loss of appetite.

3 - Stomach congestion can lead to nausea and loss of appetite
1 - Right ventricular failure causes raised venous pressure, peripheral oedema and enlarged liver
2 - Enlarged and tender liver due to congestion.This may cause jaundice

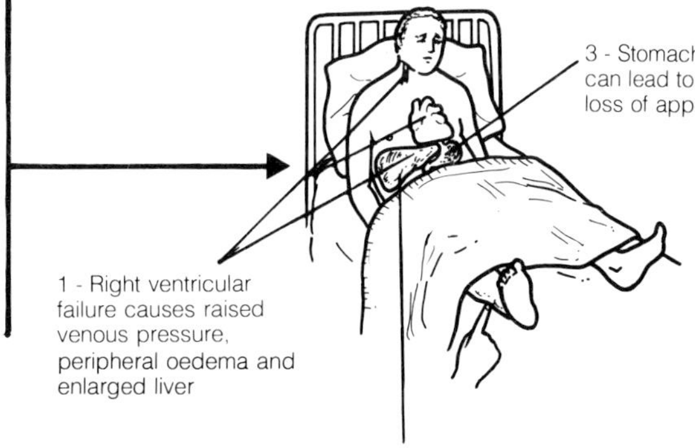

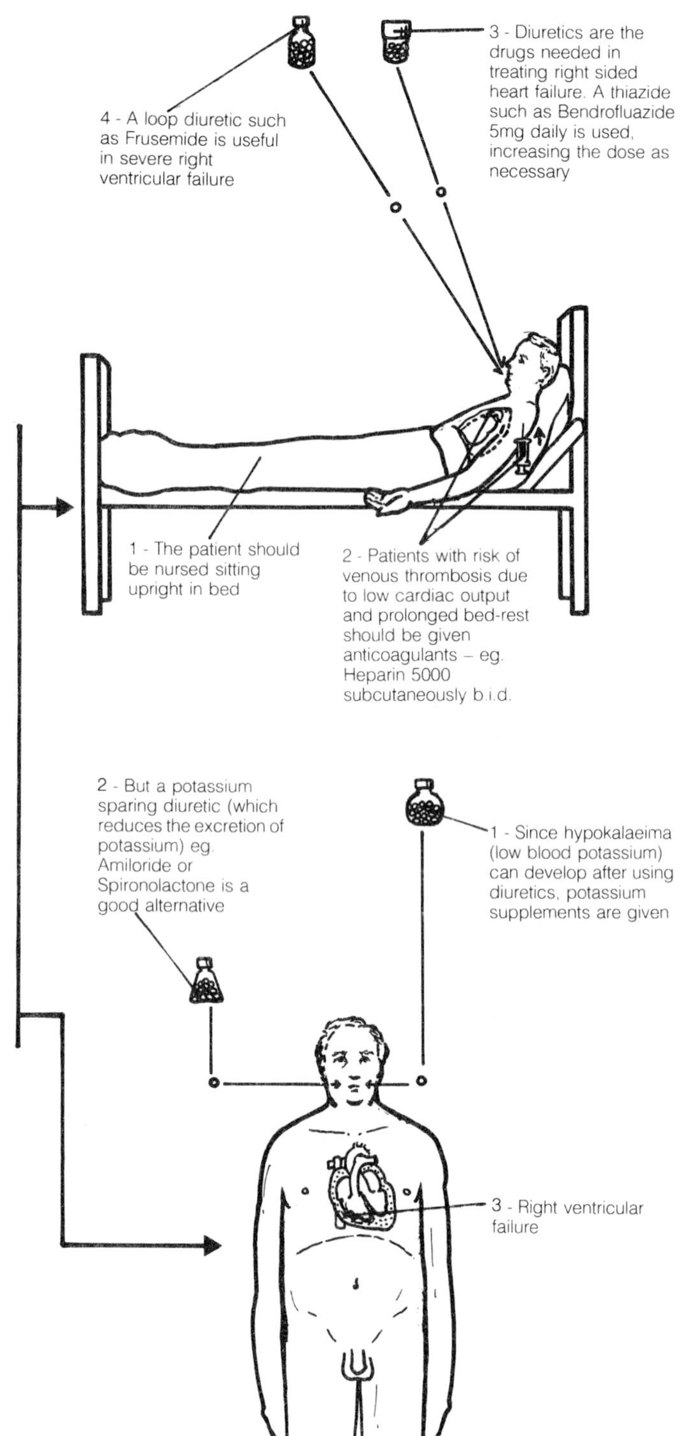
4 - A loop diuretic such as Frusemide is useful in severe right ventricular failure
3 - Diuretics are the drugs needed in treating right sided heart failure. A thiazide such as Bendrofluazide 5mg daily is used, increasing the dose as necessary
1 - The patient should be nursed sitting upright in bed
2 - Patients with risk of venous thrombosis due to low cardiac output and prolonged bed-rest should be given anticoagulants – eg. Heparin 5000 subcutaneously b.i.d.
2 - But a potassium sparing diuretic (which reduces the excretion of potassium) eg. Amiloride or Spironolactone is a good alternative
1 - Since hypokalaeima (low blood potassium) can develop after using diuretics, potassium supplements are given
3 - Right ventricular failure

Treatment of right ventricular failure

Bed rest is important and the patient should be nursed sitting upright. Patients are at increased risk of venous thrombosis because of their low cardiac output and patients on prolonged bed rest should have anti-coagulant prophylaxis (for example heparin 5000 units subcutaneously bd). **Diuretics** are the main drugs used in treating right sided heart failure. In mild heart failure, a thiazide such as bendrofluazide 5mg daily is used. The loop diuretics such as frusemide are useful in more severe cases of right ventricular failure. They also cause hypokalaemia; therefore potassium supplements usually have to be given. Alternatively they can be combined with potassium sparing diuretic such as amiloride or spironolactone.

Digoxin is no longer considered to be the first line treatment in cardiac failure. In patients in atrial fibrillation however, it is useful in slowing the heart rate and increasing ventricular contractility. In patients with more severe congestive failure, vasodilators are used. The choice of vasodilator depends on the clinical features of the patient. Where there is pulmonary congestion with dyspnoea, longacting nitrates are beneficial (such as isosorbide dinitrate slow release 20-40mg twice daily). If there is predominant forward failure, with low cardiac output and fatigue, Hydralazine usually produces a response. Critically ill patients (especially if pulmonary oedema is present) can be treated with a nitroprusside infusion (see above). Additional dopamine or dobutamine may be necessary (see myocardial infarction section).

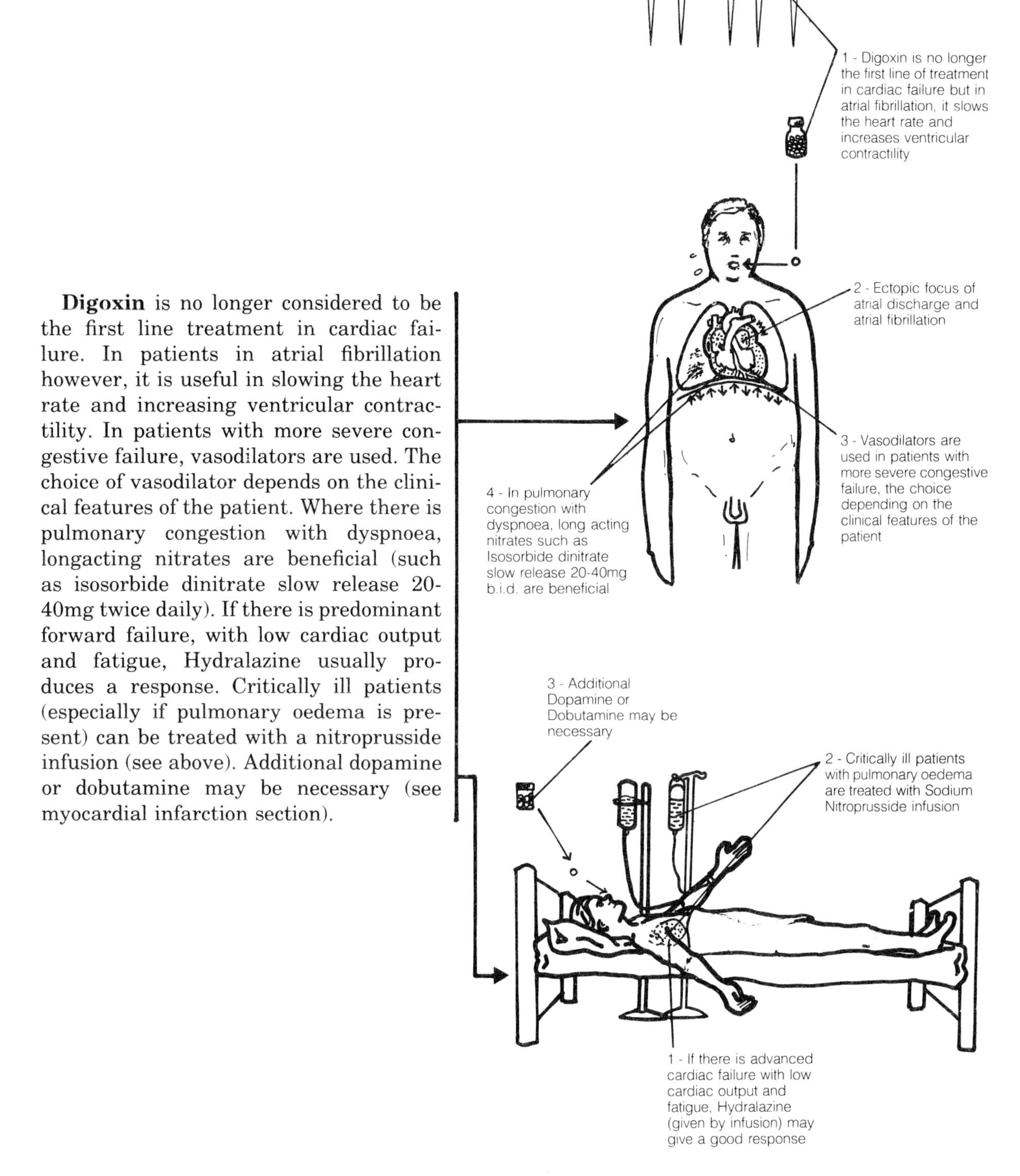
1 - Digoxin is no longer the first line of treatment in cardiac failure but in atrial fibrillation, it slows the heart rate and increases ventricular contractility
2 - Ectopic focus of atrial discharge and atrial fibrillation
3 - Vasodilators are used in patients with more severe congestive failure, the choice depending on the clinical features of the patient
4 - In pulmonary congestion with dyspnoea, long acting nitrates such as Isosorbide dinitrate slow release 20-40mg b.i.d. are beneficial
3 - Additional Dopamine or Dobutamine may be necessary
2 - Critically ill patients with pulmonary oedema are treated with Sodium Nitroprusside infusion
1 - If there is advanced cardiac failure with low cardiac output and fatigue, Hydralazine (given by infusion) may give a good response

RHEUMATIC HEART DISEASE

Rheumatic fever is a febrile illness occuring after infection with the Lancefield group A beta haemolytic streptococcus. The infective organism is usually carried in the nasopharynx, and rheumatic fever follows 0.3% of streptococcal infections. It tends to affect young people, the peak incidence being at 8 years. There are no sexual or racial differences in suseptibility, but poverty and overcrowding are more likely to produce an epidemic. Rheumatic fever is the commonest cause of heart disease in developing countries, where the prevalence is 7-33 in 1000. In countries where good health facilities exist, it has been shown that early penicillin treatment for streptococcal pharyngitis prevents rheumatic fever. The prevalence in these countries is 0.5-1 in 1000. The mechanism of damage in rheumatic fever is unknown, but certain components of the organism and host tissue are antigenically similar and it is thought that the damage is due to immunological cross reaction.

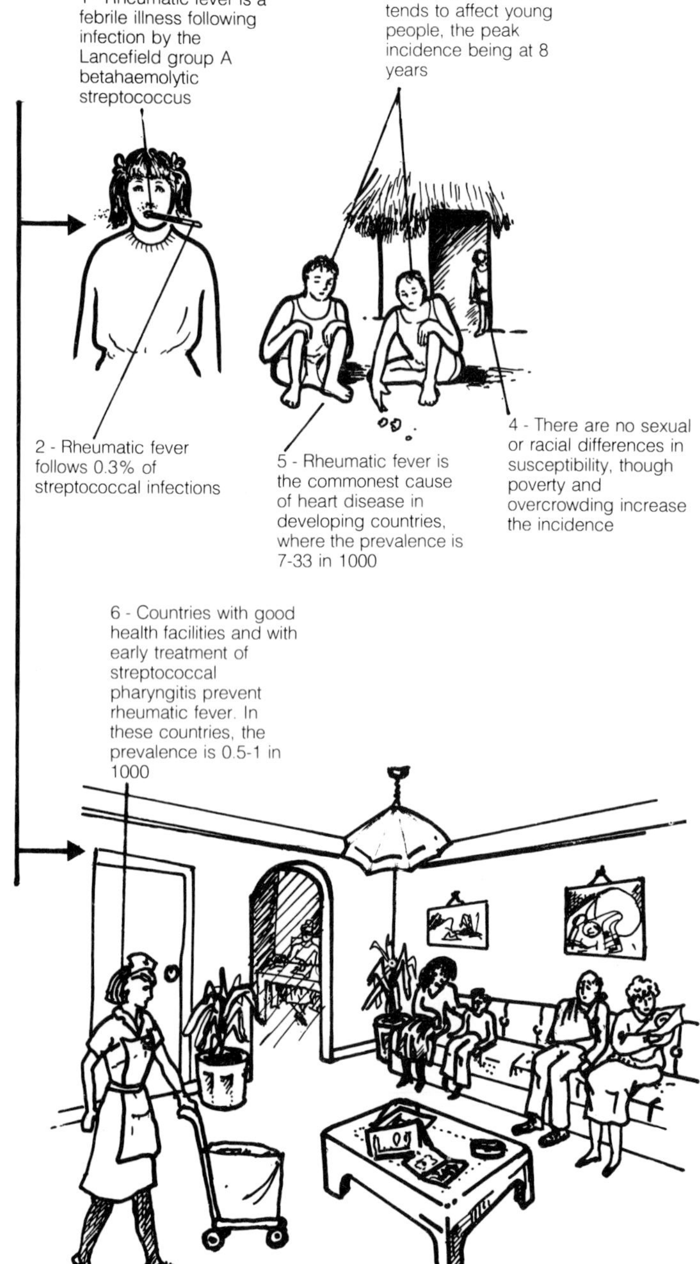
1 - Rheumatic fever is a febrile illness following infection by the Lancefield group A betahaemolytic streptococcus
2 - Rheumatic fever follows 0.3% of streptococcal infections
3 - Rheumatic fever tends to affect young people, the peak incidence being at 8 years
4 - There are no sexual or racial differences in susceptibility, though poverty and overcrowding increase the incidence
5 - Rheumatic fever is the commonest cause of heart disease in developing countries, where the prevalence is 7-33 in 1000
6 - Countries with good health facilities and with early treatment of streptococcal pharyngitis prevent rheumatic fever. In these countries, the prevalence is 0.5-1 in 1000

Clinical features

Symptoms usually develop 7-20 days after a streptococcal sore throat. The patient develops fever, malaise, and weight loss. Arthralgia (flitting joint pains affecting the large joints) occurs, and an arthritis with swelling, redness and tenderness of the affected joints occurs in three out of four patients. The large joints are usually affected one after another, the swelling lasting about 5 days. Symptoms respond dramatically to treatment with aspirin. Erythema marginatum is a migratory skin rash which is typical of rheumatic fever; it is red, scaly and serpiginous, with a paler centre. Subcutaneous nodules can occur in areas where the skin is in close contact with bone such as over the forearms, shins, wrists and ankles. Carditis occurs in the majority of cases. Patients may develop tachycardia with gallop rhythm or new heart murmurs, or there may be a change in the character of existing murmurs.

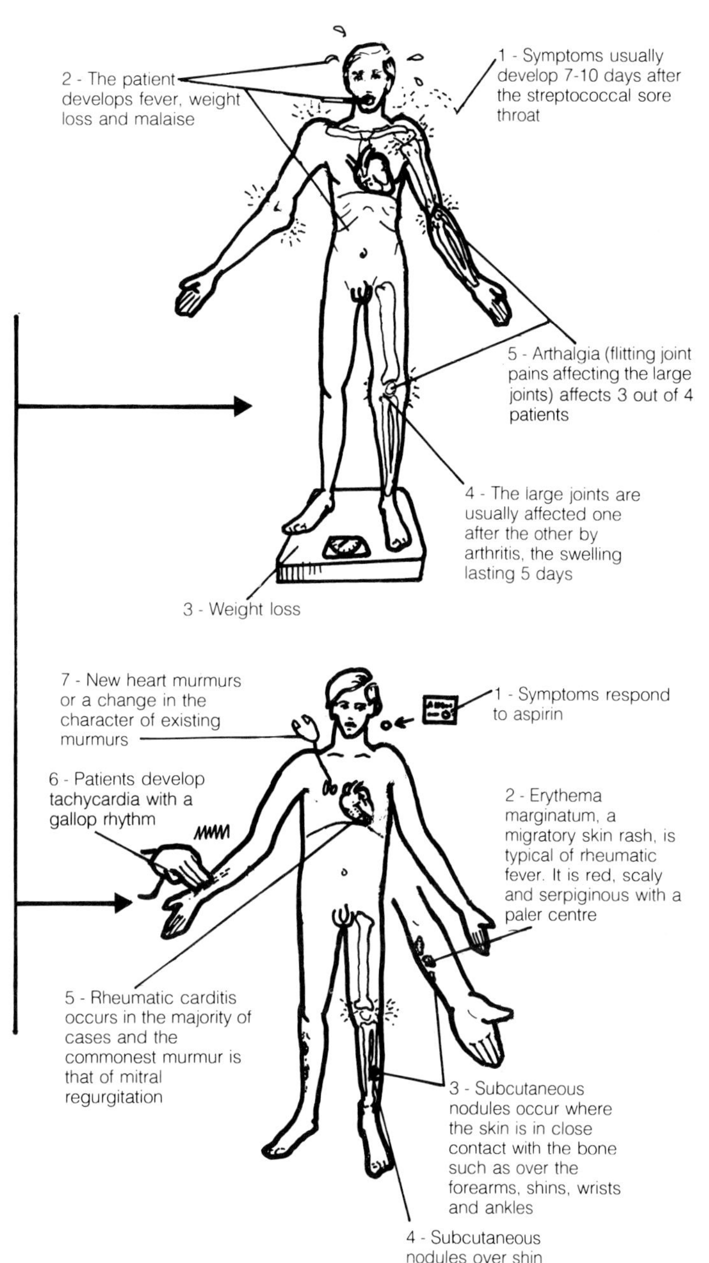
1 - Symptoms usually develop 7-10 days after the streptococcal sore throat
2 - The patient develops fever, weight loss and malaise
3 - Weight loss
4 - The large joints are usually affected one after the other by arthritis, the swelling lasting 5 days
5 - Arthalgia (flitting joint pains affecting the large joints) affects 3 out of 4 patients
1 - Symptoms respond to aspirin
2 - Erythema marginatum, a migratory skin rash, is typical of rheumatic fever. It is red, scaly and serpiginous with a paler centre
3 - Subcutaneous nodules occur where the skin is in close contact with the bone such as over the forearms, shins, wrists and ankles
4 - Subcutaneous nodules over shin
5 - Rheumatic carditis occurs in the majority of cases and the commonest murmur is that of mitral regurgitation
6 - Patients develop tachycardia with a gallop rhythm
7 - New heart murmurs or a change in the character of existing murmurs

The commonest murmur is a transient murmur of mitral regurgitation. A mitral diastolic murmur known as the **Carey-Coombs murmur** indicates mitral valve involvement and persists in two-thirds of the cases. An immediate diastolic murmur heard at the left sternal edge indicates aortic regurgitation and usually persists. Myocarditis can occur resulting in cardiomegaly or heart failure. Pericarditis is often associaed with myocarditis and is characterised by chest pain and a pericardial friction rub. Chorea is an unusual sequel of rheumatic fever. It is more common in females and occurs 3-6 months after the infection. The patient becomes clumsy and develops rapid jerky involuntary movements which are semi-purposeful and disappear during sleep.

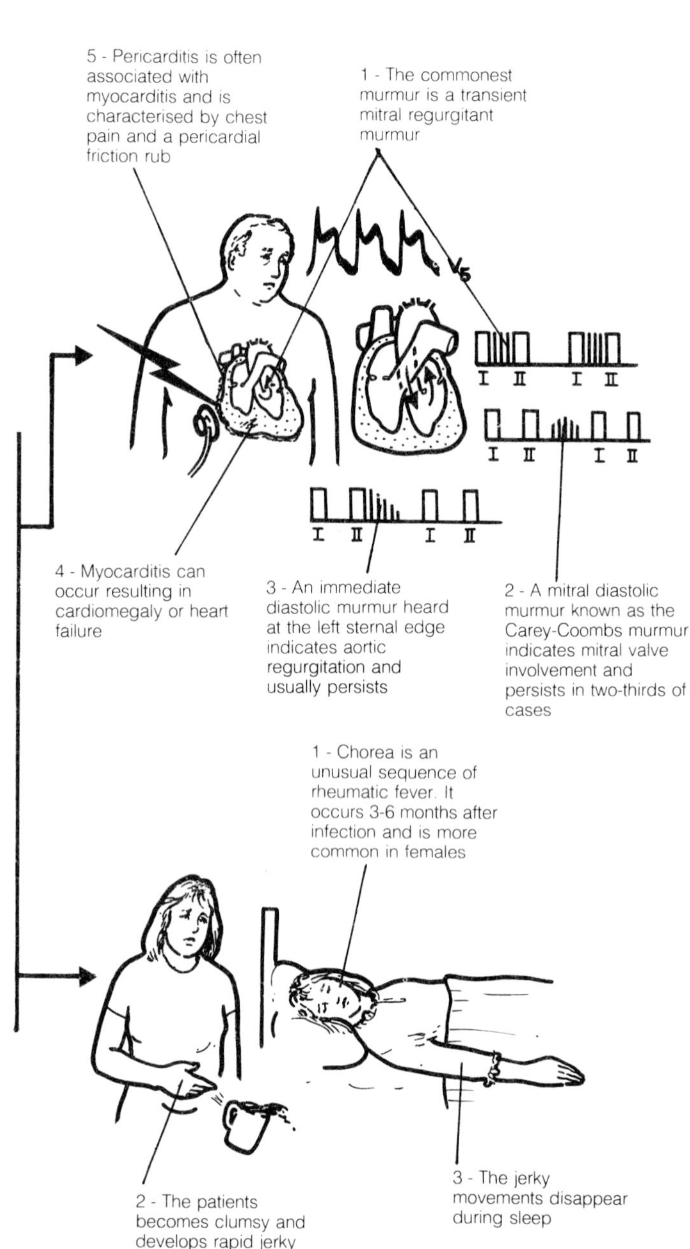
5 - Pericarditis is often associated with myocarditis and is characterised by chest pain and a pericardial friction rub
1 - The commonest murmur is a transient mitral regurgitant murmur
V5
I II I II
I II I II
I II I II
4 - Myocarditis can occur resulting in cardiomegaly or heart failure
3 - An immediate diastolic murmur heard at the left sternal edge indicates aortic regurgitation and usually persists
2 - A mitral diastolic murmur known as the Carey-Coombs murmur indicates mitral valve involvement and persists in two-thirds of cases
1 - Chorea is an unusual sequence of rheumatic fever. It occurs 3-6 months after infection and is more common in females
2 - The patients becomes clumsy and develops rapid jerky involuntary movements which are semi-purposeful
3 - The jerky movements disappear during sleep

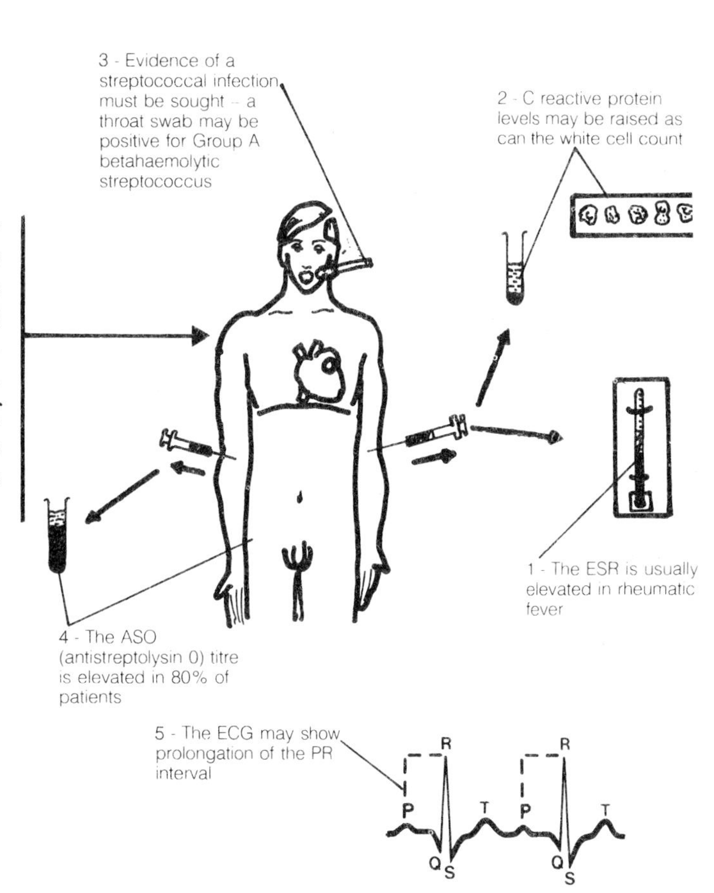
3 - Evidence of a streptococcal infection must be sought – a throat swab may be positive for Group A betahaemolytic streptococcus
2 - C reactive protein levels may be raised as can the white cell count
1 - The ESR is usually elevated in rheumatic fever
4 - The ASO (antistreptolysin 0) titre is elevated in 80% of patients
5 - The ECG may show prolongation of the PR interval
R
P
T
Q
S
R
P
T
Q
S

Investigations

The ESR is usually elevated in rheumatic fever. C reactive protein levels may be raised as is often the white cell count. The ECG may show prolongation of the PR interval. Evidence of a streptococcal infection must be sought. The ASO (antistreptolysin O) titre is elevated in 80% of patients, and a throat swab may be positive for group A beta haemolytic streptococcus.

The Duckettt Jones Criteria have been devised as a guide to the diagnosis of rheumatic fever.

Major criteria are carditis, arthritis, nodules, erythema marginatum and chorea.

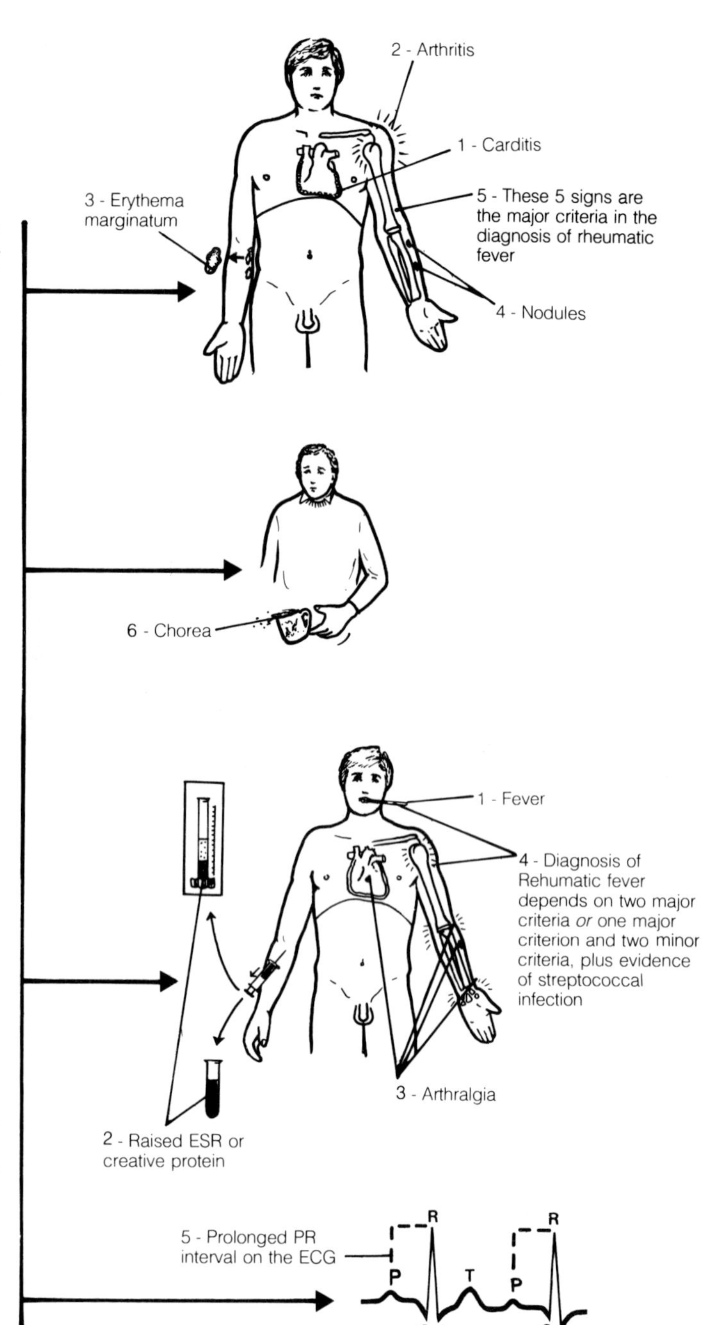
2 - Arthritis
1 - Carditis
3 - Erythema marginatum
5 - These 5 signs are the major criteria in the diagnosis of rheumatic fever
4 - Nodules
6 - Chorea
1 - Fever
4 - Diagnosis of Rehumatic fever depends on two major criteria *or* one major criterion and two minor criteria, plus evidence of streptococcal infection
3 - Arthralgia
2 - Raised ESR or creative protein
5 - Prolonged PR interval on the ECG
R
R
P
T
P
Q S
Q S

Minor criteria are fever, arthralgia, previous rheumatic fever, raised ESR or C reactive protein, and prolonged PR interval on the ECG.

To diagnose rheumatic fever, there must be two major criteria or one major and two minor criteria, plus evidence of streptococcal infection.

2 - Patients are examined daily for possible carditis and if heart failure develops, diuretic therapy may be required

1 - Patient is admitted to hospital for bed rest

1 - Digoxin may be required for atrial fibrillation

2 - Arthritis responds well to aspirin in doses of 70mg/kg/day, but if there is no response 40mg. Prednisolone daily may be used

4 - Haloperidol, 0.5g is useful in treating chorea

3 - Jerky movements of chorea

Treatment

At the onset of rheumatic fever, penicillin or erythromycin are given to eradicate any remaining streptococcal infection. All patients should be admitted to hospital and bed rest encouraged. Patients need to be examined daily for evidence of carditis, and if heart failure develops, diuretic therapy may be required. Digoxin may be necessary for atrial fibrillation. Arthritis responds well to aspirin in doses of 70mg/Kg/day. Prednisolone 40mg daily can be used if there is no response to salicylates. Haloperidol, 0.5g three times daily, is useful in treating chorea.

Prevention of rheumatic fever

If streptococcal sore throats are detected and treated early enough, rheumatic fever can be prevented. Penicillin V or erythromycin should be given (250mg four times daily for ten days). Once the attack has occured, penicillin treatment should be continued to prevent recurrence. Patients are usually kept on a maintenance dosage of penicillin V 250mg twice daily or sulphadiazine 1g daily. If there is poor compliance, benzylpenicillin can be given 1.2 munits intramuscularly/month. There is disagreement as to how long prophylaxis should be continued for, but it should be for at least 5 years.

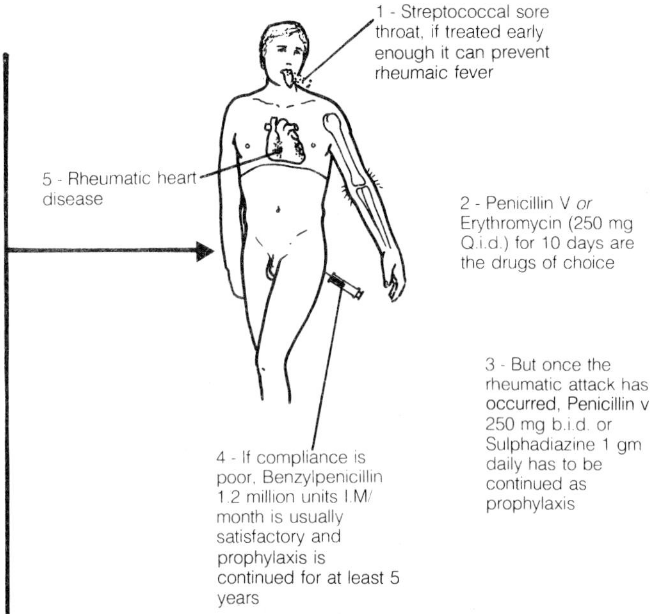

1 - Streptococcal sore throat, if treated early enough it can prevent rheumaic fever

2 - Penicillin V *or* Erythromycin (250 mg Q.i.d.) for 10 days are the drugs of choice

3 - But once the rheumatic attack has occurred, Penicillin v 250 mg b.i.d. or Sulphadiazine 1 gm daily has to be continued as prophylaxis

4 - If compliance is poor, Benzylpenicillin 1.2 million units I.M/ month is usually satisfactory and prophylaxis is continued for at least 5 years

5 - Rheumatic heart disease

Complications

Rheumatic heart disease is the major complication of rheumatic fever and occurs in a third of patients. The mitral valve is most commonly infected (70% of cases). Aortic valve disease occurs in 40% of cases, while the tricuspid and pulmonary valves are rarely involved (10% and 2% respectively).

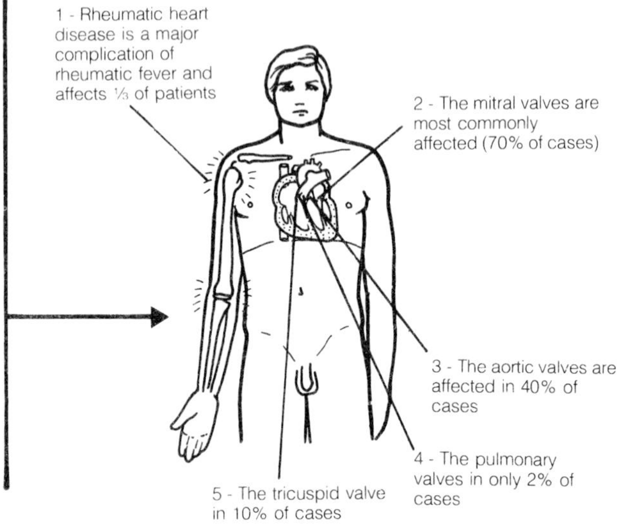

1 - Rheumatic heart disease is a major complication of rheumatic fever and affects ⅓ of patients

2 - The mitral valves are most commonly affected (70% of cases)

3 - The aortic valves are affected in 40% of cases

4 - The pulmonary valves in only 2% of cases

5 - The tricuspid valve in 10% of cases

MITRAL VALVE DISEASE

Mitral stenosis

Mitral stenosis is the commonest valve lesion following rheumatic fever although it may present 10-20 years after the attack. Mitral stenosis may be asymptomatic initially, but once the valve area has been reduced from 4-6 square cms to 2½ square cms, symptoms develop. The left atrial pressure gradually rises and this is transmitted to the pulmonary arteries and veins. As mitral stenosis progresses, pulmonary hypertension can develop with a rise in pulmonary artery pressure and thickening of the alveolar walls.

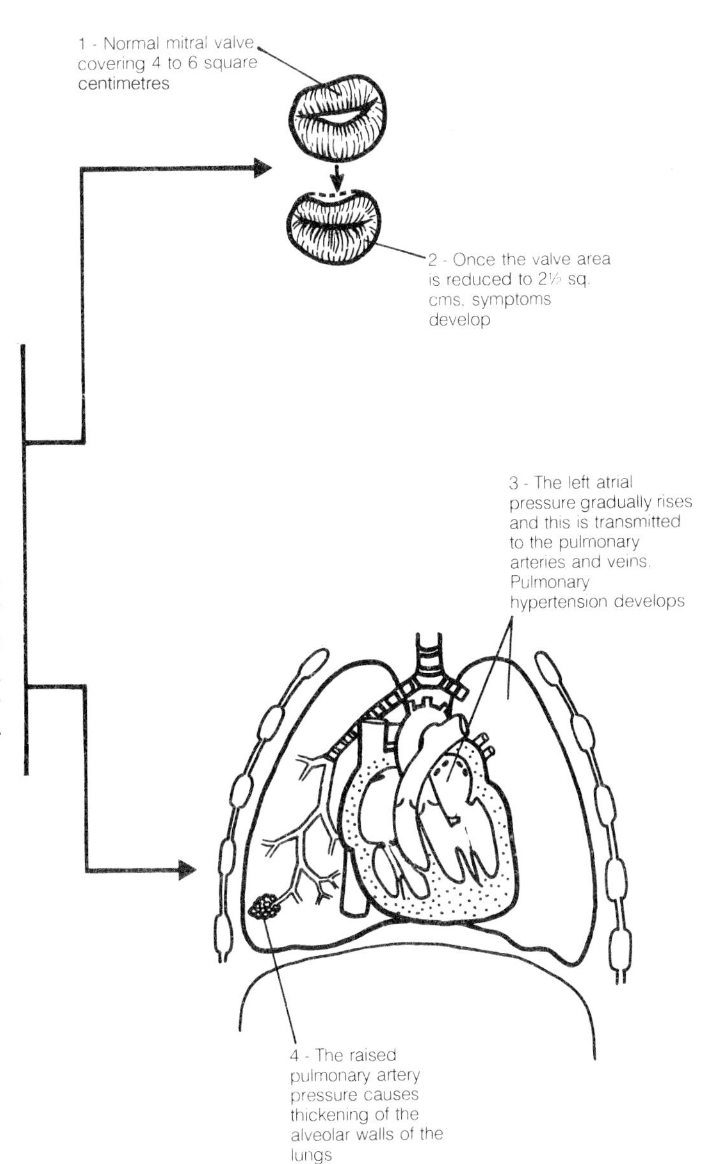
1 - Normal mitral valve covering 4 to 6 square centimetres
2 - Once the valve area is reduced to 2½ sq. cms. symptoms develop
3 - The left atrial pressure gradually rises and this is transmitted to the pulmonary arteries and veins. Pulmonary hypertension develops
4 - The raised pulmonary artery pressure causes thickening of the alveolar walls of the lungs

1 - The patient presents with breathlessness on exertion

Symptoms

The patient often presents with breathlessness on exertion, orthopnoea and paroxysmal nocturnal dyspnoea. "Bronchitis" and haemoptysis may occur due to pulmonary congestion. Atrial fibrillation is common and the patient may complain of palpitations or symptoms due to systemic emboli such as stroke or a cold pulseless leg. Because of low cardiac output, the patient may be fatigued and have cold extremities.

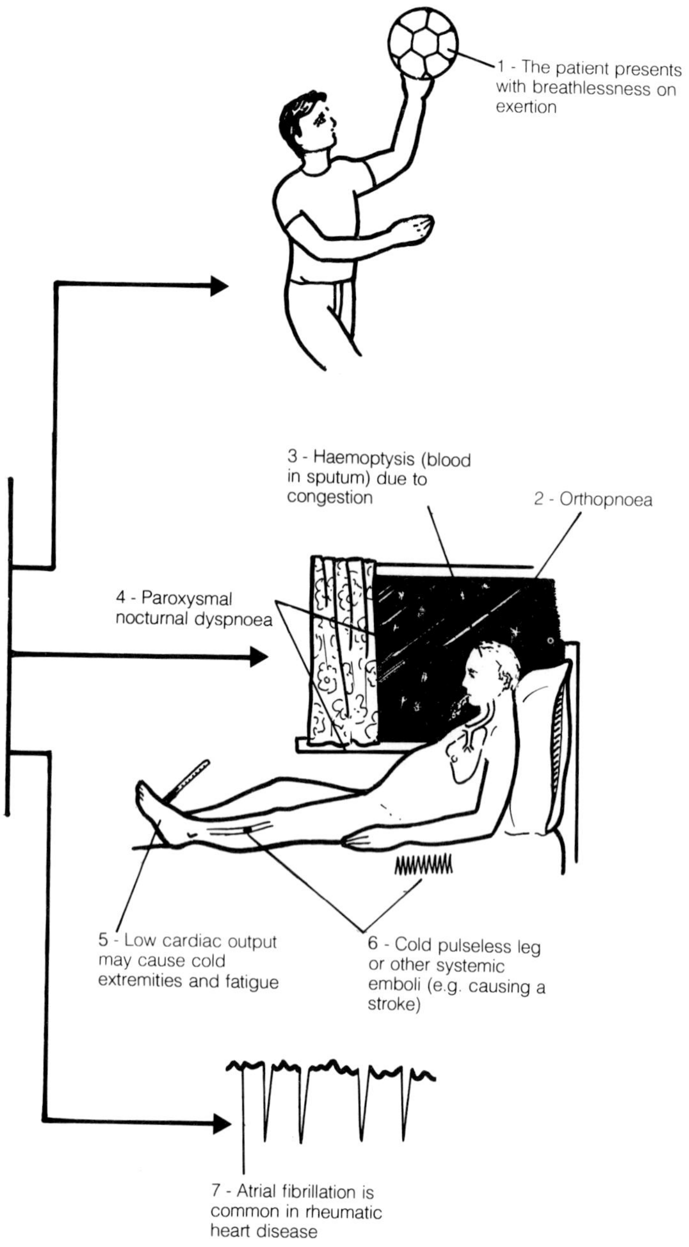

3 - Haemoptysis (blood in sputum) due to congestion

2 - Orthopnoea

4 - Paroxysmal nocturnal dyspnoea

5 - Low cardiac output may cause cold extremities and fatigue

6 - Cold pulseless leg or other systemic emboli (e.g. causing a stroke)

7 - Atrial fibrillation is common in rheumatic heart disease

Physical signs

In severe mitral stenosis with pulmonary hypertension, a malar flush may be present on the cheeks. The pulse may be irregular due to atrial fibrillation and the apex beat is "tapping" in character. There may be a left parasternal heave due to right ventricular hypertrophy. On auscultation there is a loud first heart sound, as the mitral valve is held open by high atrial pressure until slammed shut by ventricular systole. The second heart sound has a loud pulmonary component. If the valve is mobile, an opening snap occurs after the second heart sound. The diastolic murmur of mitral stenosis is classically described as low pitched and rumbling with presystolic accentuation. The length of the murmur is usually proportional to the stenosis; however if stenosis is very severe, the murmur may become almost inaudible because of low cardiac output and poor flow through the valve. If the patient is in sinus rhythm, presystolic accentuation of the murmur may occur as the atrium contracts to squeeze blood through the narrow valve. This does not occur in atrial fibrillation because of disorganised atrial activity. The **Graham Steele murmur** of pulmonary regurgitation is sometimes heard at the left sternal edge in mitral stenosis.

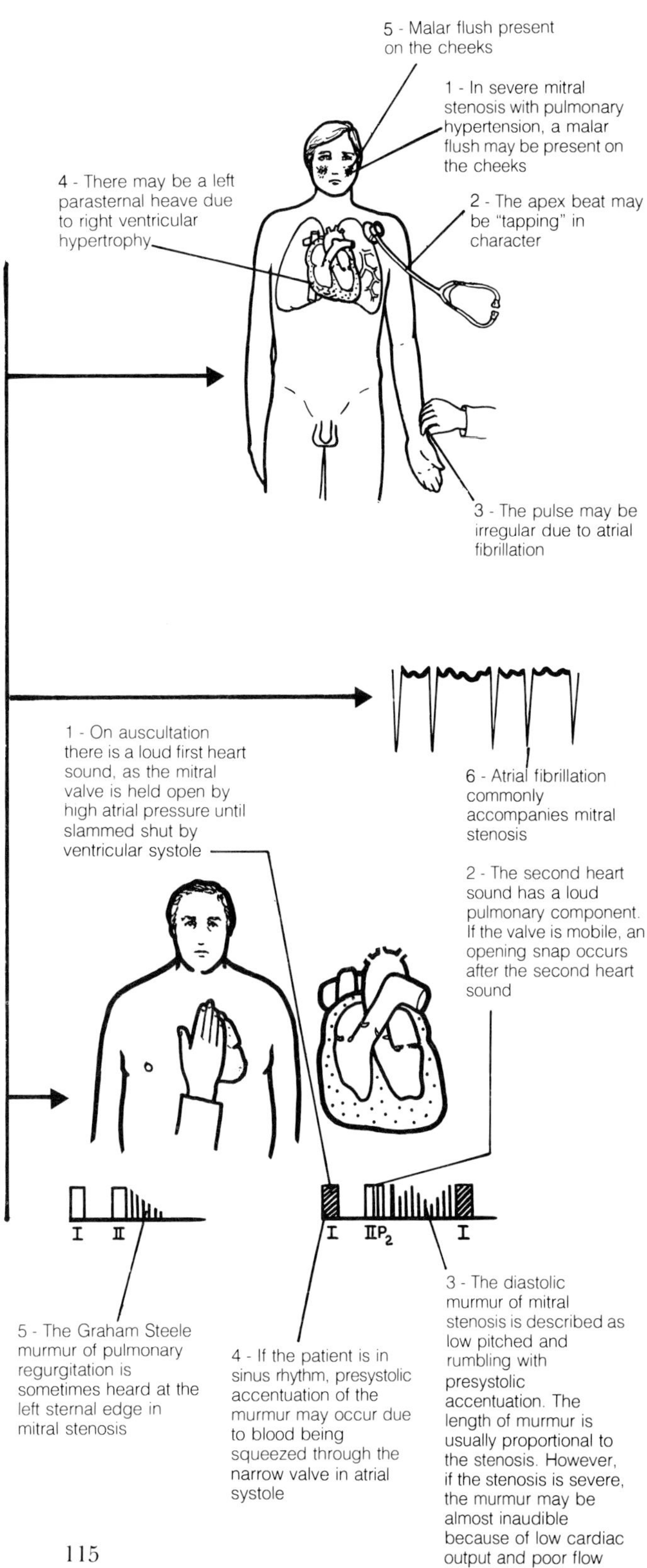
5 - Malar flush present on the cheeks
1 - In severe mitral stenosis with pulmonary hypertension, a malar flush may be present on the cheeks
4 - There may be a left parasternal heave due to right ventricular hypertrophy
2 - The apex beat may be "tapping" in character
3 - The pulse may be irregular due to atrial fibrillation
6 - Atrial fibrillation commonly accompanies mitral stenosis
1 - On auscultation there is a loud first heart sound, as the mitral valve is held open by high atrial pressure until slammed shut by ventricular systole
2 - The second heart sound has a loud pulmonary component. If the valve is mobile, an opening snap occurs after the second heart sound
I II
I IIP2 I
5 - The Graham Steele murmur of pulmonary regurgitation is sometimes heard at the left sternal edge in mitral stenosis
4 - If the patient is in sinus rhythm, presystolic accentuation of the murmur may occur due to blood being squeezed through the narrow valve in atrial systole
3 - The diastolic murmur of mitral stenosis is described as low pitched and rumbling with presystolic accentuation. The length of murmur is usually proportional to the stenosis. However, if the stenosis is severe, the murmur may be almost inaudible because of low cardiac output and poor flow through the valve

Investigations

The ECG may show atrial fibrillation. If there is sinus rhythm, wide bifid waves (P mitrale) may be seen; right ventricular hypertrophy is sometimes present.

The chest X-ray: The left heart border becomes straightened due to enlargement of the left atrial appendage. The enlarged left atrium may form a double shadow on the right side of the heart. The plain lateral X-ray may show calcification of the mitral valve. There may be evidence of pulmonary oedema, with upper lobe blood diversion, Kerley B lines and perihilar shadowing. In long standing mitral stenosis with high left atrial pressure, miliary mottling of the lung fields can appear due to haemosiderosis.

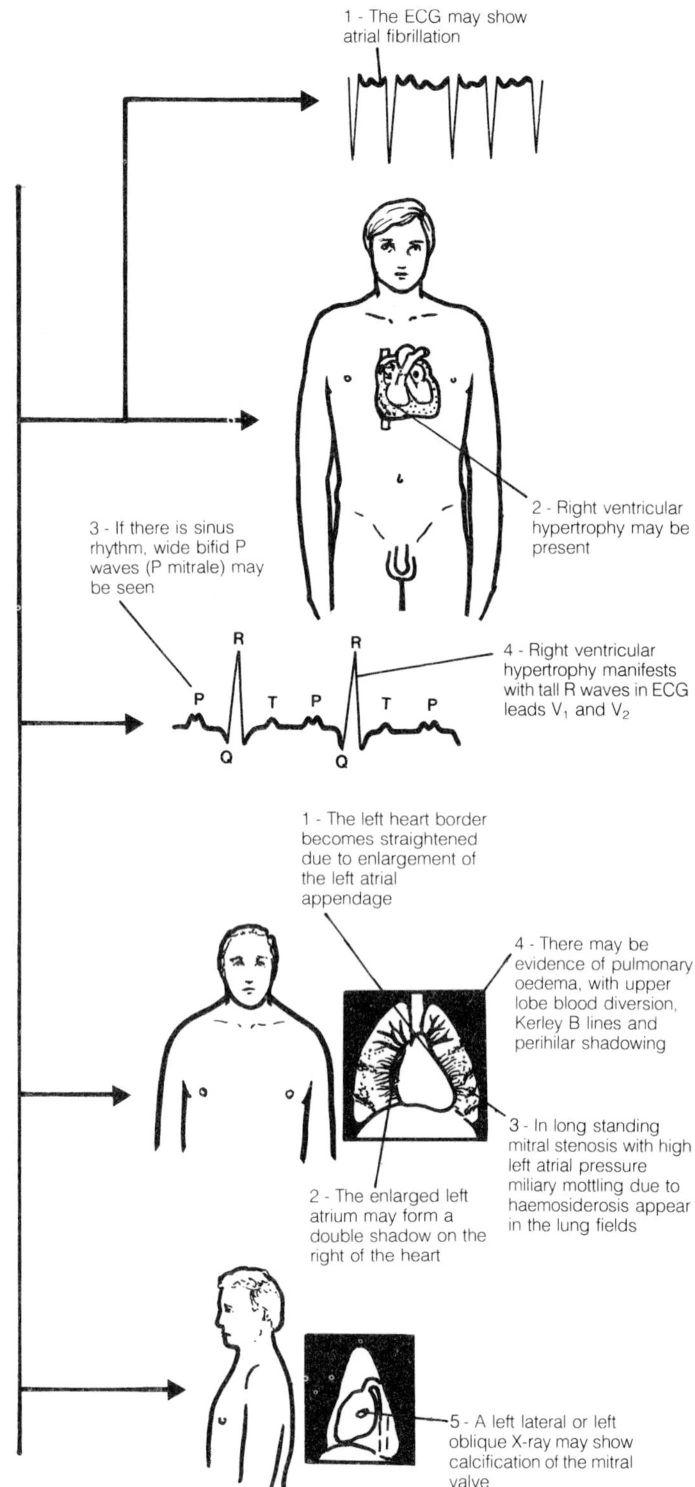
1 - The ECG may show atrial fibrillation
2 - Right ventricular hypertrophy may be present
3 - If there is sinus rhythm, wide bifid P waves (P mitrale) may be seen
R
R
P
T
P
T
P
Q
Q
4 - Right ventricular hypertrophy manifests with tall R waves in ECG leads V_1 and V_2
1 - The left heart border becomes straightened due to enlargement of the left atrial appendage
4 - There may be evidence of pulmonary oedema, with upper lobe blood diversion, Kerley B lines and perihilar shadowing
3 - In long standing mitral stenosis with high left atrial pressure miliary mottling due to haemosiderosis appear in the lung fields
2 - The enlarged left atrium may form a double shadow on the right of the heart
5 - A left lateral or left oblique X-ray may show calcification of the mitral valve

Echocardiography shows a characteristic picture with the reduced closure rate of the anterior mitral valve leaflet and no significant reopening of the valve in atrial systole. The posterior leaflet of the valve moves anteriorly with the anterior leaflet during diastole because of fusion of the commissures.

Cardiac catheterisation is useful in measuring pressures across the mitral valve, which indicate the severity of the stenosis.

1 - Echocardiography shows reduced closure rate of the anterior mitral valve leaflet

2 - There is no significant reopening of the valve in atrial systole. The posterior leaflet of the valve moves anteriorly with the anterior leaflet during diastole because of fusion of the commissures

3 - Mitral valve with fusion of the anterior and posterior leaflets

Management of mitral stenosis

Pulmonary oedema is treated with diuretics. Digoxin is used to treat atrial fibrillation and any patient who has had systemic emboli should be anticoagulated. Antibiotic cover should be given for any dental or surgical procedure (see section on infective endocarditis).

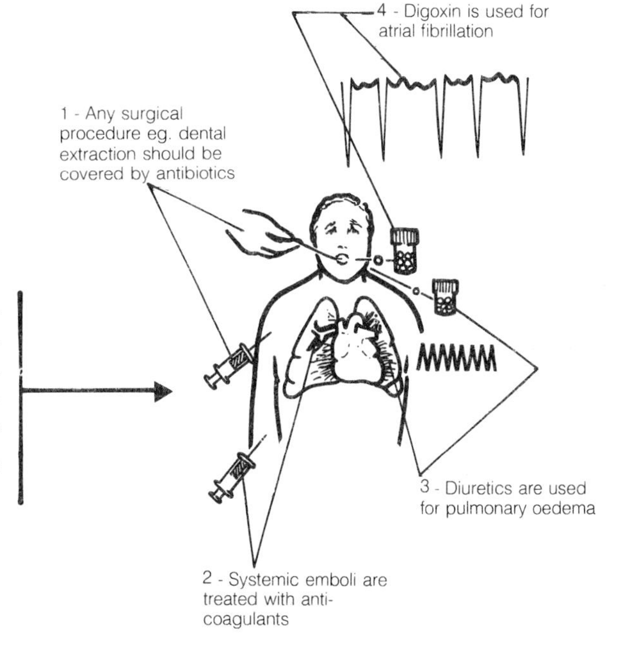
4 - Digoxin is used for atrial fibrillation
1 - Any surgical procedure eg. dental extraction should be covered by antibiotics
3 - Diuretics are used for pulmonary oedema
2 - Systemic emboli are treated with anti-coagulants

Surgery

Mitral valvotomy can be performed if the patient's symptoms severely limit his activity. Sudden attacks of pulmonary oedema without any precipitating cause, attacks of pulmonary oedema in pregnancy, and systemic emboli are all indications for surgery. Closed mitral valvotomy can be performed if the valve is mobile, there is no evidence of calcification, and no evidence of mitral regurgitation. The mortality rate of closed operation is less than 5% and restenosis occurs in 5% of patients per annum. If the above criteria cannot be met, an open operation possibly with valve replacement should be performed.

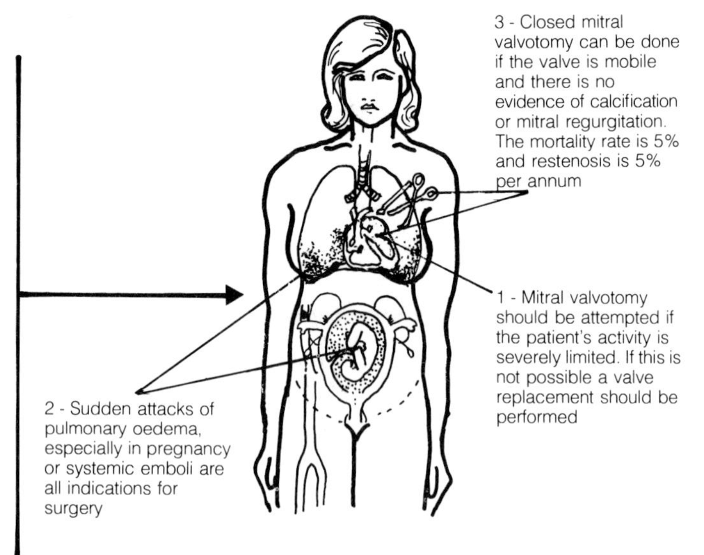
3 - Closed mitral valvotomy can be done if the valve is mobile and there is no evidence of calcification or mitral regurgitation. The mortality rate is 5% and restenosis is 5% per annum
1 - Mitral valvotomy should be attempted if the patient's activity is severely limited. If this is not possible a valve replacement should be performed
2 - Sudden attacks of pulmonary oedema, especially in pregnancy or systemic emboli are all indications for surgery

Differential diagnosis

The main differential diagnosis of mitral stenosis is **left atrial myxoma**. This is a benign mobile tumour usually arising from the fossa ovalis. The patient can present with fevers, weight loss, joint pains, systemic emboli and a high ESR. The echocardiogram shows multiple dense echoes within the mitral valve in diastole. Once the diagnosis is made, surgical treatment is important as the tumour can sometimes completely obstruct the valve opening.

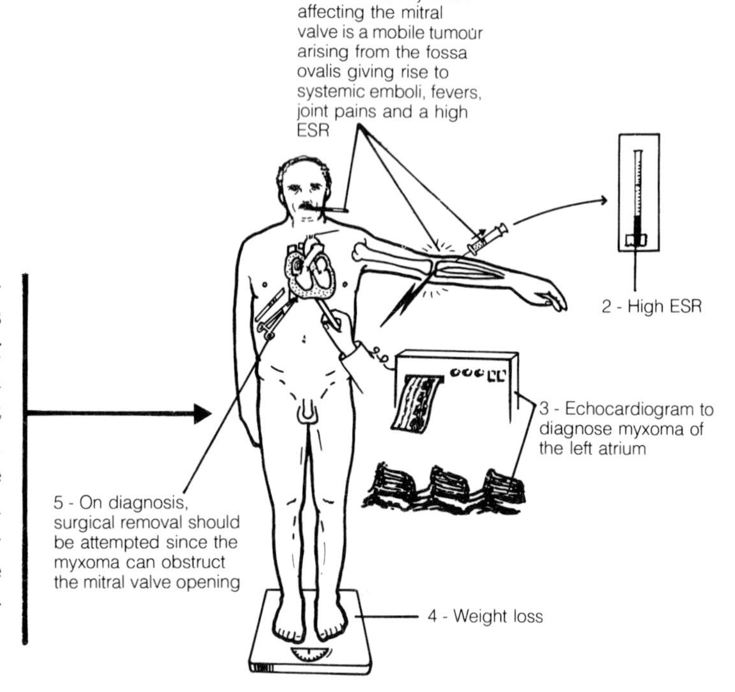
1 - Left atrial myxoma affecting the mitral valve is a mobile tumour arising from the fossa ovalis giving rise to systemic emboli, fevers, joint pains and a high ESR
2 - High ESR
3 - Echocardiogram to diagnose myxoma of the left atrium
4 - Weight loss
5 - On diagnosis, surgical removal should be attempted since the myxoma can obstruct the mitral valve opening

Mitral regurgitation

Mitral regurgitation is usually the result of previous rheumatic fever. Other causes include mitral valve prolapse, papillary muscle dysfunction or ruptured chordae tendinae, hypertrophic cardiomyopathy, and connective tissue disorders.

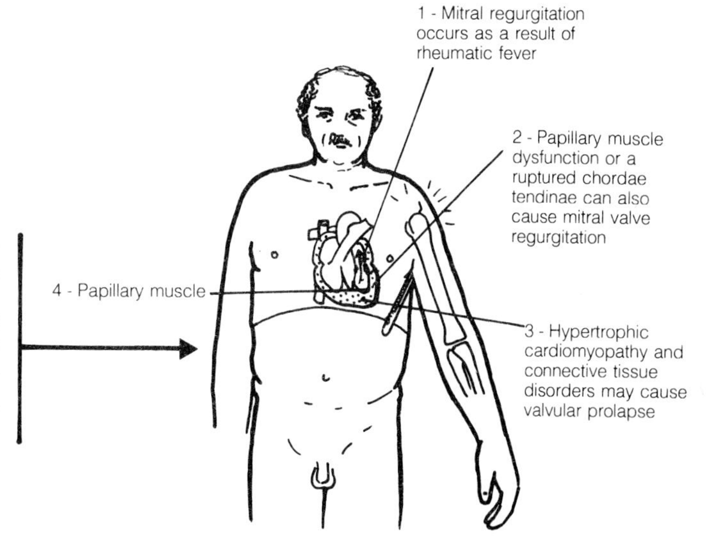
1 - Mitral regurgitation occurs as a result of rheumatic fever
2 - Papillary muscle dysfunction or a ruptured chordae tendinae can also cause mitral valve regurgitation
4 - Papillary muscle
3 - Hypertrophic cardiomyopathy and connective tissue disorders may cause valvular prolapse

Rheumatic mitral regurgitation

Mitral regurgitation can occur transiently during an acute attack of rheumatic fever. Thirty per cent of patients do not lose the murmur and the valve leaflets become contracted and fibrosed. The leaflets do not close completely during ventricular contraction and blood is regurgitated into the left atrium. Left atrial and ventricular enlargement occur because of the increased stroke volume.

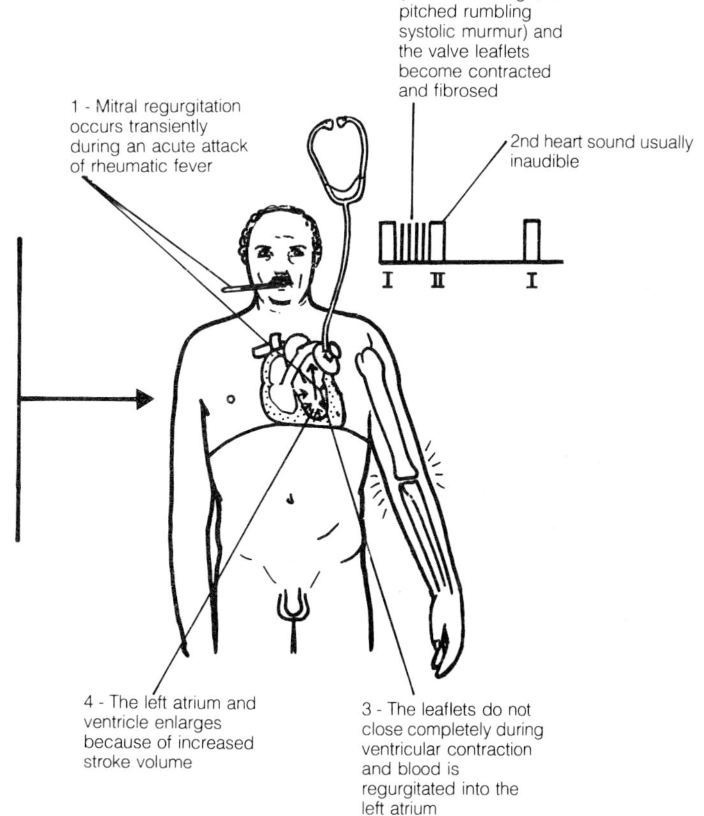
2 - 30% of patients do not lose the murmur (which is a rough, low-pitched rumbling systolic murmur) and the valve leaflets become contracted and fibrosed
1 - Mitral regurgitation occurs transiently during an acute attack of rheumatic fever
2nd heart sound usually inaudible
I II I
4 - The left atrium and ventricle enlarges because of increased stroke volume
3 - The leaflets do not close completely during ventricular contraction and blood is regurgitated into the left atrium

Symptoms of mitral regurgitation

Breathlessness on exertion, tiredness and congestive cardiac failure occur. Occasionally the patient may complain of dysphagia due to pressure on the oesophagus by the enlarging left atrium. **Ewart's sign** (collapse of the left lower lobe) for the same reason occasionally occurs.

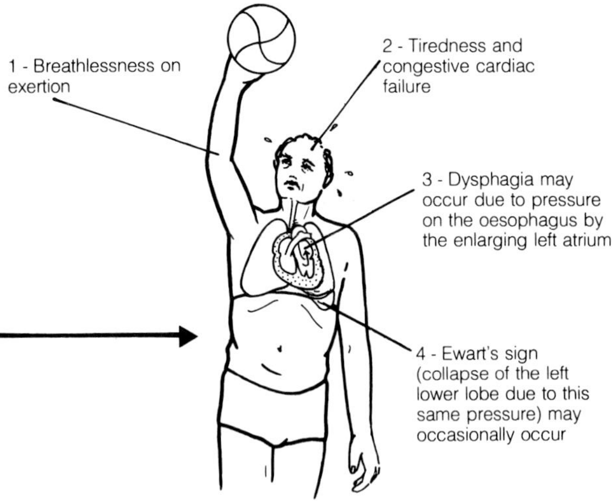

1 - Breathlessness on exertion

2 - Tiredness and congestive cardiac failure

3 - Dysphagia may occur due to pressure on the oesophagus by the enlarging left atrium

4 - Ewart's sign (collapse of the left lower lobe due to this same pressure) may occasionally occur

2 - The apex beat is displaced because of left ventricular enlargement and is hyperkinetic. There may be a thrill

1 - Atrial fibrillation is often present

3 - On auscultation there is a pan-systolic murmur radiating to the axilla

4 - A third heart sound due to rapid ventricular filling and a brief diastolic murmur sometimes occur

Physical signs of mitral regurgitation

Atrial fibrillation is often present. The apex beat is usually displaced because of left ventricular enlargement, and hyperkinetic. There may be a thrill. On auscultation there is an apical pansystolic murmur radiating to the axilla. A third heart sound may be present due to rapid ventricular filling, and a brief mid-diastolic murmur sometimes occurs.

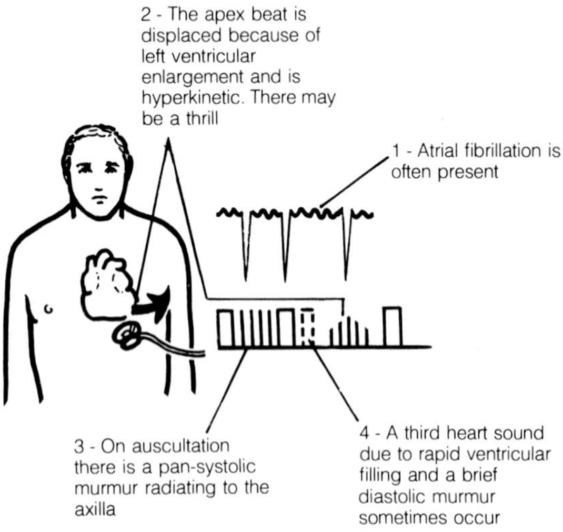

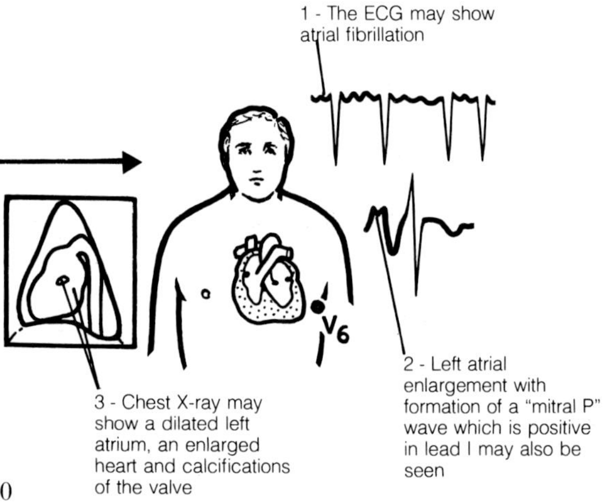

1 - The ECG may show atrial fibrillation

V6

2 - Left atrial enlargement with formation of a "mitral P" wave which is positive in lead I may also be seen

3 - Chest X-ray may show a dilated left atrium, an enlarged heart and calcifications of the valve

Investigations

The **ECG** may show atrial fibrillation or left atrial enlargement.

Chest X-ray: The heart may be enlarged with a dilated left atrium, and calcification of the valve may be visible.

The echocardiogram may show valvular calcification and left ventricular dilation. Other causes of mitral regurgitation such as mitral valve prolapse may be visible.

3 - Mitral regurgitation

2 - Other causes of mitral regurgitation such as mitral valve prolapse may be visible

1 - The echocardiogram shows left ventricular dilation and valvular calcification

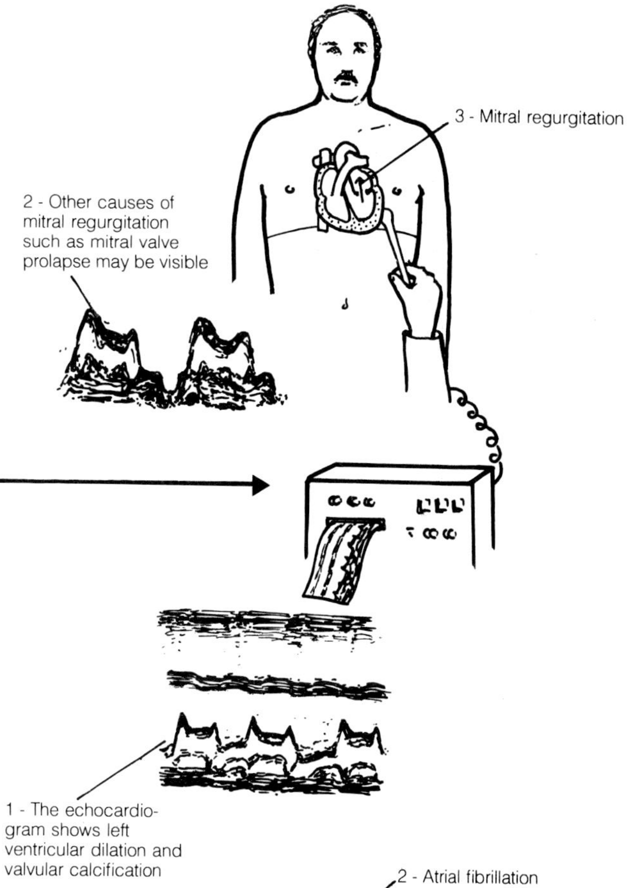

Treatment

Heart failure should be treated when it occurs; diuretics and vasodilators are useful. Atrial fibrillation is treated with digoxin. Valve replacement may be required when the patient has significant symptoms. Acute mitral regurgitation often requires urgent surgery.

2 - Atrial fibrillation treated with Digoxin

3 - Vasodilators are used to reduce overload on the heart

1 - Heart failure is treated when it occurs

6 - Valve replacement is used when the patient has significant symptoms

5 - Acute mitral regurgitation often requires urgent surgery

4 - Diuretics are used in pulmonary congestion

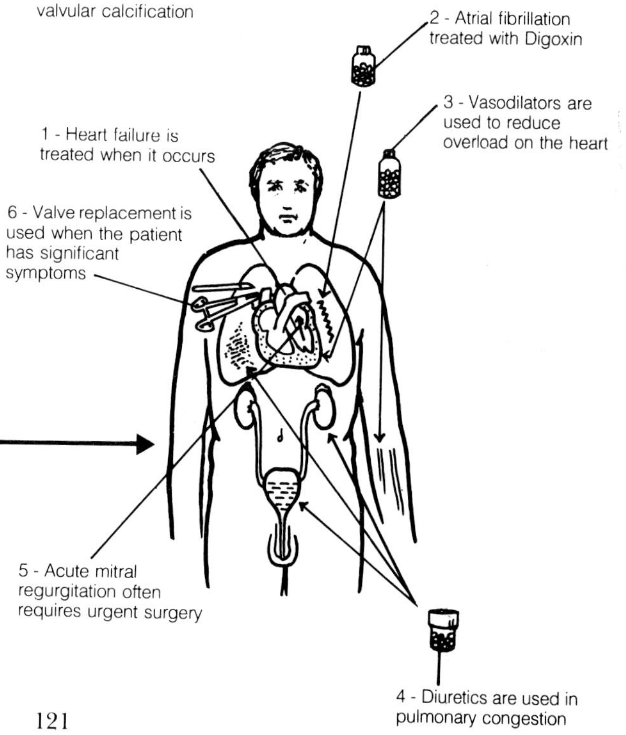

Problems of valve replacement

Patients with mechanical valves need to be treated with anticoagulants for life, as the rate of systemic emboli is otherwise high (5-10%). Tissue valves (xenografts) do not require anticoagulant therapy, but their life span is shorter than that of the mechanical valves.

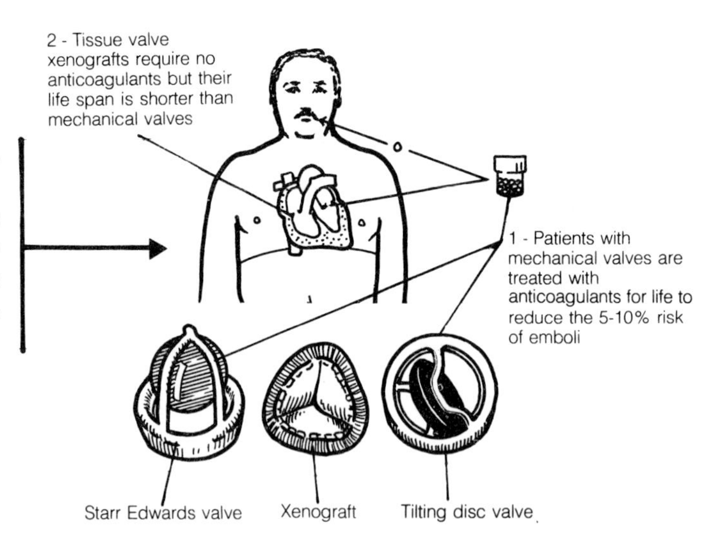
2 - Tissue valve xenografts require no anticoagulants but their life span is shorter than mechanical valves
1 - Patients with mechanical valves are treated with anticoagulants for life to reduce the 5-10% risk of emboli
Starr Edwards valve
Xenograft
Tilting disc valve

Mitral valve prolapse

In this condition myxomatous degeneration of the mitral valve leaflets occurs and the chordae tendinae become stretched. The condition is often asymptomatic and discovered at routine medical examinations. Some patients complain of chest pain, palpitations and exertional dyspnoea. On examination there is usually a mid-systolic click followed by a late systolic murmur. Echocardiography shows

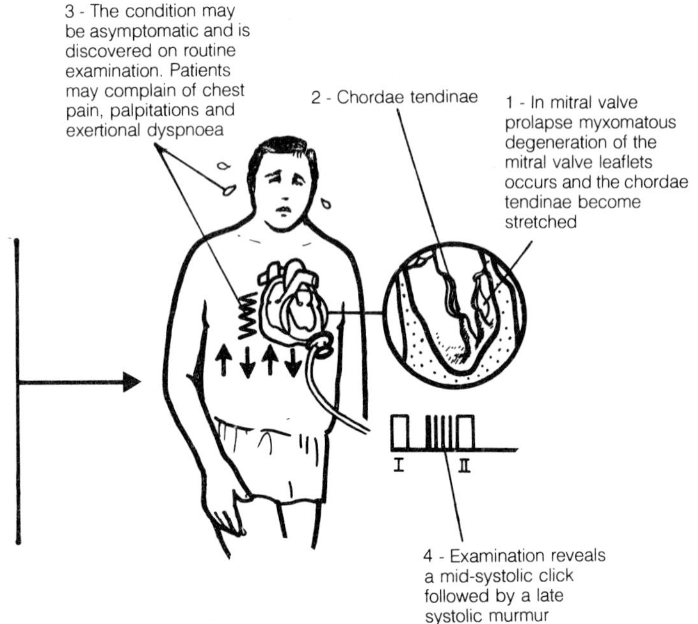
3 - The condition may be asymptomatic and is discovered on routine examination. Patients may complain of chest pain, palpitations and exertional dyspnoea
2 - Chordae tendinae
1 - In mitral valve prolapse myxomatous degeneration of the mitral valve leaflets occurs and the chordae tendinae become stretched
I
II
4 - Examination reveals a mid-systolic click followed by a late systolic murmur

sagging of the valve posteriorly in late systole. Prognosis of the condition is good, although there is increased risk of infective endocarditis and ventricular arrhythmias

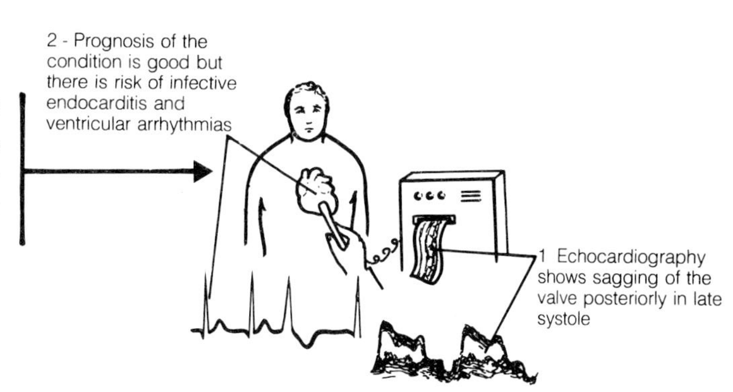
2 - Prognosis of the condition is good but there is risk of infective endocarditis and ventricular arrhythmias
1 Echocardiography shows sagging of the valve posteriorly in late systole

Papillary muscle dysfunction

Ischaemia of the papillary muscle or left ventricular dilatation can cause the papillary muscle to function poorly with resulting regurgitation of blood through the mitral valve during systole. **Papillary muscle rupture** can occur after myocardial infarction resulting in acute onset of mitral regurgitation and severe pulmonary oedema.

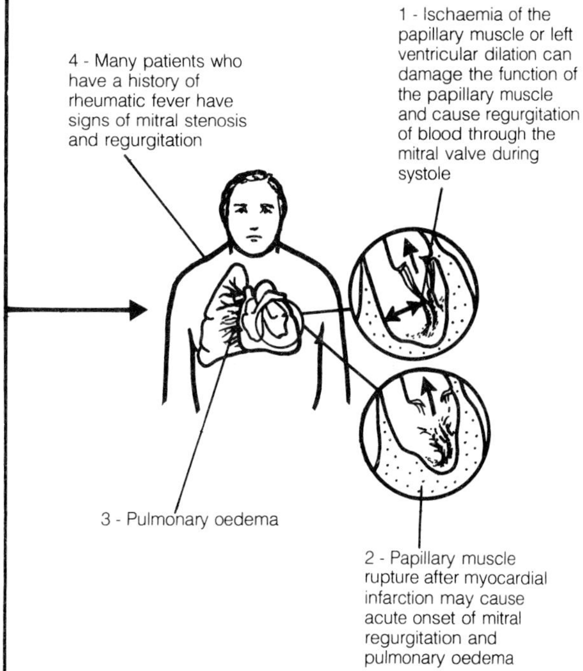

4 - Many patients who have a history of rheumatic fever have signs of mitral stenosis and regurgitation

1 - Ischaemia of the papillary muscle or left ventricular dilation can damage the function of the papillary muscle and cause regurgitation of blood through the mitral valve during systole

3 - Pulmonary oedema

2 - Papillary muscle rupture after myocardial infarction may cause acute onset of mitral regurgitation and pulmonary oedema

Mixed mitral valve disease

Many patients who have a history of rheumatic fever have signs both of mitral stenosis and regurgitation.

TRICUSPID VALVE DISEASE

It is unusual to find isolated tricuspid valve disease in a patient who has had previous rheumatic fever; other valves especially the mitral are usually affected. **Tricuspid stenosis** is usually due to previous rheumatic heart disease. The symptoms and signs are those of right sided heart failure with low cardiac output. The patient complains of weakness and swelling of the legs and abdomen. The venous pressure is raised with a prominent "a" wave if there is sinus rhythm. This is due to atrial contraction against a stenosed tricuspid valve. On auscultation a diastolic murmur which increases on inspiration can be heard. There is peripheral oedema and ascites may be present.

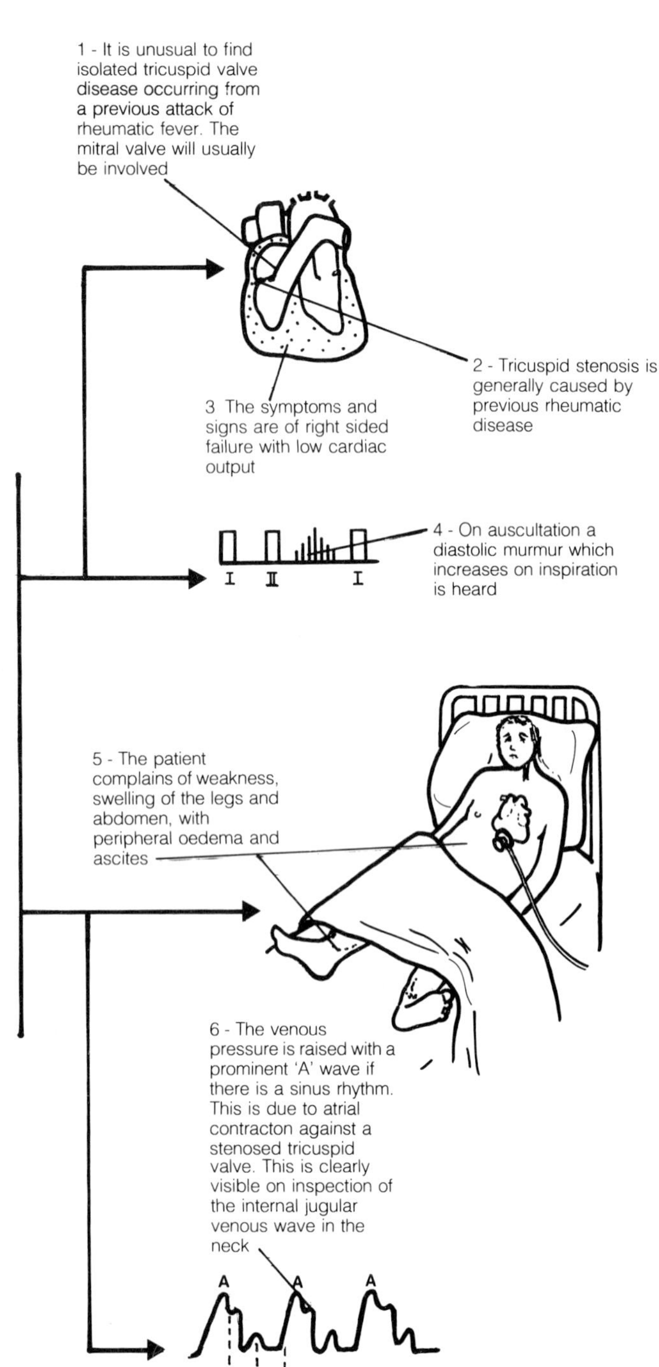
1 - It is unusual to find isolated tricuspid valve disease occurring from a previous attack of rheumatic fever. The mitral valve will usually be involved
2 - Tricuspid stenosis is generally caused by previous rheumatic disease
3 The symptoms and signs are of right sided failure with low cardiac output
4 - On auscultation a diastolic murmur which increases on inspiration is heard
I
II
I
5 - The patient complains of weakness, swelling of the legs and abdomen, with peripheral oedema and ascites
6 - The venous pressure is raised with a prominent 'A' wave if there is a sinus rhythm. This is due to atrial contracton against a stenosed tricuspid valve. This is clearly visible on inspection of the internal jugular venous wave in the neck
A
A
A

Tricuspid regurgitation

This may be congenital or acquired. **Acquired** tricuspid regurgitation can occur after rheumatic fever, after infective endocarditis in a "main-lining" drug addict, or in carcinoid syndrome. **Functional** tricuspid regurgitation occurs with dilatation of the right ventricle secondary to pulmonary hypertension. On examination the patient has signs of congestive cardiac failure. The venous pressure is raised with a prominent systolic "V" wave. There is a right ventricular heave, and a systolic murmur can be heard at the lower end of the sternum which becomes

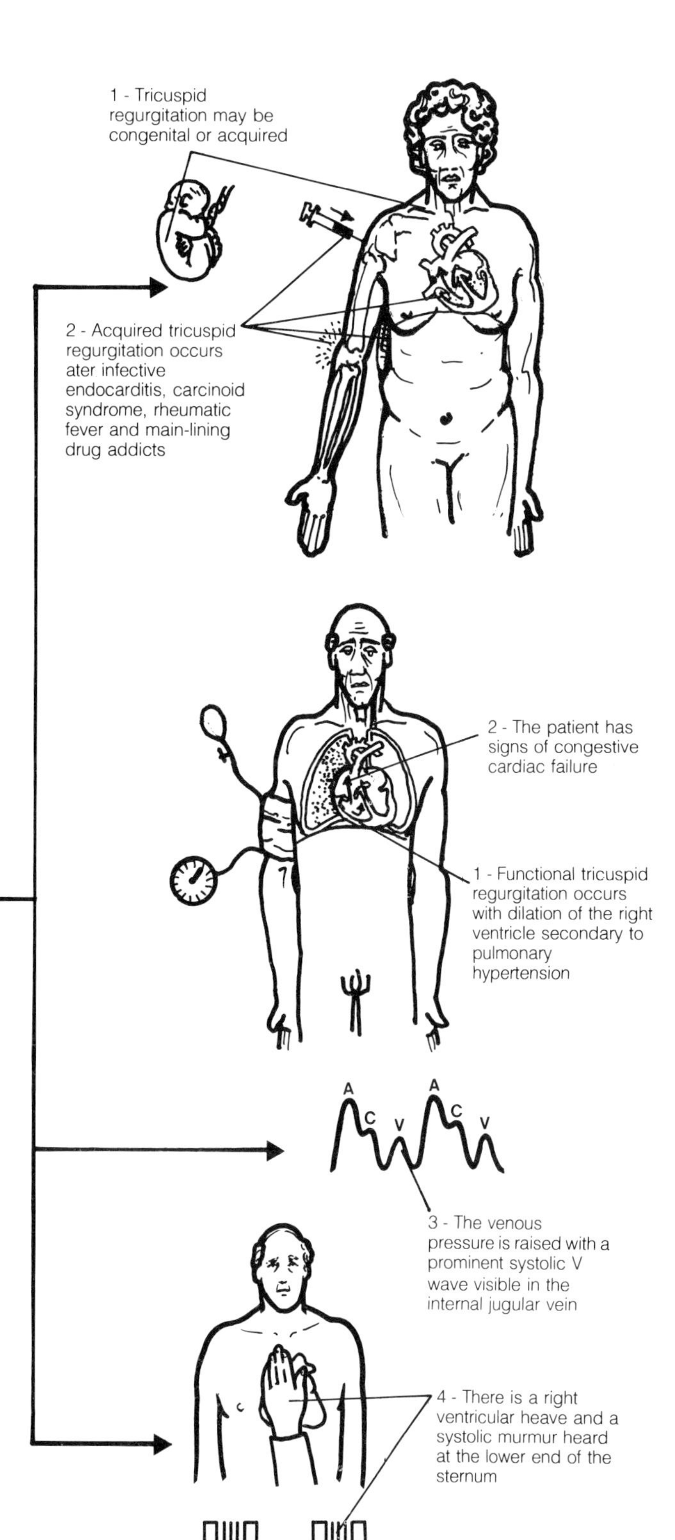
1 - Tricuspid regurgitation may be congenital or acquired
2 - Acquired tricuspid regurgitation occurs ater infective endocarditis, carcinoid syndrome, rheumatic fever and main-lining drug addicts
2 - The patient has signs of congestive cardiac failure
1 - Functional tricuspid regurgitation occurs with dilation of the right ventricle secondary to pulmonary hypertension
A
C
V
A
C
V
3 - The venous pressure is raised with a prominent systolic V wave visible in the internal jugular vein
4 - There is a right ventricular heave and a systolic murmur heard at the lower end of the sternum
I
II
I
II

louder on inspiration. The liver is enlarged and pulsatile and there is peripheral oedema. Ascites may be present and proteinuria sometimes occurs because of renal congestion.

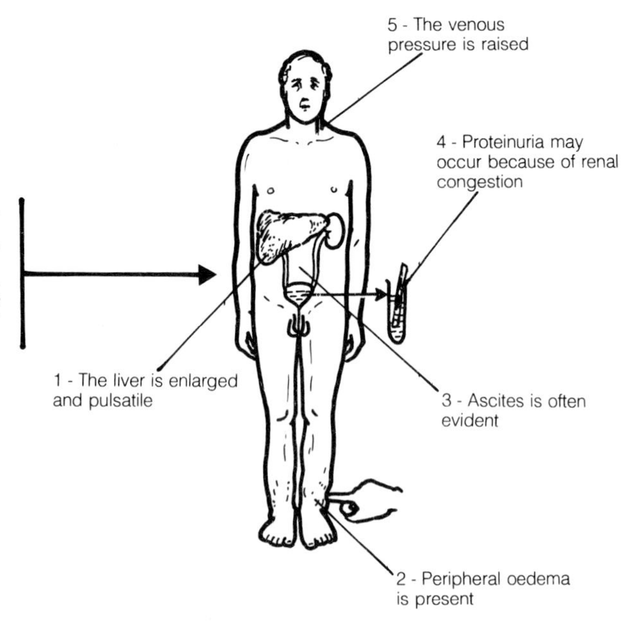
5 - The venous pressure is raised
4 - Proteinuria may occur because of renal congestion
1 - The liver is enlarged and pulsatile
3 - Ascites is often evident
2 - Peripheral oedema is present

Treatment

Congestive cardiac failure is treated with diuretics. If there is organic tricuspid regurgitation an annuloplasty or tricuspid valve replacement can be performed.

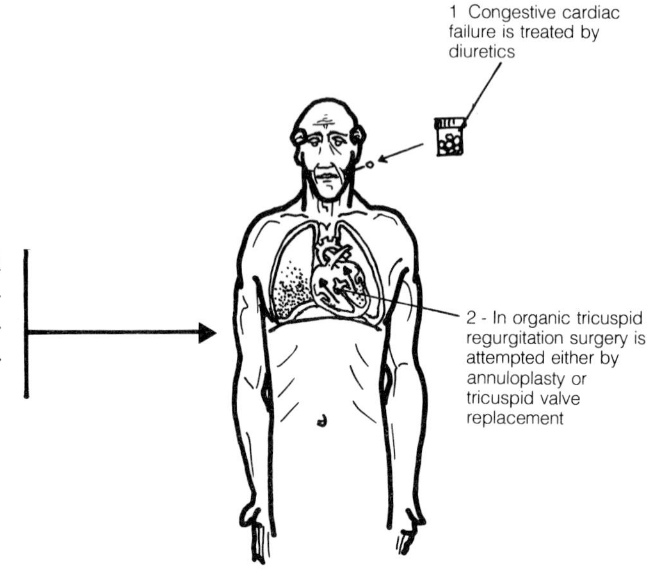
1 Congestive cardiac failure is treated by diuretics
2 - In organic tricuspid regurgitation surgery is attempted either by annuloplasty or tricuspid valve replacement

AORTIC VALVE DISEASE

Aortic stenosis may be congenital or acquired. Congenital aortic stenosis is usually due to fusion of the commissures of a bicuspid valve. In adults with bicuspid valves, calcification can produce significant narrowing. Sometimes, coarctation of the aorta, Turner's syndrome or Marfan's syndrome are accompanied by a bicuspid aortic valve.

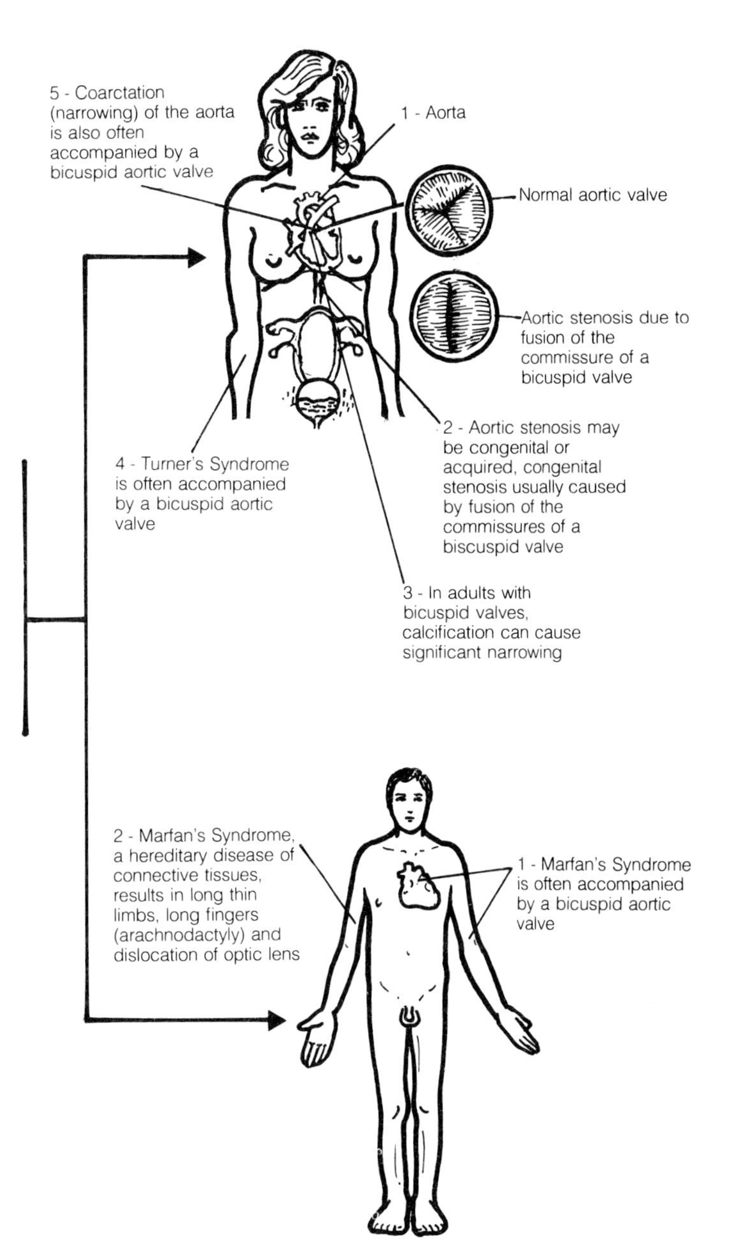

5 - Coarctation (narrowing) of the aorta is also often accompanied by a bicuspid aortic valve
1 - Aorta
Normal aortic valve
Aortic stenosis due to fusion of the commissure of a bicuspid valve
2 - Aortic stenosis may be congenital or acquired, congenital stenosis usually caused by fusion of the commissures of a biscuspid valve
4 - Turner's Syndrome is often accompanied by a bicuspid aortic valve
3 - In adults with bicuspid valves, calcification can cause significant narrowing
2 - Marfan's Syndrome, a hereditary disease of connective tissues, results in long thin limbs, long fingers (arachnodactyly) and dislocation of optic lens
1 - Marfan's Syndrome is often accompanied by a bicuspid aortic valve

Rheumatic heart disease

Is the commonest acquired cause of aortic valve disease. Aortic stenosis is less common than regurgitation, and often both are present.

Aortic stenosis

As the aortic valve diameter is reduced in size, the left ventricular pressure increases to overcome the obstruction; this results in hypertrophy of the ventricular muscle. Eventually, left ventricular failure and pulmonary oedema occur.

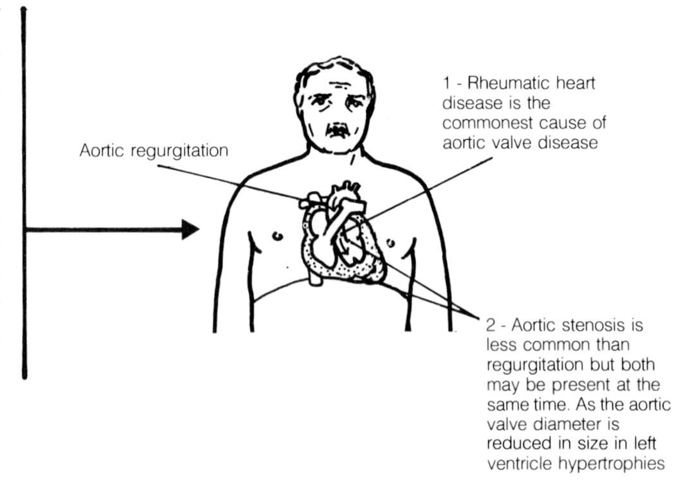
Aortic regurgitation
1 - Rheumatic heart disease is the commonest cause of aortic valve disease
2 - Aortic stenosis is less common than regurgitation but both may be present at the same time. As the aortic valve diameter is reduced in size in left ventricle hypertrophies

Symptoms

The patient may present with exertional dyspnoea, and left ventricular failure may cause orthopnoea and paroxysmal noctural dyspnoea. **Effort syncope** and angina may be related to reduced cardiac output during exercise and co-existing coronary artery disease. Cardiac arrhythmias are common, and may also cause syncope.

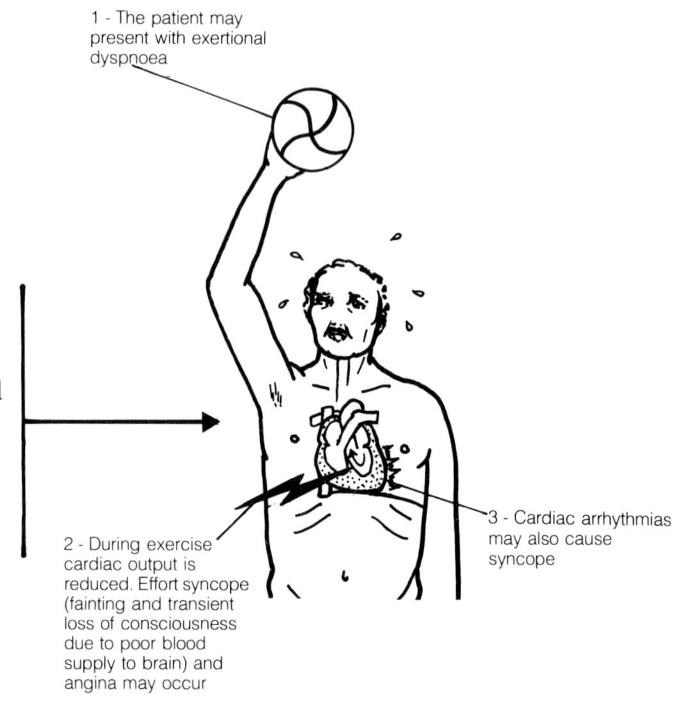
1 - The patient may present with exertional dyspnoea
2 - During exercise cardiac output is reduced. Effort syncope (fainting and transient loss of consciousness due to poor blood supply to brain) and angina may occur
3 - Cardiac arrhythmias may also cause syncope

Signs

The pulse in aortic stenosis is slow rising and best felt at the carotid. The apex beat is heaving in nature, sometimes with a palpable systolic thrill. On auscultation, a harsh mid systolic murmur is heard at the left sternal edge, in the aortic area, and is transmitted up to the carotids. If the valve is mobile as in young people, an ejection click may be heard. In older patients with an immobile calcified valve, there is no ejection click, and the aortic component of the second heart sound becomes delayed, producing a single second sound. Some aortic regurgitation may be present, resulting in an immediate, flowing diastolic murmur, best heard in expiration at the left sternal edge.

4 - Orthopnoea (difficulty in breathing on lying flat) and paroxysmal nocturnal dyspnoea may occur
1 - The pulse in aortic stenosis is slow rising and best felt at the carotid. The apex beat is heaving in nature, sometimes with a palpable systolic thrill
2 - On auscultation a harsh mid-systolic murmur is heard at the left sternal edge in the aortic area and transmitted to the carotid
I II I
I II
I II
4 - In calcified valves, aortic regurgitation occurs. This results in an immediate flowing diastolic murmur best heard at the left sternal edge
3 - If the valve is mobile as in young people, an ejection click is heard, but in older patients with calcified valves, the "click" is absent

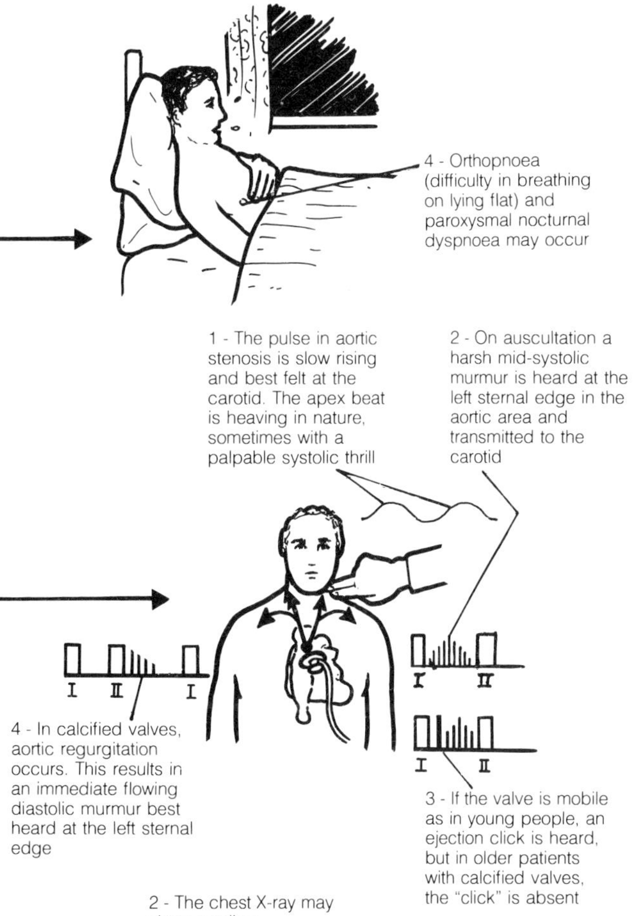

Investigations

The ECG may show left ventricular hypertrophy. The **chest X-ray** may show cardiac enlargement, and the calcified aortic valve can sometimes be seen on a penetrated lateral film. There may also be post stenotic dilation of the ascending aorta. **Echocardiography** is useful in assessing valvular calcification and the degree of stenosis as well as left ventricular function. **Cardiac catheterization** is

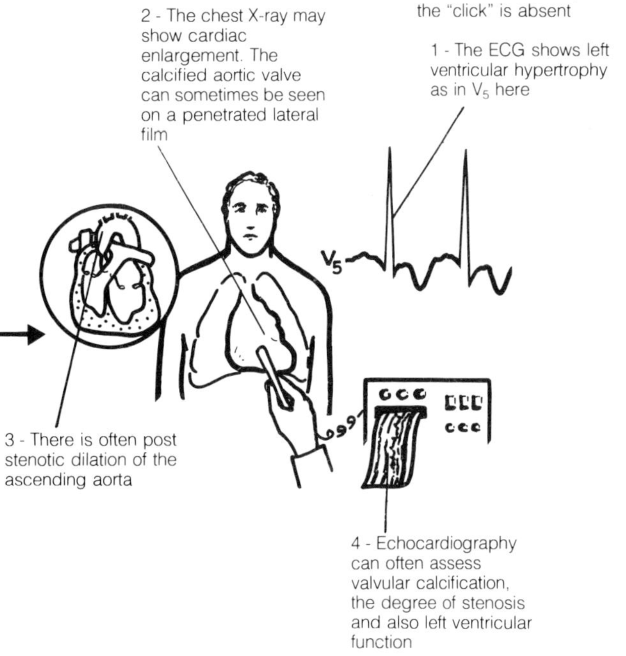

2 - The chest X-ray may show cardiac enlargement. The calcified aortic valve can sometimes be seen on a penetrated lateral film
1 - The ECG shows left ventricular hypertrophy as in V_5 here
V_5
3 - There is often post stenotic dilation of the ascending aorta
4 - Echocardiography can often assess valvular calcification, the degree of stenosis and also left ventricular function

useful in assessing the severity of aortic stenosis, as well as the state of the coronary arteries. The pressure gradient across the aortic valve can be measured; if this is more than 70 mmHg surgery is usually necessary. The left ventricular end diastolic pressure and ventriculogram will give a measure of left ventricular function.

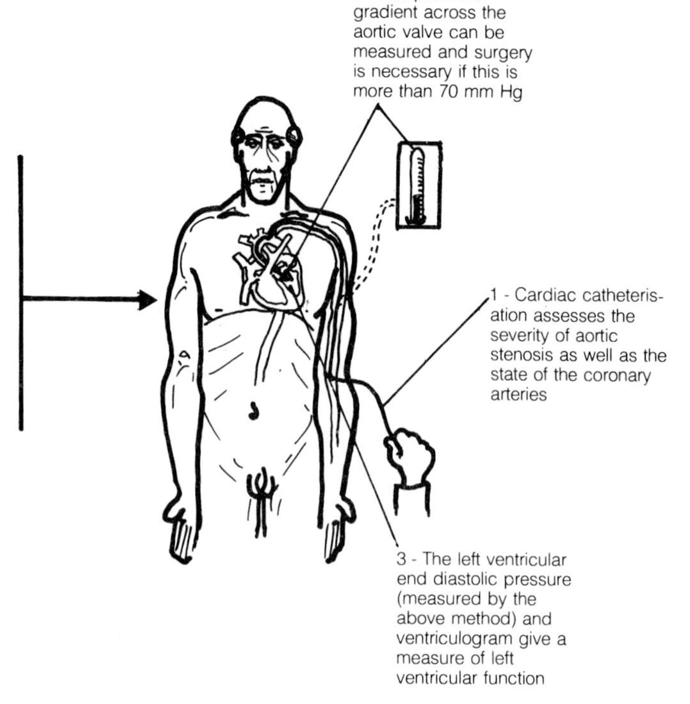
2 - The pressure gradient across the aortic valve can be measured and surgery is necessary if this is more than 70 mm Hg
1 - Cardiac catheterisation assesses the severity of aortic stenosis as well as the state of the coronary arteries
3 - The left ventricular end diastolic pressure (measured by the above method) and ventriculogram give a measure of left ventricular function

Management of aortic stenosis
1. Medical

Asymptomatic patients, with no evidence of left ventricular hypertrophy, should be followed up carefully. They should avoid strenuous physical exertion, and any dental procedures should be covered with prophylactic antibiotics.

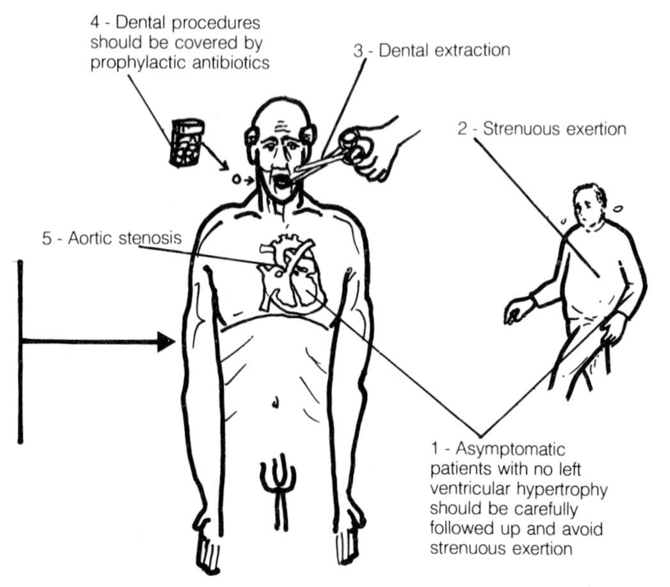
4 - Dental procedures should be covered by prophylactic antibiotics
3 - Dental extraction
2 - Strenuous exertion
5 - Aortic stenosis
1 - Asymptomatic patients with no left ventricular hypertrophy should be carefully followed up and avoid strenuous exertion

2. Surgical

Once a patient with aortic stenosis develops symptoms such as exertional dyspnoea, angina pectoris or syncopal episodes, surgical intervention is indicated. In young patients with mobile valves, a valvotomy is sometimes possible. When the valve is calcified or aortic regurgitation is present, valve replacement is necessary. If significant coronary artery disease is present, coronary artery bypass grafting is performed at the same time.

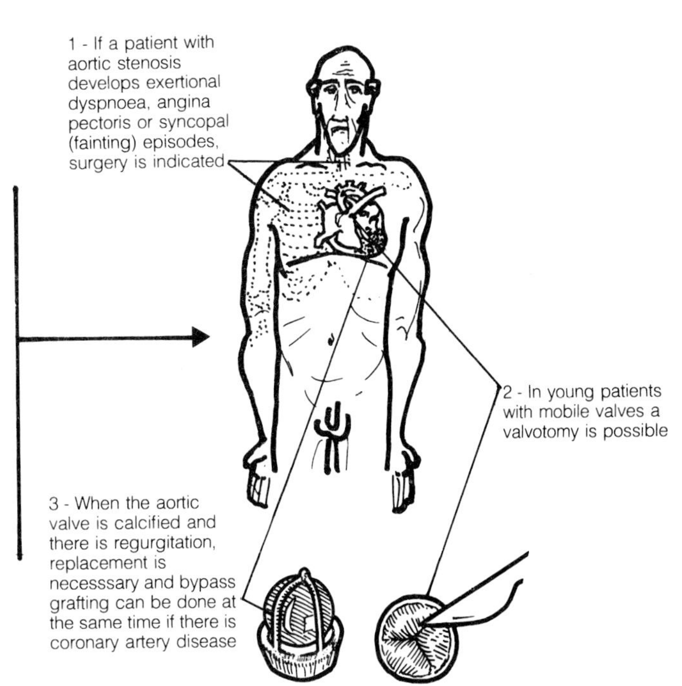
1 - If a patient with aortic stenosis develops exertional dyspnoea, angina pectoris or syncopal (fainting) episodes, surgery is indicated
2 - In young patients with mobile valves a valvotomy is possible
3 - When the aortic valve is calcified and there is regurgitation, replacement is necesssary and bypass grafting can be done at the same time if there is coronary artery disease

Aortic regurgitation

Rheumatic heart disease is the commonest cause of aortic regurgitation, and usually gives rise to mixed valvular disease. Other causes include syphilitic aortitis, aortic aneurysm, infective endocarditis, ankylosing spondylitis, and Marfan's syndrome. In aortic regurgitation, blood leaks back through the aortic valve in diastole. This results in increasing volume load on the left ventricle, and increase in stroke volume with resulting dilatation and hypertrophy. The characteristic pulse produced has a sharp upstroke and falls away rapidly (sometimes called collapsing pulse). Ultimately left ventricular failure can develop.

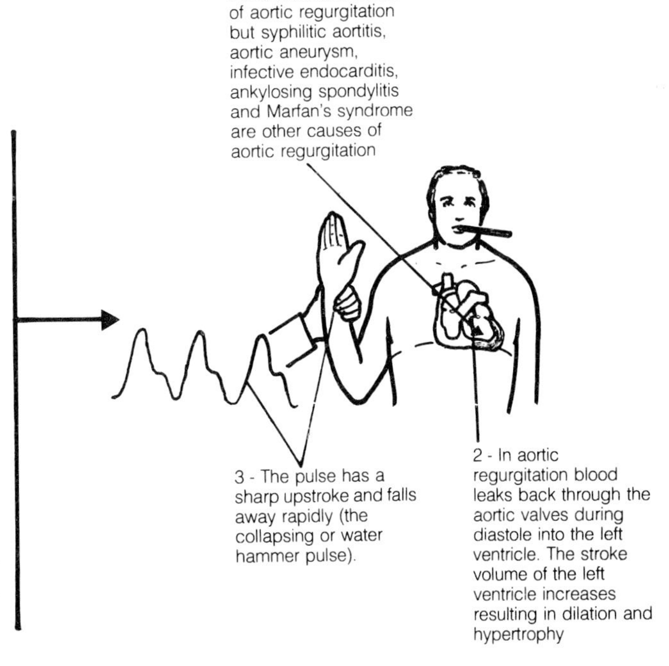
1 - Rheumatic fever is the commonest cause of aortic regurgitation but syphilitic aortitis, aortic aneurysm, infective endocarditis, ankylosing spondylitis and Marfan's syndrome are other causes of aortic regurgitation
2 - In aortic regurgitation blood leaks back through the aortic valves during diastole into the left ventricle. The stroke volume of the left ventricle increases resulting in dilation and hypertrophy
3 - The pulse has a sharp upstroke and falls away rapidly (the collapsing or water hammer pulse).

1 - The patient may be asymptomatic for years until heart failure develops. He may present with exertional dyspnoea and paroxysmal nocturnal dyspnoea

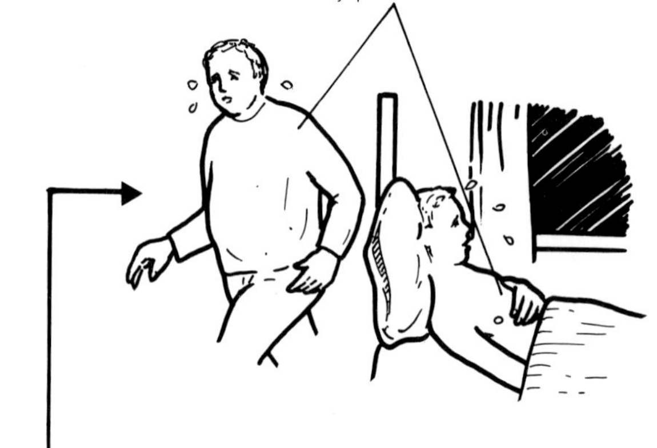

Symptoms

The patient may be asymptomatic for years, until heart failure develops. He may present with exertional dyspnoea and paroxysmal nocturnal dyspnoea. Angina can occcur in patients with syphilitic aortitis, where the coronary ostia are often involved, or may be due to co-existent coronary artery disease.

2 - Angina can occur in patients with syphilitic aortitis, where the coronary ostia are often involved, or may be due to co-existent coronary artery disease

1 - Collapsing pulse with sharp upstroke

2 - The pulse in aortic regurgitation is the collapsing or water hammer type with a sharp upstroke

3 - The carotid pulsation is prominent

4 - There is often a large difference between the systolic and diastolic pressures

5 - The left ventricle is enlarged and hyperdynamic and the apex beat displaced outwards and thrusting in nature

Signs

The pulse in aortic regurgitation has a sharp upstroke (collapsing pulse). Prominent carotid pulsation may be seen, and on measuring the blood pressure, there is often a wide pulse pressure (that is a large difference between systolic and diastolic pressures). The left ventricle is usually enlarged and hyperdynamic; the apex beat is displaced outwards and is thrusting in nature. On auscultation, there is an immediate blowing diastolic murmur, best heard in expiration at the left sternal edge and in the aortic area.

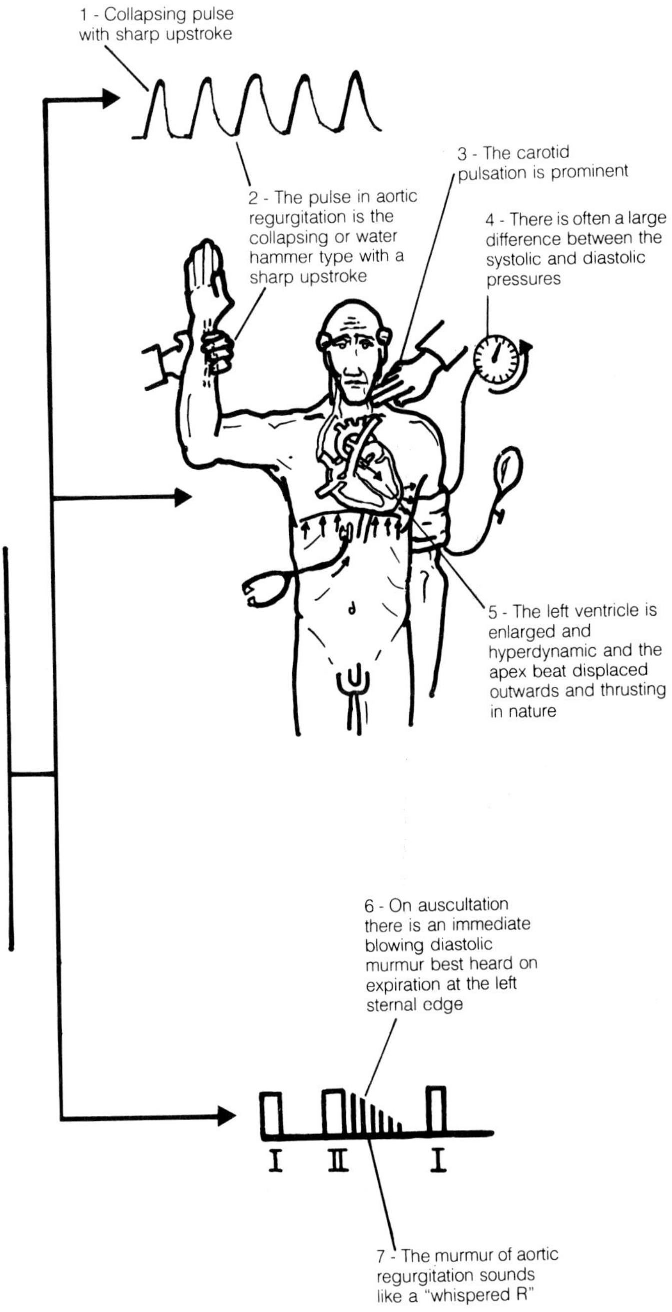

6 - On auscultation there is an immediate blowing diastolic murmur best heard on expiration at the left sternal edge

I II I

7 - The murmur of aortic regurgitation sounds like a "whispered R"

An aortic systolic murmur is commonly present due to increased blood flow through the valve. The **Austin Flint** murmur is a low pitched, mid-diastolic murmur, sometimes heard in aortic regurgitation. It is caused by the regurgitant jet of blood producing vibration of the mitral valve leaflet.

Investigations

The **ECG** may show left ventricular hypertrophy while the **chest X-ray** shows left ventricular enlargement. **Echocardiography** in aortic regurgitation, a characteristic flutter pattern occurs on the anterior leaflet of the mitral valve. Separation of the aortic valve cusps in diastole can be visualised, and left ventricular size and contractility can be assessed. **Cardiac catheterization** with aortography gives a good indication of the degree of regurgitation. In the presence of multiple valve lesions, it is useful in assessing the severity of each.

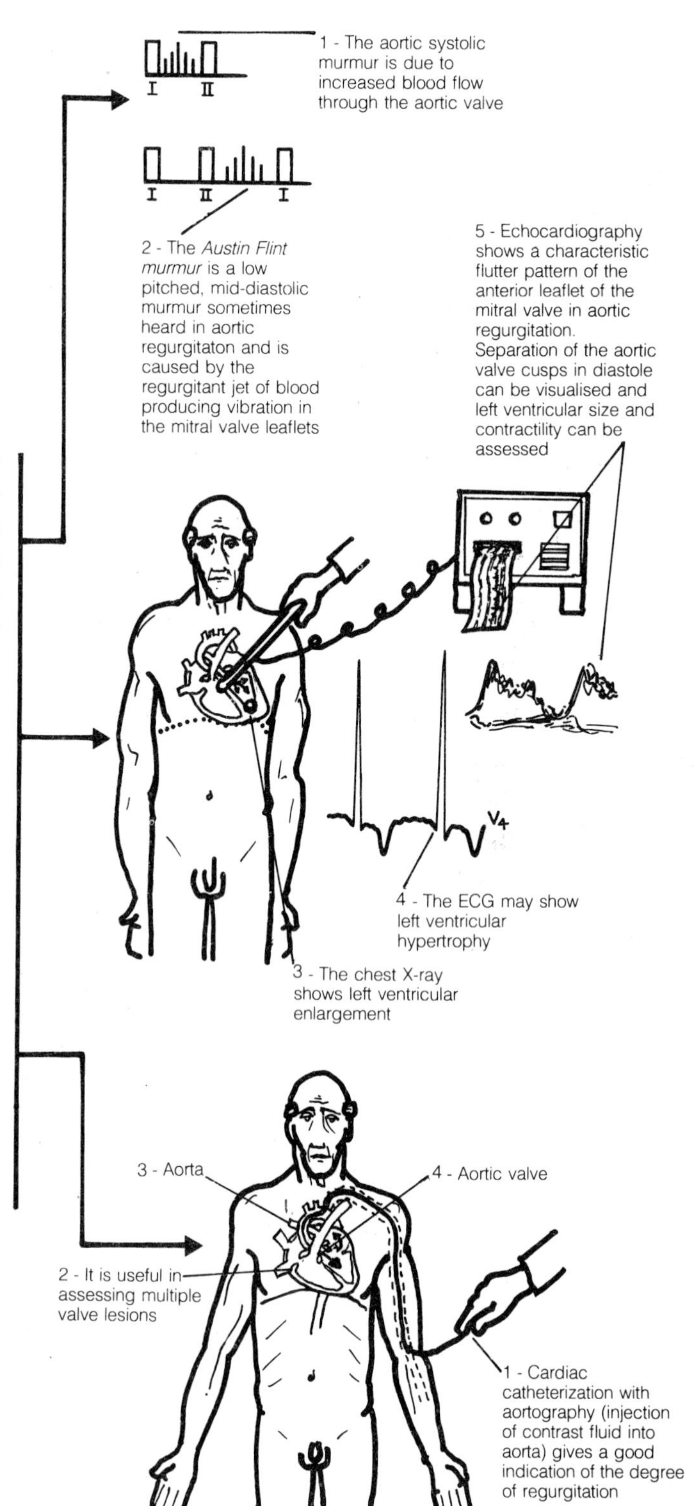
I II
1 - The aortic systolic murmur is due to increased blood flow through the aortic valve
I II I
2 - The *Austin Flint murmur* is a low pitched, mid-diastolic murmur sometimes heard in aortic regurgitaton and is caused by the regurgitant jet of blood producing vibration in the mitral valve leaflets
5 - Echocardiography shows a characteristic flutter pattern of the anterior leaflet of the mitral valve in aortic regurgitation. Separation of the aortic valve cusps in diastole can be visualised and left ventricular size and contractility can be assessed
V4
4 - The ECG may show left ventricular hypertrophy
3 - The chest X-ray shows left ventricular enlargement
3 - Aorta
4 - Aortic valve
2 - It is useful in assessing multiple valve lesions
1 - Cardiac catheterization with aortography (injection of contrast fluid into aorta) gives a good indication of the degree of regurgitation

Management

1. Medical

Antibiotic cover should be given for any surgical or dental procedure. Asymptomatic patients should be closely followed up, and any deterioration in signs or development of symptoms are an indication for surgery.

2. Surgical

Aortic valve replacment should be carried out before heart failure develops. Any co-existing mitral valve or coronary artery disease is usually operated on at the same time.

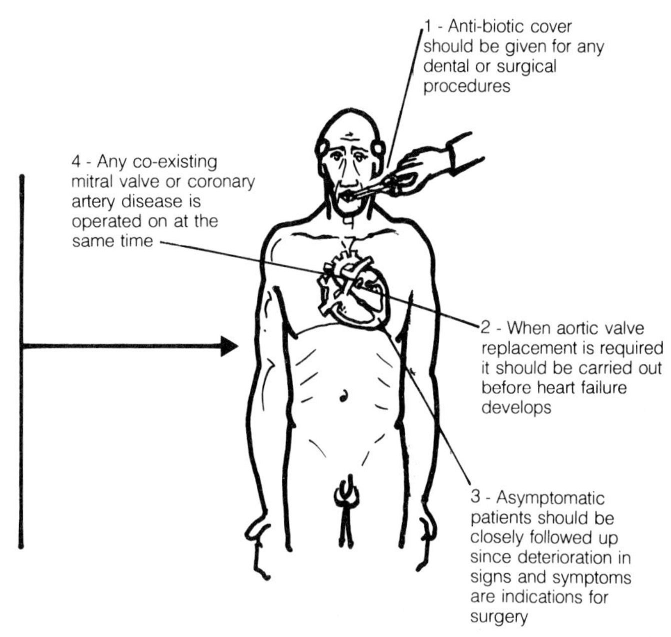
1 - Anti-biotic cover should be given for any dental or surgical procedures
4 - Any co-existing mitral valve or coronary artery disease is operated on at the same time
2 - When aortic valve replacement is required it should be carried out before heart failure develops
3 - Asymptomatic patients should be closely followed up since deterioration in signs and symptoms are indications for surgery

Acute aortic regurgitation

This can occur in infective endocarditis or dissecting aneurysms. Acute left ventricular failure occurs and shock may ensue. Urgent cardiac investigation is required and emergency valve replacement is sometimes necessary.

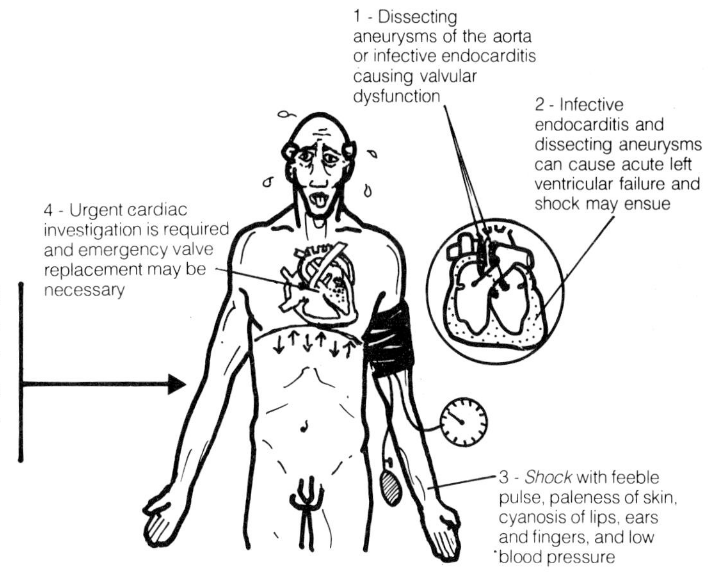
1 - Dissecting aneurysms of the aorta or infective endocarditis causing valvular dysfunction
2 - Infective endocarditis and dissecting aneurysms can cause acute left ventricular failure and shock may ensue
4 - Urgent cardiac investigation is required and emergency valve replacement may be necessary
3 - *Shock* with feeble pulse, paleness of skin, cyanosis of lips, ears and fingers, and low blood pressure

CONGENITAL HEART DISEASE

There are two groups of congenital heart disease; acyanotic and cyanotic.

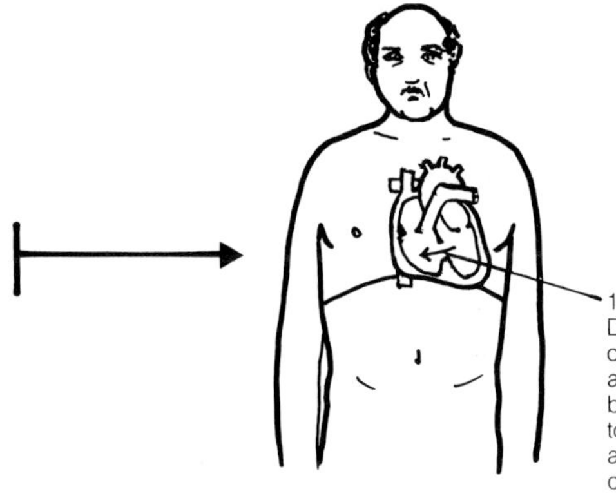
1 - *Acyanotic* Heart Disease can occur with or without a shunt. With a shunt as in this case, blood will flow from left to right side through an abnormal communication provided there is no pulmonary hypertension which would abnormally increase the pressure in the right ventricle

ACYANOTIC HEART DISEASE

Can occur with or without a shunt. With a shunt, blood will flow from the left side of the heart to the right side through any abnormal communication, providing that there is no pulmonary hypertension.

1 - Patent ductus arteriosus occurs normally in intra-uterine life, to by-pass the non-functioning lungs of the foetus

Patent ductus arteriosus

The ductus arteriosus, which forms a connection between the aorta and pulmonary artery, usually closes at birth. If it remains patent, blood is shunted from the aorta to the pulmonary artery, producing a high output cardiac state.

Symptoms

The patient may be asymptomatic. Heart failure or infective endocarditis are common complications.

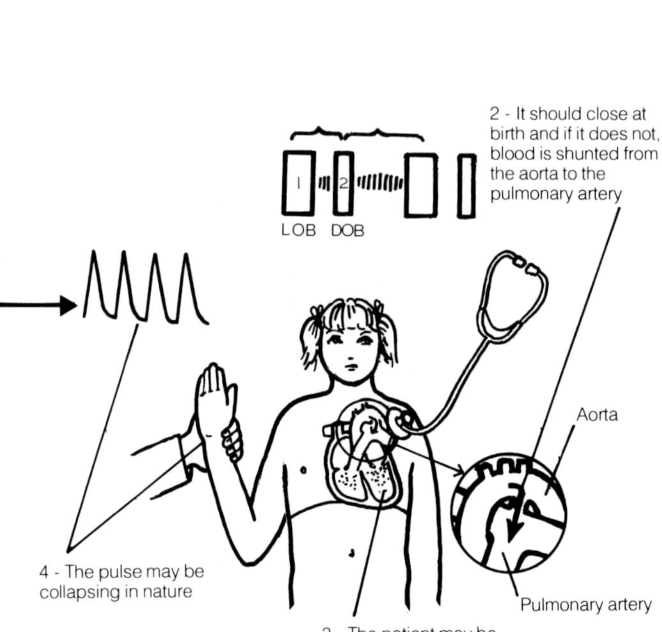
2 - It should close at birth and if it does not, blood is shunted from the aorta to the pulmonary artery
1
2
LOB DOB
Aorta
4 - The pulse may be collapsing in nature
Pulmonary artery
3 - The patient may be asymptomatic but heart failure or infective endocarditis are common complications

Signs

The pulse may be collapsing in nature, and on auscultation a murmur which is continuous throughout systole and diastole can be heard in the pulmonary area (**machinery murmur**). It is loudest around the second heart sound, and a thrill may be present. Sometimes a short mitral diastolic flow murmur can be heard.

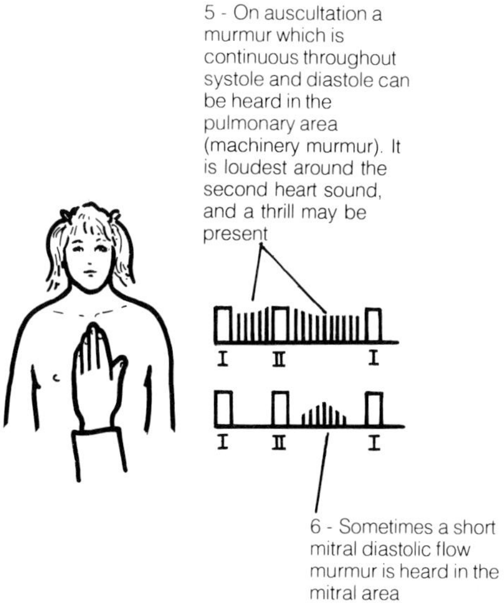

5 - On auscultation a murmur which is continuous throughout systole and diastole can be heard in the pulmonary area (machinery murmur). It is loudest around the second heart sound, and a thrill may be present

6 - Sometimes a short mitral diastolic flow murmur is heard in the mitral area

Management

A patent ductus arteriosus should be surgically closed, unless pulmonary hypertension with reversed flow through the shunt has developed. However, in infancy prostaglandin inhibitors (eg. indomethacin) may promote closure and now form a medical means of therapy.

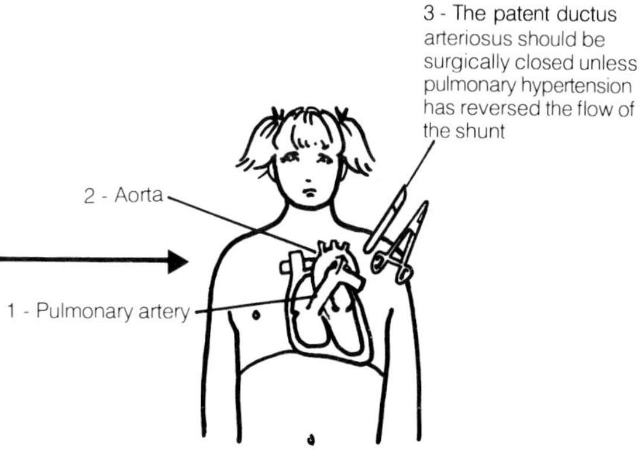

3 - The patent ductus arteriosus should be surgically closed unless pulmonary hypertension has reversed the flow of the shunt

2 - Aorta

1 - Pulmonary artery

Atrial septal defect

Atrial septal defects (**ASD**) are usually due to failure of development of the septum secundum in the fossa ovalis. More rarely, incomplete fusion of the septum primum with the endocardial cushions occurs, associated with a defect in the mitral valve – the ostium primum **ASD**. In **ASD** blood flows from the left atrium to the right, and the blood flow through the defect is often twice that through the left ventricle. A pulmonary systolic flow murmur is characteristic of **ASD** and is due to this larger blood flow.

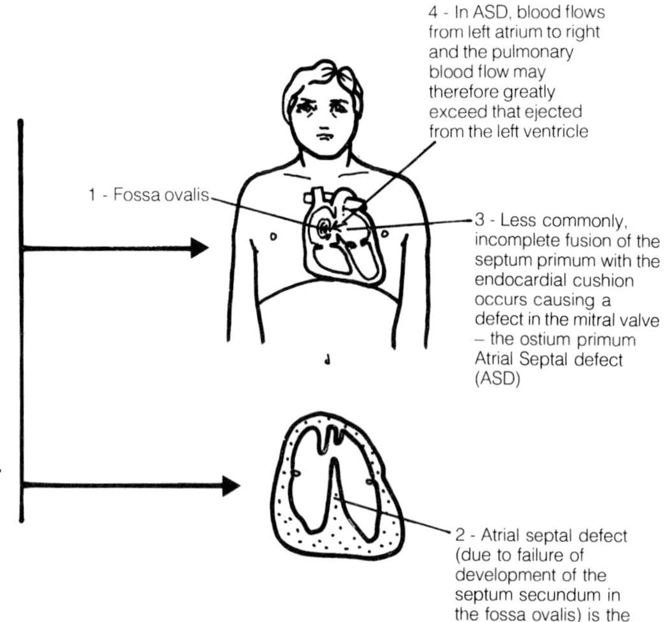
4 - In ASD, blood flows from left atrium to right and the pulmonary blood flow may therefore greatly exceed that ejected from the left ventricle
1 - Fossa ovalis
3 - Less commonly, incomplete fusion of the septum primum with the endocardial cushion occurs causing a defect in the mitral valve – the ostium primum Atrial Septal defect (ASD)
2 - Atrial septal defect (due to failure of development of the septum secundum in the fossa ovalis) is the most common and usually less serious form of ASD

Symptoms

Most patients are asymptomatic though ostium primum defects with significant mitral regurgitation can result in heart failure.

Signs

There may be left parasternal heave due to right ventricular hypertrophy. There is fixed splitting of the second heart sound as the right ventricular stroke volume does not alter with respiration. A systolic pulmonary flow murmur is present. In ostium primum defects the pansystolic murmur of mitral regurgitation may also be present. The **ECG** shows right bundle branch block and in ostium primum defects left axis deviation is present. **Chest X-ray** shows prominent pulmonary arteries and a prominent right heart border.

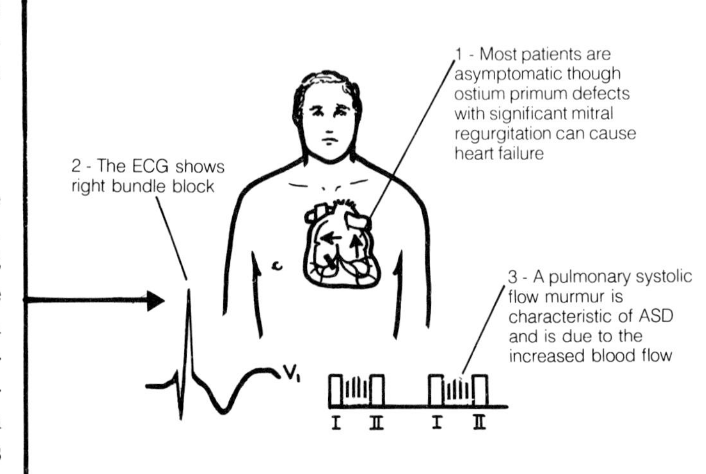
1 - Most patients are asymptomatic though ostium primum defects with significant mitral regurgitation can cause heart failure
2 - The ECG shows right bundle block
3 - A pulmonary systolic flow murmur is characteristic of ASD and is due to the increased blood flow
V1
I II I II

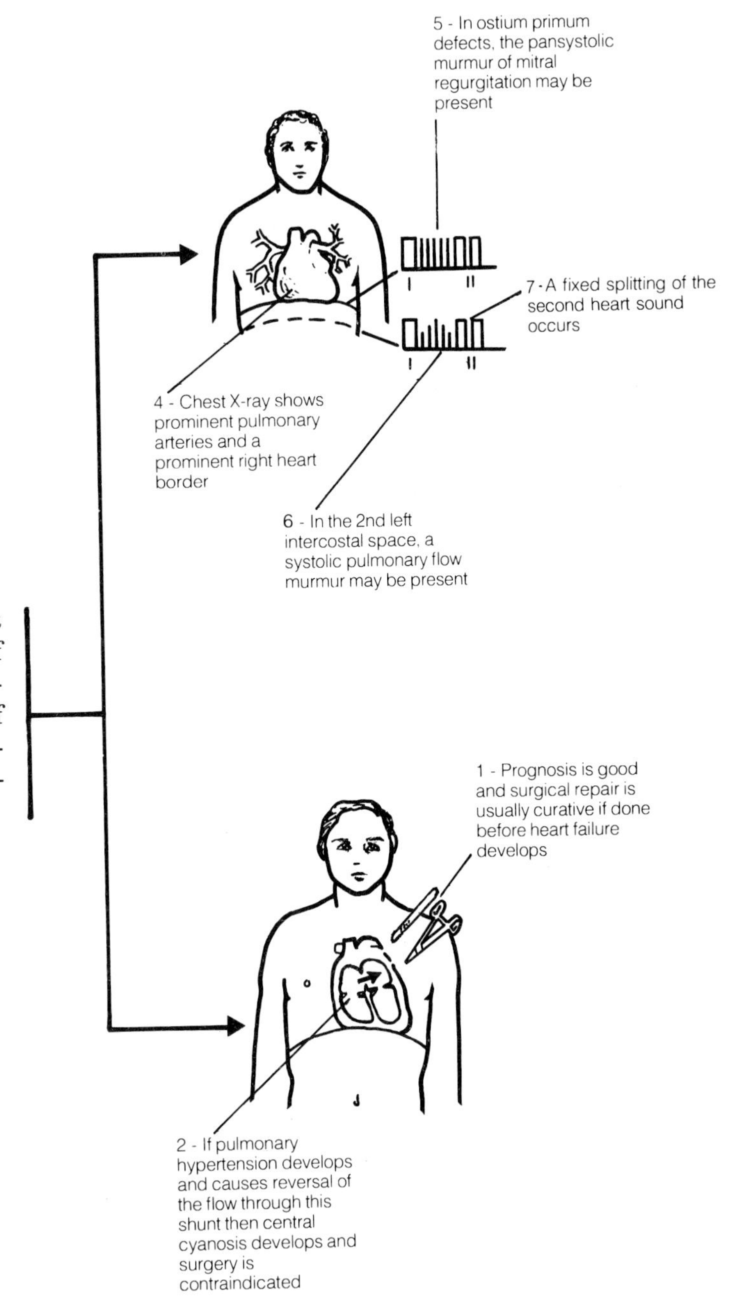
5 - In ostium primum defects, the pansystolic murmur of mitral regurgitation may be present
I
II
7 - A fixed splitting of the second heart sound occurs
I
II
4 - Chest X-ray shows prominent pulmonary arteries and a prominent right heart border
6 - In the 2nd left intercostal space, a systolic pulmonary flow murmur may be present
1 - Prognosis is good and surgical repair is usually curative if done before heart failure develops
2 - If pulmonary hypertension develops and causes reversal of the flow through this shunt then central cyanosis develops and surgery is contraindicated

Prognosis

Is usually good but worsens if heart failure has developed. Surgical repair of the defect is usually curative. If pulmonary hypertension develops, reversal of flow through the shunt and central cyanosis occur. Surgery is then contraindicated.

Ventricular septal defect (VSD)

The clinical features depend on the size of the defect. With a small defect (**maladie de Roger**) there is restricted flow from the left to the right ventricle and the condition is asymptomatic. A loud pansystolic murmur best heard at the lower end of the sternum often accompanied by a thrill occurs. Larger defects result in a large left to right shunt, with increased pulmonary blood flow. Cardiac failure can result, and if pulmonary hypertension becomes severe, reversal of the shunt with central cyanosis can occur.

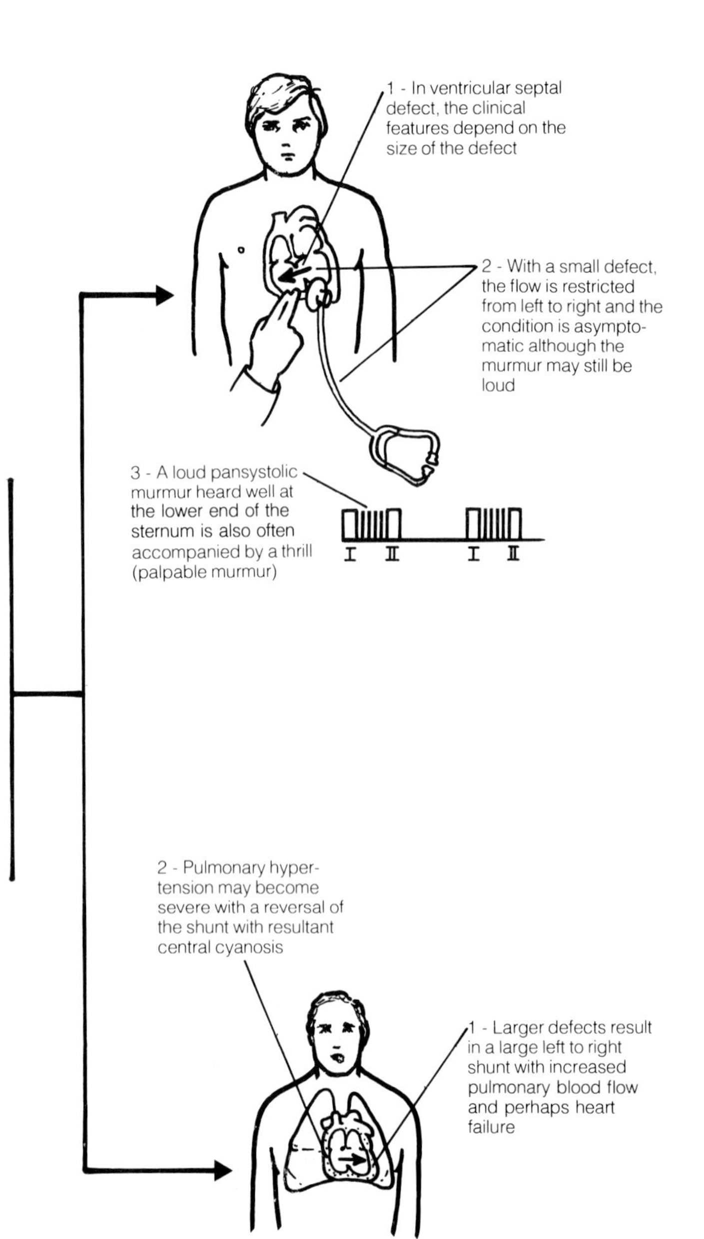
1 - In ventricular septal defect, the clinical features depend on the size of the defect
2 - With a small defect, the flow is restricted from left to right and the condition is asymptomatic although the murmur may still be loud
3 - A loud pansystolic murmur heard well at the lower end of the sternum is also often accompanied by a thrill (palpable murmur)
I II I II
2 - Pulmonary hypertension may become severe with a reversal of the shunt with resultant central cyanosis
1 - Larger defects result in a large left to right shunt with increased pulmonary blood flow and perhaps heart failure

Management

Small defects need close observation; sixty per cent will close spontaneously. Antibiotic prophylaxis is required for surgical and dental procedures. Larger defects require surgical repair.

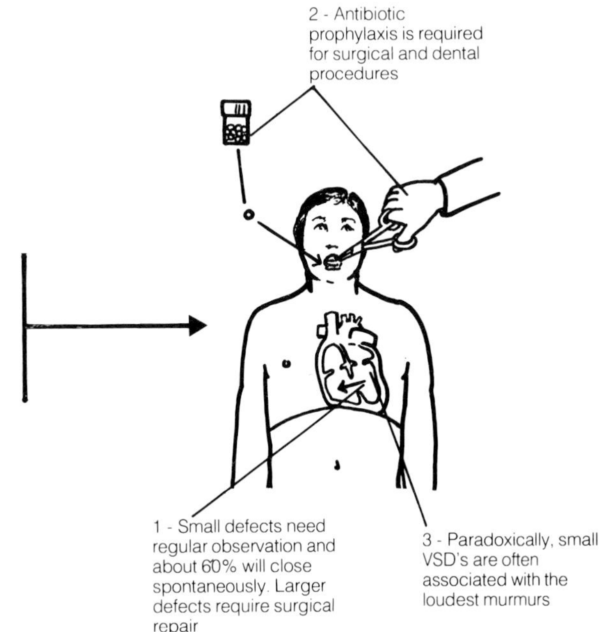
2 - Antibiotic prophylaxis is required for surgical and dental procedures
1 - Small defects need regular observation and about 60% will close spontaneously. Larger defects require surgical repair
3 - Paradoxically, small VSD's are often associated with the loudest murmurs

Acyanotic congenital heart diseases without a shunt:–

In **coarctation of the aorta** there is stricture of the aorta near the insertion of the ligamentum arteriosum, just below the origin of the left subclavian artery. A collateral circulation develops linking the subclavian arteries above the stricture with the intercostal arteries below. Patients may present with heart failure or hypertension in the upper limbs; however they may be asymptomatic for years.

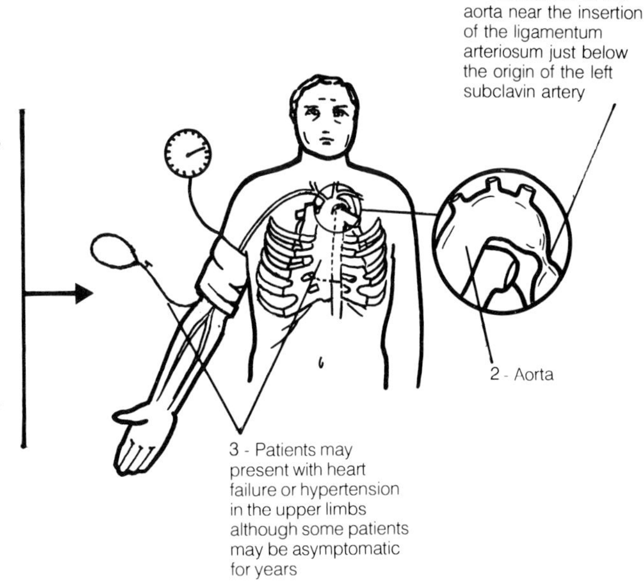
1 - Coarctation of the aorta near the insertion of the ligamentum arteriosum just below the origin of the left subclavin artery
2 - Aorta
3 - Patients may present with heart failure or hypertension in the upper limbs although some patients may be asymptomatic for years

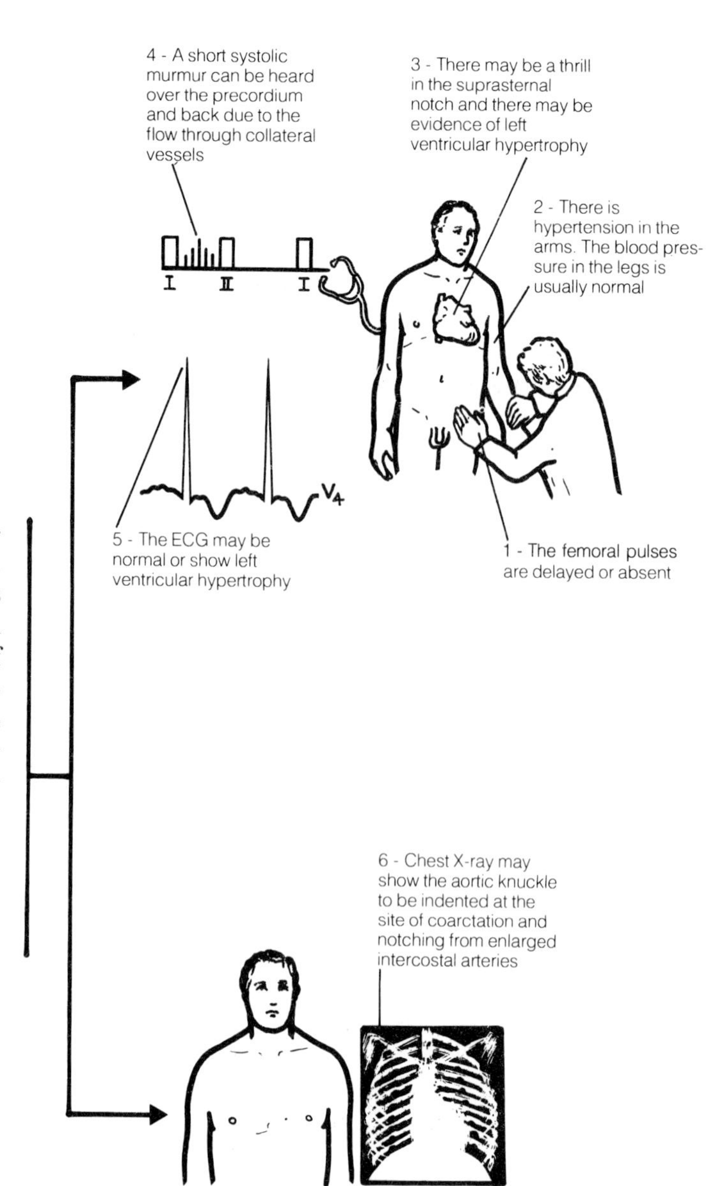
4 - A short systolic murmur can be heard over the precordium and back due to the flow through collateral vessels
3 - There may be a thrill in the suprasternal notch and there may be evidence of left ventricular hypertrophy
2 - There is hypertension in the arms. The blood pressure in the legs is usually normal
I
II
I
V4
5 - The ECG may be normal or show left ventricular hypertrophy
1 - The femoral pulses are delayed or absent
6 - Chest X-ray may show the aortic knuckle to be indented at the site of coarctation and notching from enlarged intercostal arteries

Signs

The femoral pulses are delayed or absent. There is hypertension in the arms, while the blood pressure in the legs is usually normal. There may be a thrill in the suprasternal notch, and evidence of left ventricular hypertrophy. A short systolic murmur can be heard over the precordium and back, due to blood flow through collateral vessels. The **ECG** may be normal or show left ventricular hypertrophy. **Chest X-ray** The aortic knuckle may be indented at the site of coarctation, and rib notching occurs from enlarged intercostal arteries.

Treatment

The coarctation should be resected as soon as possible. After the age of four, hypertension may persist even though the coarctation has been corrected. Long-term complications of coarctation include cerebral haemorrhage, due to ruptured aneurysms in the circle of Willis, and heart failure.

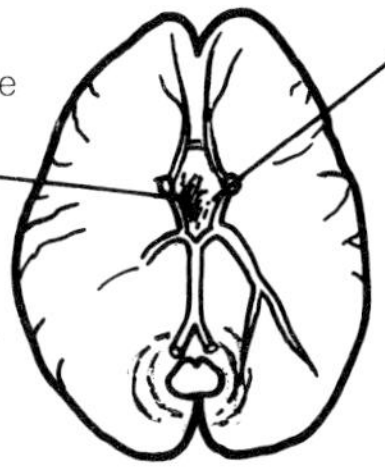

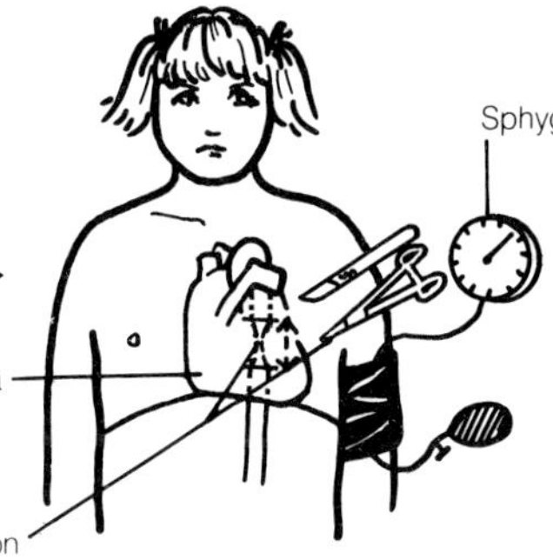

1 - Haemorrhage of aneurysms of the circle of Willis due to hypertension
2 - Circle of Willis
Sphygmomanometer
3 - Heart failure is a complication
4 - The coarctation should be resected as soon as possible but after the age of 4, hypertension may persist despite the correction

Congenital aortic stenosis

This may be valvular, subvalvular or supravalvular. Valvular aortic stenosis is usually due to fusion of the commissures of a bicuspid aortic valve. This may be asymptomatic, but there is increased risk of infective endocarditis. Subvalvular stenosis may be due to muscle hypertrophy or a fibrous ring. In supravalvular stenosis, a constriction occurs just distal to the origin of the coronary artery; this may be associated with an abnormal facial appearance.

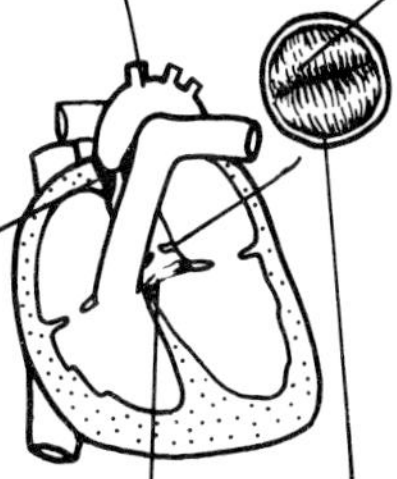

1 - Aorta
2 - Valvular aortic stenosis is usually due to fusion of the commissures of a bicuspid aortic valve
3 - This may be asymptomatic but there is increased risk of infective endocarditis
4 - Subvalvular stenosis may be due to muscle hypertrophy or a fibrous ring
5 - In supravalvular stenosis a constriction occurs just distal to the origin of the coronary artery

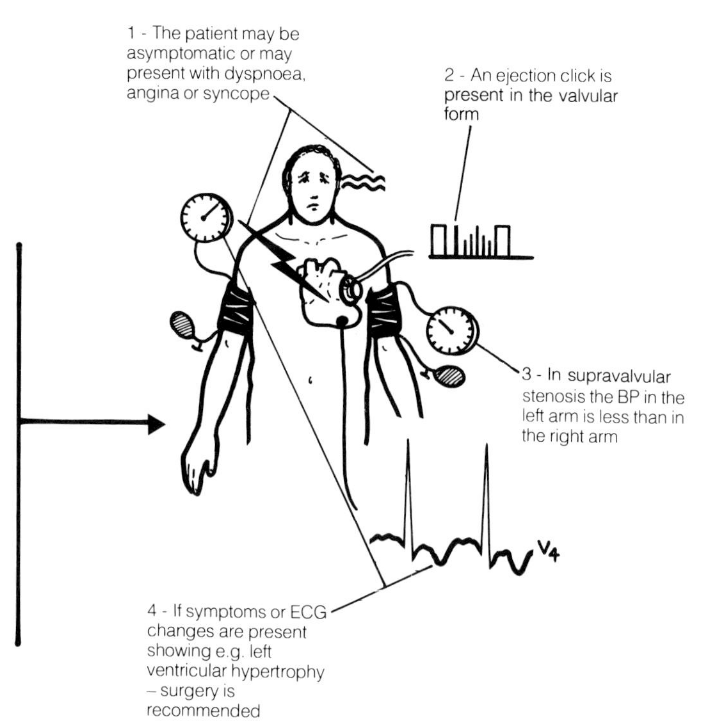
1 - The patient may be asymptomatic or may present with dyspnoea, angina or syncope
2 - An ejection click is present in the valvular form
3 - In supravalvular stenosis the BP in the left arm is less than in the right arm
V4
4 - If symptoms or ECG changes are present showing e.g. left ventricular hypertrophy – surgery is recommended

Symptoms
The patient may be asymptomatic or may present with dyspnoea, angina or syncope

Signs
On examination, an ejection click is present in the valvular form. In supravalvular stenosis, the blood pressure in the left arm is less than that in the right.

Management
If symptoms or ECG changes are present, surgery is recommended.

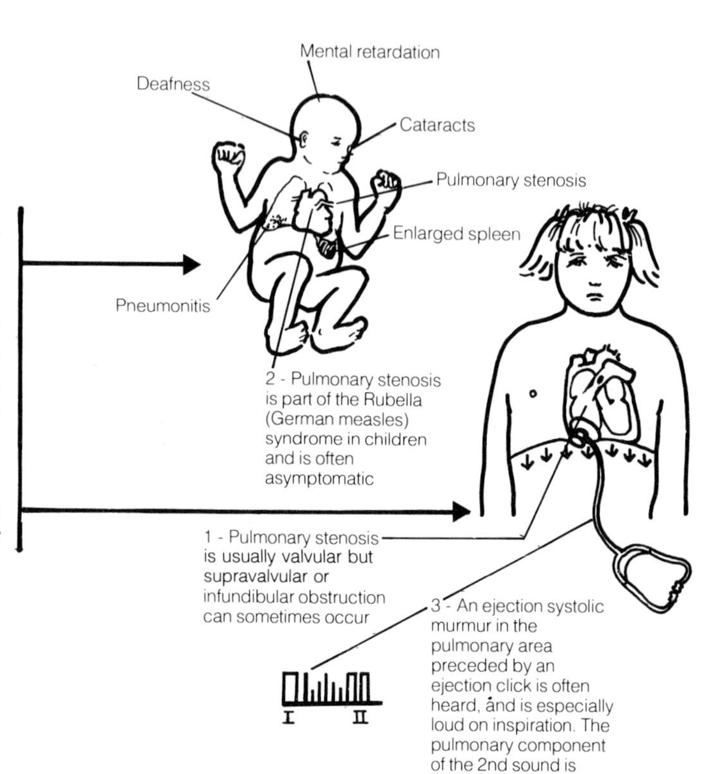
Mental retardation
Deafness
Cataracts
Pulmonary stenosis
Enlarged spleen
Pneumonitis
2 - Pulmonary stenosis is part of the Rubella (German measles) syndrome in children and is often asymptomatic
1 - Pulmonary stenosis is usually valvular but supravalvular or infundibular obstruction can sometimes occur
3 - An ejection systolic murmur in the pulmonary area preceded by an ejection click is often heard, and is especially loud on inspiration. The pulmonary component of the 2nd sound is delayed
I
II

Pulmonary stenosis
Pulmonary stenosis is usually valvular, although supravalvular or infundibular obstruction can occasionally occur. Pulmonary stenosis is found as part of the rubella syndrome in children. It is often asymptomatic, and characteristic findings are an ejection systolic murmur in the pulmonary area, preceded by an ejection click. The murmur is louder in inspiration, and the pulmonary component of the second sound is delayed. If

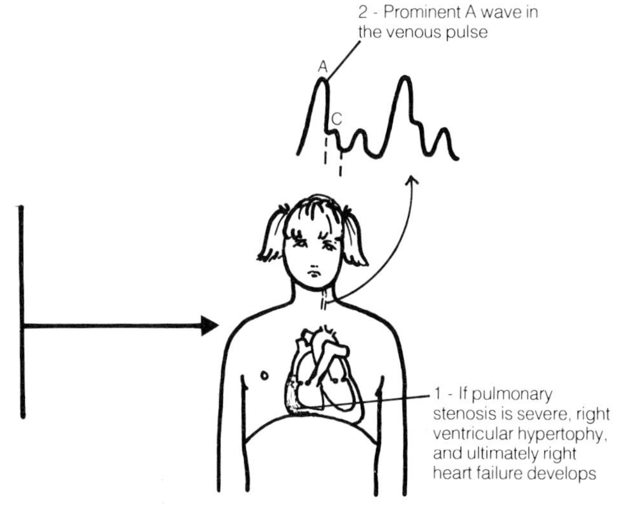
2 - Prominent A wave in the venous pulse
A
C
1 - If pulmonary stenosis is severe, right ventricular hypertophy, and ultimately right heart failure develops

pulmonary stenosis is severe, right ventricular hypertrophy and ultimately right heart failure develop, with a prominent "a" wave in the venous pulse, a palpable right ventricular impulse at the left sternal edge and a soft delayed pulmonary second sound.

CYANOTIC HEART DISEASE

Fallot's tetralogy

Fallot's tetralogy consists of pulmonary stenosis, ventricular septal defect, right ventricular hypertrophy and over-riding of the interventricular septum by the aorta. Venous blood is partly ejected into the aorta during systole; this results in cyanosis.

Symptoms

The child usually presents with cyanosis, sometimes associated with loss of consciousness and dyspnoea on exertion. The child often squats after exertion; this reduces the right to left shunt by increasing arterial resistance.

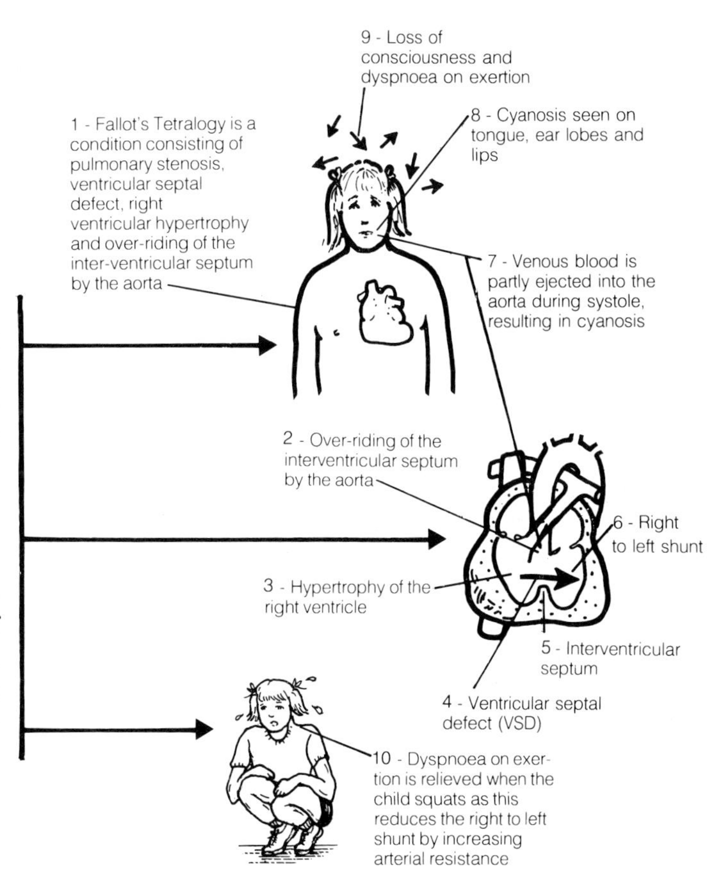
9 - Loss of consciousness and dyspnoea on exertion
1 - Fallot's Tetralogy is a condition consisting of pulmonary stenosis, ventricular septal defect, right ventricular hypertrophy and over-riding of the inter-ventricular septum by the aorta
8 - Cyanosis seen on tongue, ear lobes and lips
7 - Venous blood is partly ejected into the aorta during systole, resulting in cyanosis
2 - Over-riding of the interventricular septum by the aorta
6 - Right to left shunt
3 - Hypertrophy of the right ventricle
5 - Interventricular septum
4 - Ventricular septal defect (VSD)
10 - Dyspnoea on exertion is relieved when the child squats as this reduces the right to left shunt by increasing arterial resistance

Signs

Polycythaemia and finger clubbing may be present. There is an ejection systolic murmur in the pulmonary area and a single second heart sound. The **ECG** shows right ventricular hypertrophy with right axis deviation. The **chest X-ray** shows a hollow pulmonary artery bay and oligaemic lung fields. The appearance is described as "boot shaped" (coeur en sabot). **Echocardiography** may show over-riding of the septum by the aorta. **Cardiac catheterization:** The right and left ventricular pressures are equal, and oxygen saturation in the aorta is reduced.

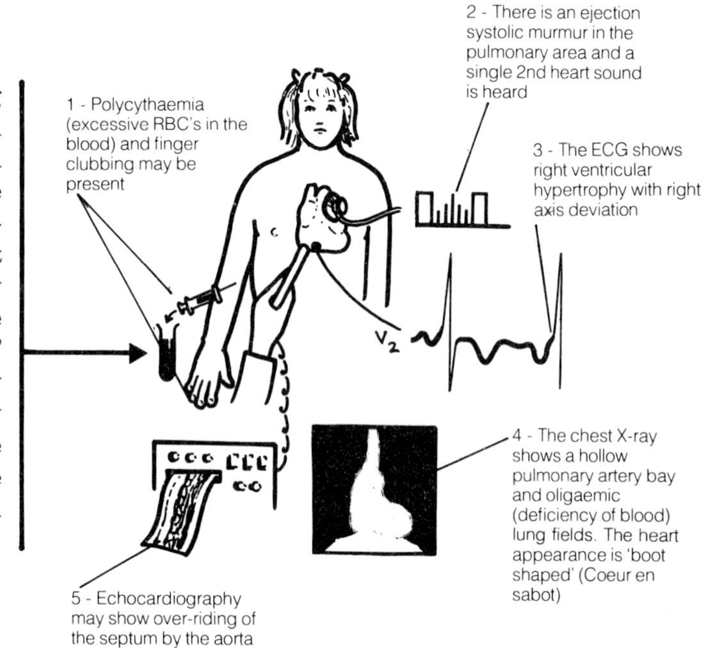

1 - Polycythaemia (excessive RBC's in the blood) and finger clubbing may be present
2 - There is an ejection systolic murmur in the pulmonary area and a single 2nd heart sound is heard
3 - The ECG shows right ventricular hypertrophy with right axis deviation
V2
4 - The chest X-ray shows a hollow pulmonary artery bay and oligaemic (deficiency of blood) lung fields. The heart appearance is 'boot shaped' (Coeur en sabot)
5 - Echocardiography may show over-riding of the septum by the aorta

Management

Without treatment, 85% of patients die before adolescence. Surgical repair of the tetralogy should be performed as soon as possible. This can be done as a one stage procedure on cardiopulmonary bypass; the ventricular septal defect is patched and pulmonary stenosis is relieved. The Blalock-Taussig procedure is sometimes used in small infants; here a systemic artery (often subclavian) is anastomosed to the pulmonary artery. Definitive repair is then performed at 2-4 years.

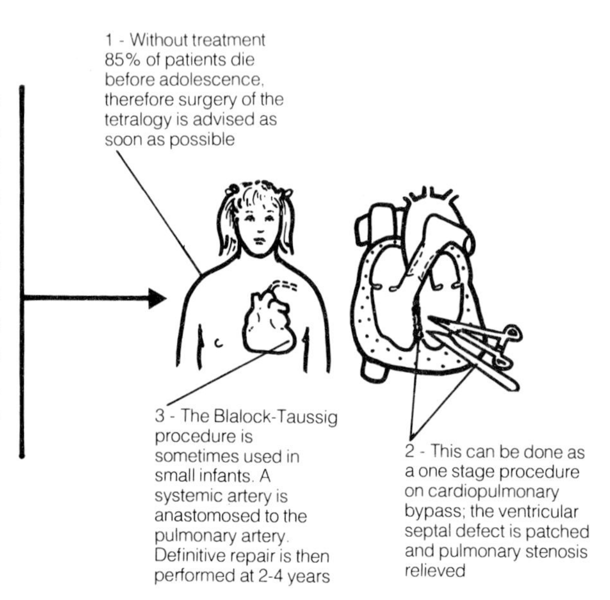

1 - Without treatment 85% of patients die before adolescence, therefore surgery of the tetralogy is advised as soon as possible
2 - This can be done as a one stage procedure on cardiopulmonary bypass; the ventricular septal defect is patched and pulmonary stenosis relieved
3 - The Blalock-Taussig procedure is sometimes used in small infants. A systemic artery is anastomosed to the pulmonary artery. Definitive repair is then performed at 2-4 years

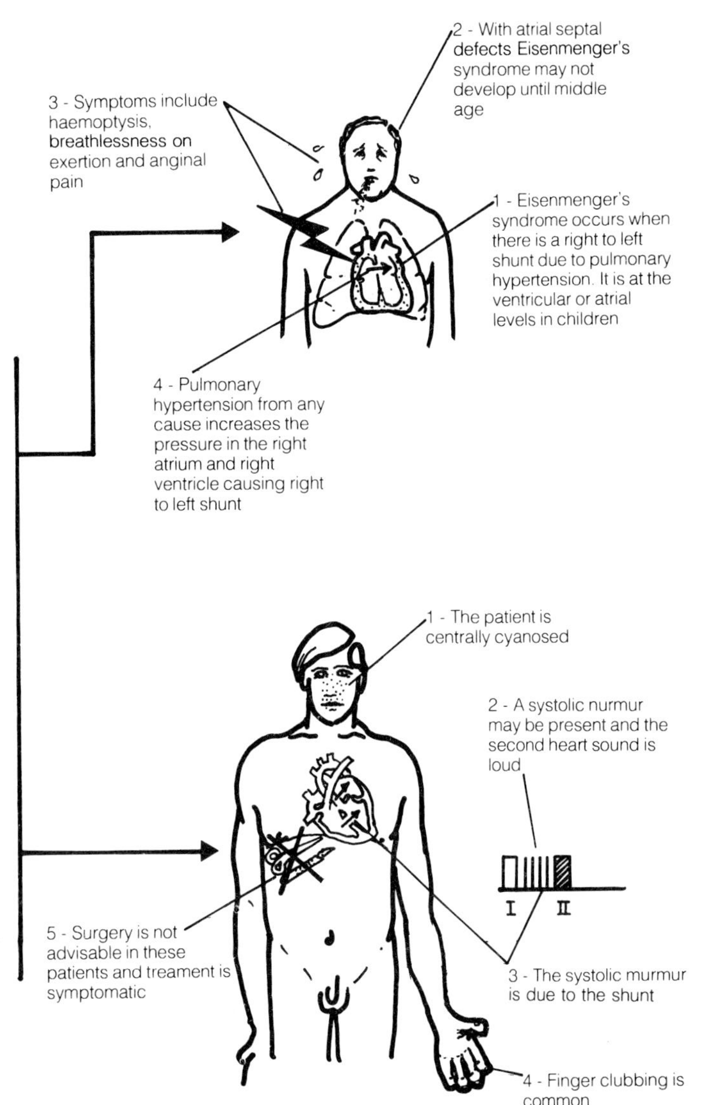
2 - With atrial septal defects Eisenmenger's syndrome may not develop until middle age
3 - Symptoms include haemoptysis, breathlessness on exertion and anginal pain
1 - Eisenmenger's syndrome occurs when there is a right to left shunt due to pulmonary hypertension. It is at the ventricular or atrial levels in children
4 - Pulmonary hypertension from any cause increases the pressure in the right atrium and right ventricle causing right to left shunt
1 - The patient is centrally cyanosed
2 - A systolic nurmur may be present and the second heart sound is loud
I
II
5 - Surgery is not advisable in these patients and treament is symptomatic
3 - The systolic murmur is due to the shunt
4 - Finger clubbing is common

Eisenmenger's syndrome

This occurs when there is a right to left shunt due to pulmonary hypertension. Shunting usually occurs at ventricular or atraial levels in children. With atrial septal defects, Eisenmenger's syndrome may not develop until middle age.

Symptoms

Include breathlessness on exertion, angina and haemoptysis.

Signs

The patient is centrally cyanosed and has finger clubbing. The second heart sound is loud, and a systolic murmur may be present due to the shunt.

Treatment

Surgery is not advisable in these patients, and treatment is symptomatic.

INFECTIVE ENDOCARDITIS

In the United Kingdom there are approximately 1,000 cases of infective endocarditis per year. This number has not decreased since the advent of antibiotics, but the pattern of disease and infective organism have changed dramatically.

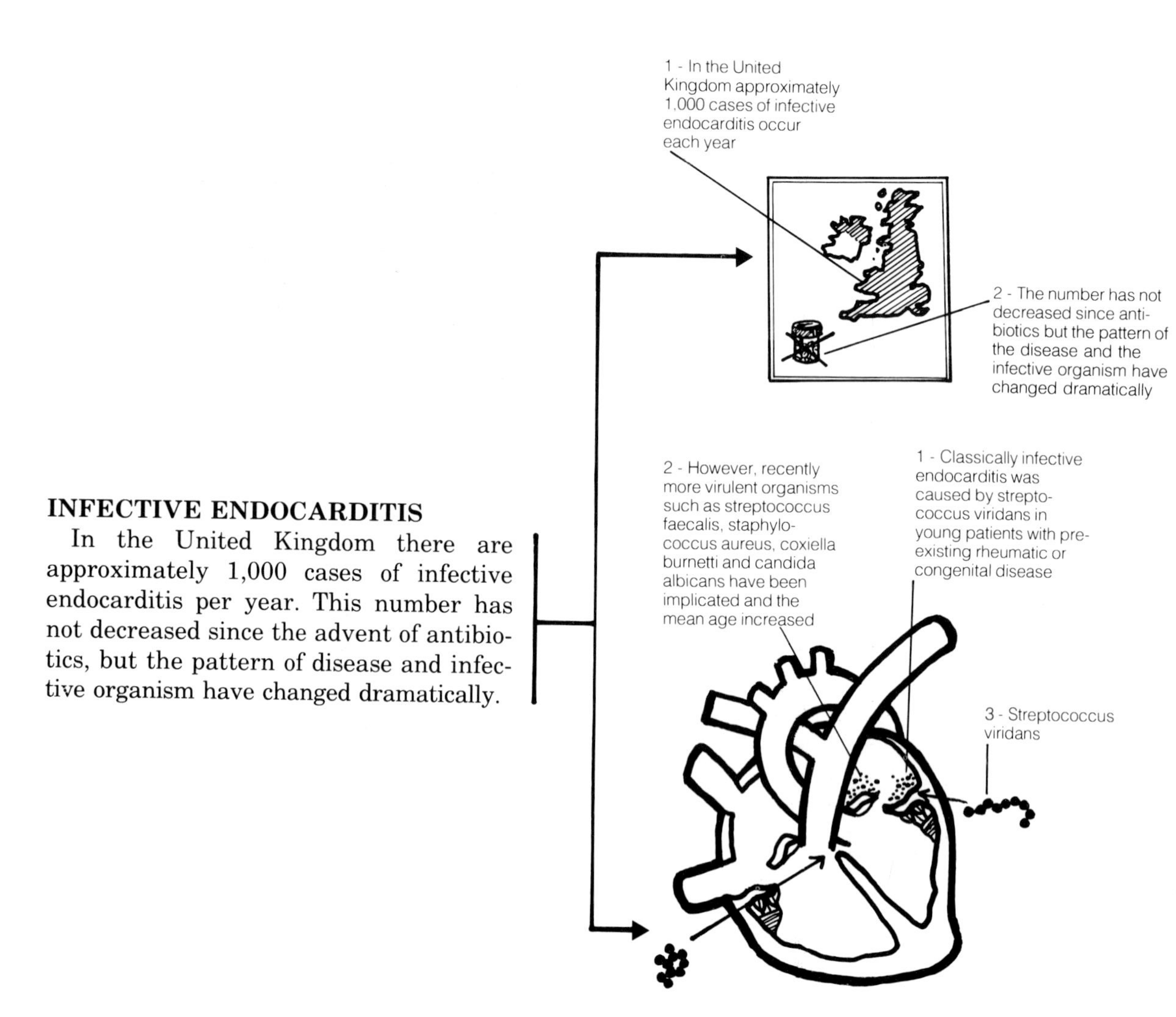
1 - In the United Kingdom approximately 1,000 cases of infective endocarditis occur each year
2 - The number has not decreased since anti-biotics but the pattern of the disease and the infective organism have changed dramatically
1 - Classically infective endocarditis was caused by strepto-coccus viridans in young patients with pre-existing rheumatic or congenital disease
2 - However, recently more virulent organisms such as streptococcus faecalis, staphylo-coccus aureus, coxiella burnetti and candida albicans have been implicated and the mean age increased
3 - Streptococcus viridans

Classically infective endocarditis was caused by *Streptococcus viridans* in young patients with pre-existing rheumatic or congenital heart disease. However, more recently, the mean age of infection has risen, and more virulent organisms such as *Staphylococcus aureus, Streptococcus faecalis, Coxiella burnetti,* and fungi especially candida have been involved. Other pre-disposing factors include prosthetic heart valves, haemodialysis, and intravenous drug abuse, which can cause right sided endocarditis. In the elderly, genitourinary surgery, colonic disease, and gallbladder disease predispose to infective endocarditis.

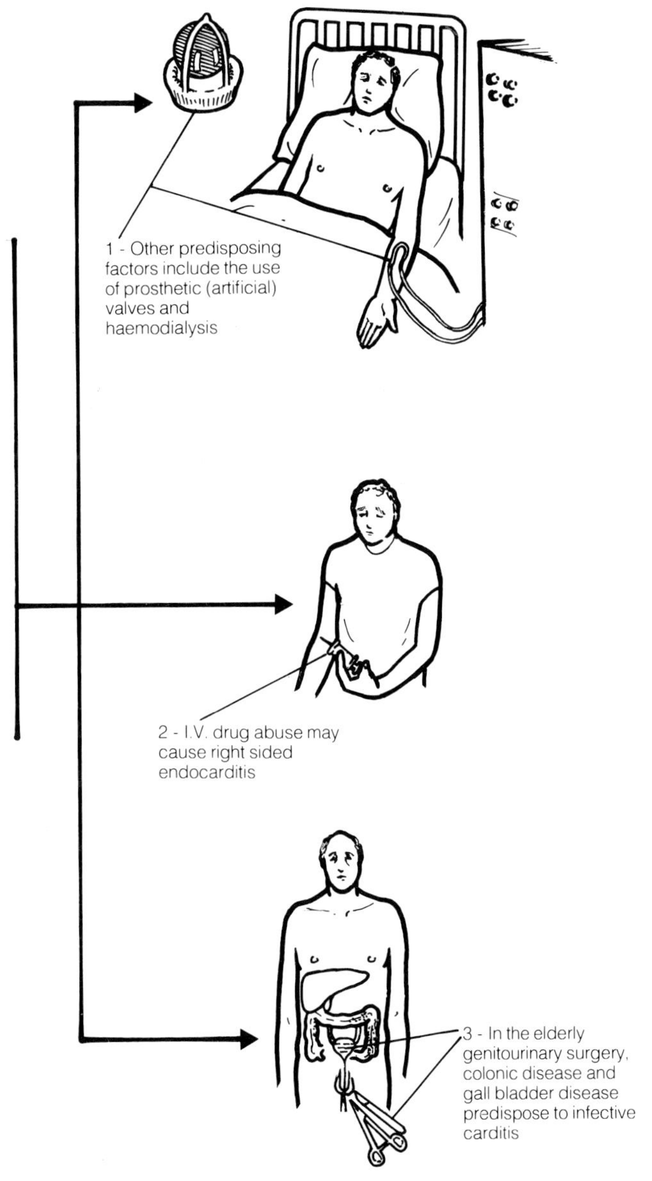
1 - Other predisposing factors include the use of prosthetic (artificial) valves and haemodialysis
2 - I.V. drug abuse may cause right sided endocarditis
3 - In the elderly genitourinary surgery, colonic disease and gall bladder disease predispose to infective carditis

Where the infecting organism is virulent (for example *Streptococcus faecalis*) normal heart valves can be affected. Less virulent organisms tend to affect patients with pre-existing valvular disease, especially where there is a "jet lesion" with rapid blood flow through a regurgitant valve or intracardiac shunt. A transient bacteraemia, often induced by dental manipulation, is usually the starting point for infection. A platelet/fibrin clot forms on the heart valve, and is colonized by circulating organisms to produce an infected vegetation. Infected emboli can be released into the circulation and immune complexes are produced, which can lodge in the kidneys and skin.

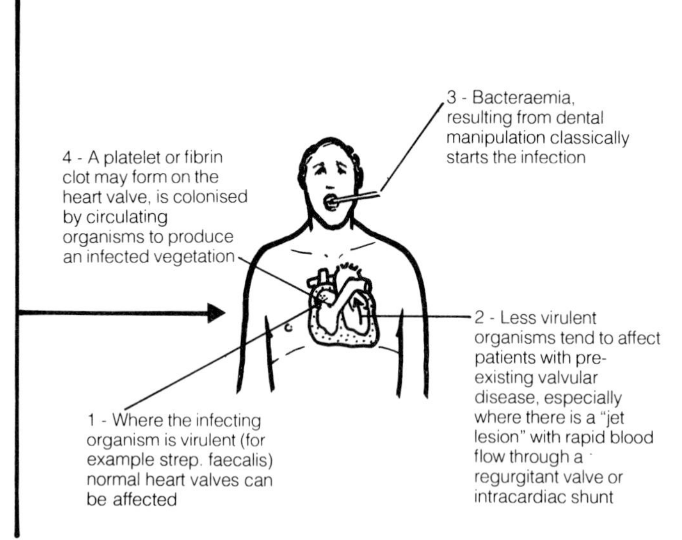
3 - Bacteraemia, resulting from dental manipulation classically starts the infection
4 - A platelet or fibrin clot may form on the heart valve, is colonised by circulating organisms to produce an infected vegetation
2 - Less virulent organisms tend to affect patients with pre-existing valvular disease, especially where there is a "jet lesion" with rapid blood flow through a regurgitant valve or intracardiac shunt
1 - Where the infecting organism is virulent (for example strep. faecalis) normal heart valves can be affected

Clinical features

Patients may develop malaise, weight loss, tiredness and pyrexia. On examination the patient may be anaemic with finger clubbing. Features of immune complex disease may be seen. These include splinter haemorrhages found in the nails, Osler's nodes (tender transient erythematous areas on the finger pads), Roth's spots (haemorrhagic areas on the retina) and vasculitic skin rashes. Systemic or pulmonary emboli can occur, and patients may develop hemiparesis, ischaemic limbs, splenic infarction or sudden blindness. On auscultation of the heart there

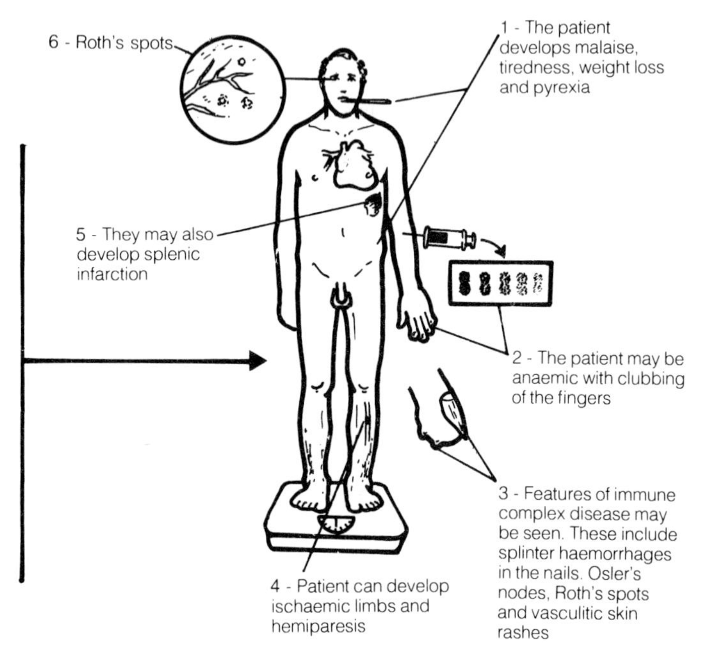
6 - Roth's spots
1 - The patient develops malaise, tiredness, weight loss and pyrexia
5 - They may also develop splenic infarction
2 - The patient may be anaemic with clubbing of the fingers
3 - Features of immune complex disease may be seen. These include splinter haemorrhages in the nails. Osler's nodes, Roth's spots and vasculitic skin rashes
4 - Patient can develop ischaemic limbs and hemiparesis

may be a change in the character of the heart murmur, or the development of new murmurs. Heart failure or conduction defects may develop. Occasionally infected valves may rupture. The spleen is usually enlarged and the kidney is affected. Glomerulonephritis is common and gives rise to microscopic haematuria.

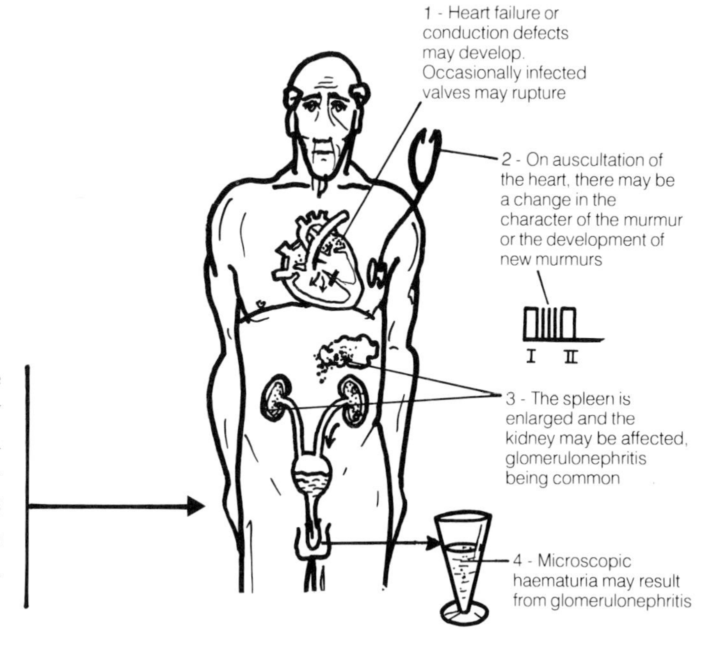
1 - Heart failure or conduction defects may develop. Occasionally infected valves may rupture
2 - On auscultation of the heart, there may be a change in the character of the murmur or the development of new murmurs
I II
3 - The spleen is enlarged and the kidney may be affected, glomerulonephritis being common
4 - Microscopic haematuria may result from glomerulonephritis

Investigations

There is usually anaemia with a raised ESR. Examination of the urine will reveal microscopic haematuria. Blood cultures should be performed whenever the diagnosis is suspected, and six such specimens of blood should be sent for culture before commencing antibiotic treatment. An **echocardiogram** should be performed; this is very useful in demonstrating vegetations, which will appear as echodense areas attached to the leaflets of affected valves. In right-sided endocarditis or left-to-right shunts, infected emboli occur in the lungs, producing areas of infarction which can be seen on the **chest X-ray.**

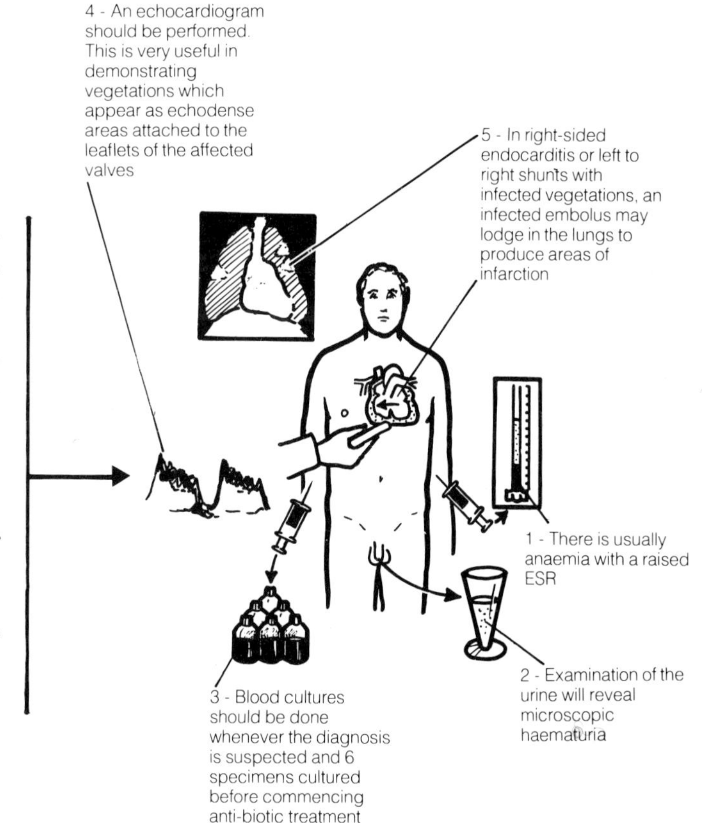
4 - An echocardiogram should be performed. This is very useful in demonstrating vegetations which appear as echodense areas attached to the leaflets of the affected valves
5 - In right-sided endocarditis or left to right shunts with infected vegetations, an infected embolus may lodge in the lungs to produce areas of infarction
1 - There is usually anaemia with a raised ESR
2 - Examination of the urine will reveal microscopic haematuria
3 - Blood cultures should be done whenever the diagnosis is suspected and 6 specimens cultured before commencing anti-biotic treatment

Treatment

This depends on the infecting organism, but bactericidal rather than bacteristatic drugs should be used, and intravenous therapy is required initially. Antibiotic treatment is continued for at least 4 weeks and some organisms such as *Staphlococcus aureus* require treatment for at least 6 weeks. Serum concentrations of the antibiotic should be measured during treatment, and related to the MIC (minimal inhibitory concentration) of the infecting organism. *Streptococcus viridans, Streptococcus pyogenes* and *Streptoccus pneumoniae* are treated with benzyl penicillin, 2 mega units intravenously, 6 hourly. Gentamicin is added if the patient is very ill and is given in divided doses 8 hourly (5mg per kg body weight per 24 hours if renal function is normal). Once there is a clinical response, the patient can be given oral amoxycillin with probenecid. *Staphlococcus aureus* is treated with flucloxicillin and fusidic acid. If there is no response, vancomycin or rifampicin can be added, in consultation with microbiologists. *Streptoccus faecalis* is treated with amoxycillin and gentamicin, while *E.coli* infection is treated with gentamicin. Fungal infections are treated with amphotericin B intravenously.

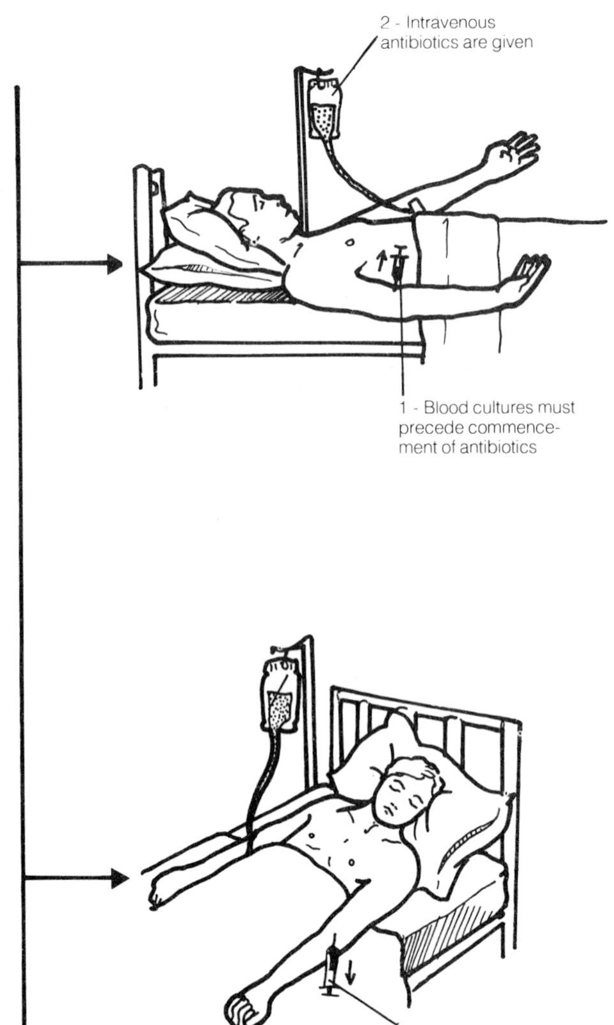
2 - Intravenous antibiotics are given
1 - Blood cultures must precede commencement of antibiotics
2 - Blood drawn for antibiotic serum levels

Culture negative endocarditis

Treatment often has to begin before the results of blood cultures are known, and cultures may sometimes be sterile. In these cases treatment is based on which infecting organism is suspected. Penicillin and gentamicin are given if *Streptococcus viridans* is suspected, for example after recent dental surgery. If the patient has recently had genitourinary surgery or instrumentation of the bowel and *Streptococcus faecalis* is suspected, Amoxycillin and gentamicin are given, while flucloxacillin would be given to cover any staphylococcal infection.

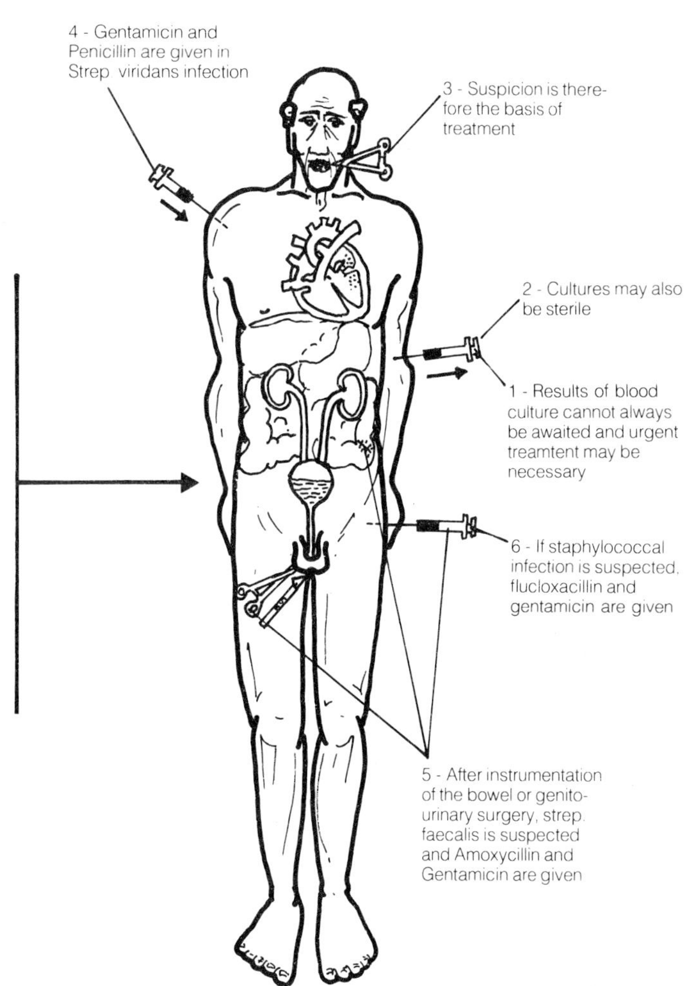
4 - Gentamicin and Penicillin are given in Strep viridans infection
3 - Suspicion is therefore the basis of treatment
2 - Cultures may also be sterile
1 - Results of blood culture cannot always be awaited and urgent treamtent may be necessary
6 - If staphylococcal infection is suspected, flucloxacillin and gentamicin are given
5 - After instrumentation of the bowel or genito-urinary surgery, strep. faecalis is suspected and Amoxycillin and Gentamicin are given

Prevention of infective endocarditis

In dental patients at risk of endocarditis, oral amoxycillin 3g taken 1 hour before the procedure is recommended, followed by 3g at 8 hours after the procedure. If the patient is allergic to penicillin, erythromycin stearate, 1.5g orally should be given 1-2 hours before dental treatment, followed by two oral doses of erythromycin 500mg at 6 hourly intervals. If the patient is receiving a general anaesthetic, benzyl penicillin 2 mega units intra-muscularly should be given just before operation. If the patient is allergic to penicillin, vancomycin 1g intravenously can be given half an hour before operation. Patients with prosthetic heart valves are at especially high risk, and need to be given a combination of Benzyl penicillin 3 mega units intravenously plus gentamicin intravenously just before operation. In penicillin allergy, vancomycin can be substituted. If colonic or genitourinary procedures are to be performed, gentamicin plus ampicillin 1g should be given intravenously half an hour preoperatively, and repeated at 8 and 16 hours post-operatively.

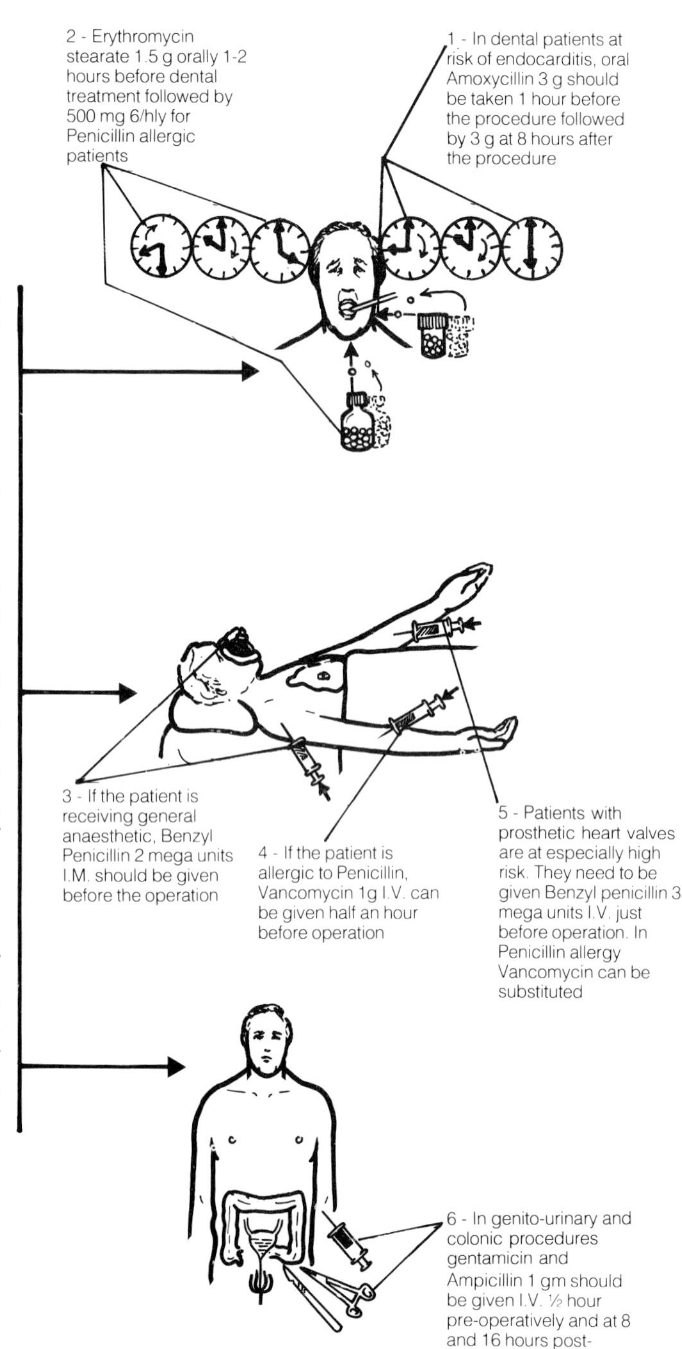
2 - Erythromycin stearate 1.5 g orally 1-2 hours before dental treatment followed by 500 mg 6/hly for Penicillin allergic patients
1 - In dental patients at risk of endocarditis, oral Amoxycillin 3 g should be taken 1 hour before the procedure followed by 3 g at 8 hours after the procedure
3 - If the patient is receiving general anaesthetic, Benzyl Penicillin 2 mega units I.M. should be given before the operation
4 - If the patient is allergic to Penicillin, Vancomycin 1g I.V. can be given half an hour before operation
5 - Patients with prosthetic heart valves are at especially high risk. They need to be given Benzyl penicillin 3 mega units I.V. just before operation. In Penicillin allergy Vancomycin can be substituted
6 - In genito-urinary and colonic procedures gentamicin and Ampicillin 1 gm should be given I.V. ½ hour pre-operatively and at 8 and 16 hours post-operatively

HEART MUSCLE DISEASE

Acute myocarditis

Inflammation of the myocardium can occur in infections and in connective tissue disorders. Viral or parasitic infections can often produce myocarditis, especially the Cocksackie and echo viruses, Chagas disease and toxoplasmosis. Certain drugs such as cytotoxic agents, and severe bacterial infections, for example with *Clostridium* species or *Corynebacterium diphtheriae* can produce a toxic myocarditis.

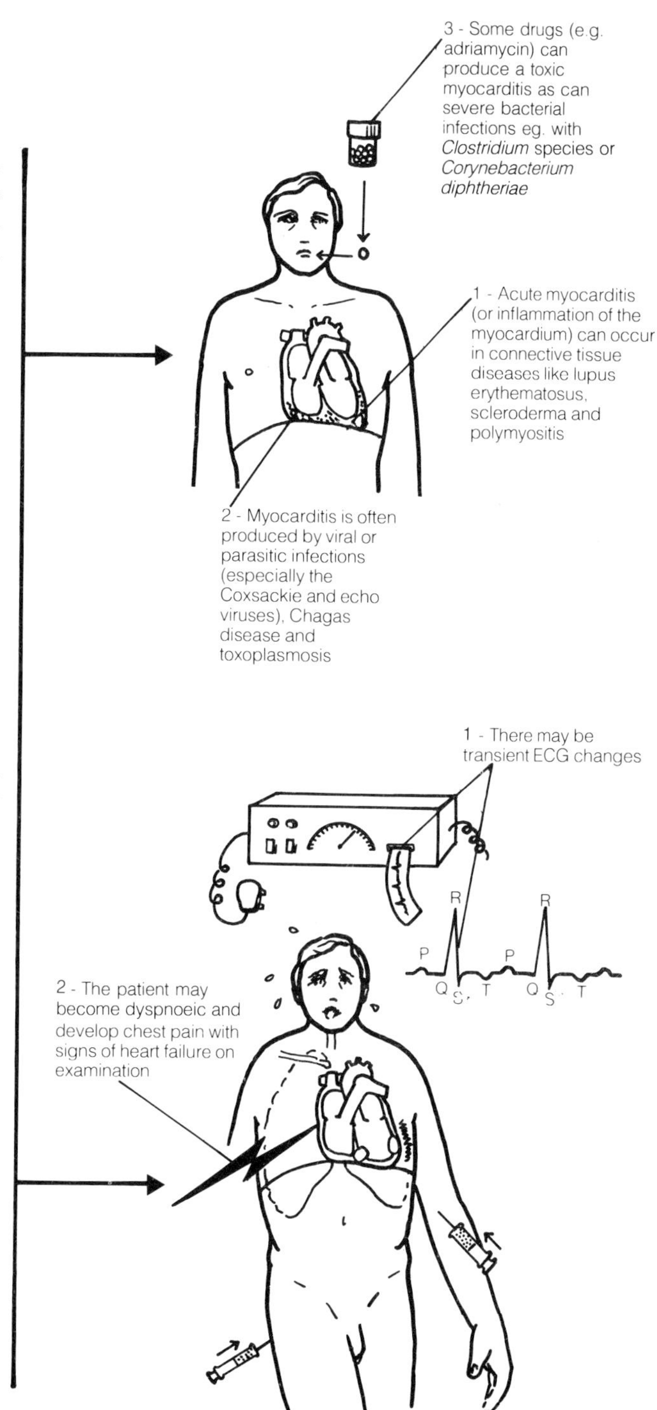
3 - Some drugs (e.g. adriamycin) can produce a toxic myocarditis as can severe bacterial infections eg. with *Clostridium* species or *Corynebacterium diphtheriae*
1 - Acute myocarditis (or inflammation of the myocardium) can occur in connective tissue diseases like lupus erythematosus, scleroderma and polymyositis
2 - Myocarditis is often produced by viral or parasitic infections (especially the Coxsackie and echo viruses), Chagas disease and toxoplasmosis
1 - There may be transient ECG changes
R
R
P
P
Q
S
T
Q
S
T
2 - The patient may become dyspnoeic and develop chest pain with signs of heart failure on examination

Clinical features

There may be transient ECG changes with no symptoms. The patient may however become dyspnoeic and develop chest pain, with signs of heart failure on examination.

Treatment

Is usually supportive, unless a treatable organism such as *C.diphtheriae* is present. Heart failure is treated and specific antiarrhythmic drugs may be required. In diphtheritic myocarditis, a demand pacemaker may be required for atrioventricular block.

CARDIOMYOPATHIES

A cardiomyopathy is a disorder of heart muscle of unknown aetiology

Congestive cardiomyopathy

The heart is dilated, with flabby cardiac muscle and poor contractility. The patient usually presents with breathlessness, cardiac failure, cardiomegaly and arrhythmias. On examination, the heart is enlarged with gallop rhythm, and functional tricuspid and mitral regurgitation may be present. **Echocardiography** shows a dilated poorly contracting left ventricle.

Treatment

As for cardiac failure. Diuretics and vasodilators are often required, with digoxin if atrial fibrillation is present. Cardiac transplantation is the only definitive treatment.

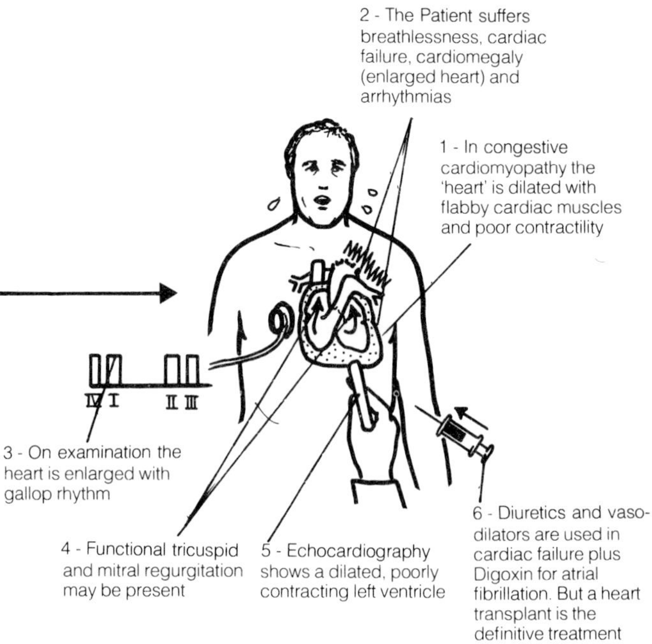

1 - In congestive cardiomyopathy the 'heart' is dilated with flabby cardiac muscles and poor contractility

2 - The Patient suffers breathlessness, cardiac failure, cardiomegaly (enlarged heart) and arrhythmias

IV I II III

3 - On examination the heart is enlarged with gallop rhythm

4 - Functional tricuspid and mitral regurgitation may be present

5 - Echocardiography shows a dilated, poorly contracting left ventricle

6 - Diuretics and vasodilators are used in cardiac failure plus Digoxin for atrial fibrillation. But a heart transplant is the definitive treatment

Alcoholic cardiomyopathy

Heavy alcohol consumption can result in cardiomyopathy, with exertional dyspnoea, cardiac failure and atrial fibrillation. If the patient abstains from alcohol, recovery can occur. If not, the condition is progressive.

1 - Heavy alcohol consumption can cause cardiomyopathy

2 - Exertional dyspnoea, atrial fibrillation and cardiac failure result from cardiomyopathy

3 - If the patient abstains from alcohol, recovery can occur, otherwise the condition is progressive

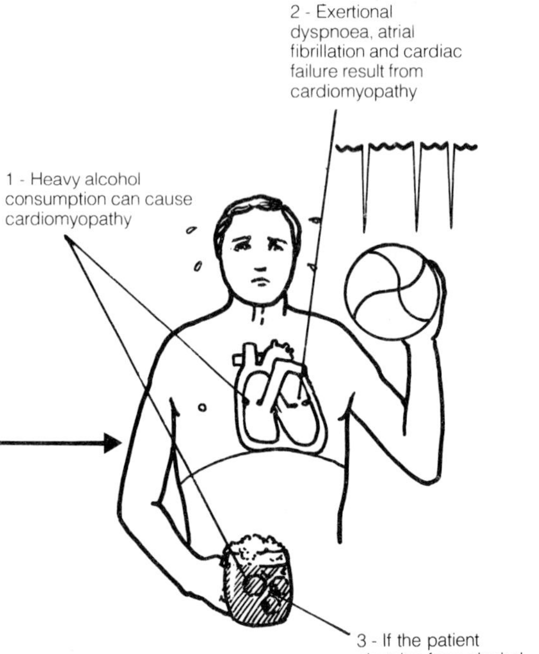

Hypertrophic cardiomyopathy

There is marked hypertrophy of ventricular muscle without ventricular dilation. There may also be outflow tract obstruction because of muscle hypertrophy. The patient may present with breathlessness on exertion, angina, syncopal episodes, or palpitations. On examination, a jerky pulse is present, and a third heart sound may be audible. A late systolic murmur, due to outflow tract obstruction and mitral regurgitation may be present. The **ECG** may show left ventricular hypertrophy with left bundle branch block. Echocardiography shows asymmetric septal hypertrophy, with marked thickening of the interventricular septum (ASH). Systolic anterior movement of the mitral valve cusps (SAM) occurs and mid-systolic closure of the aortic valve is also characteristic.

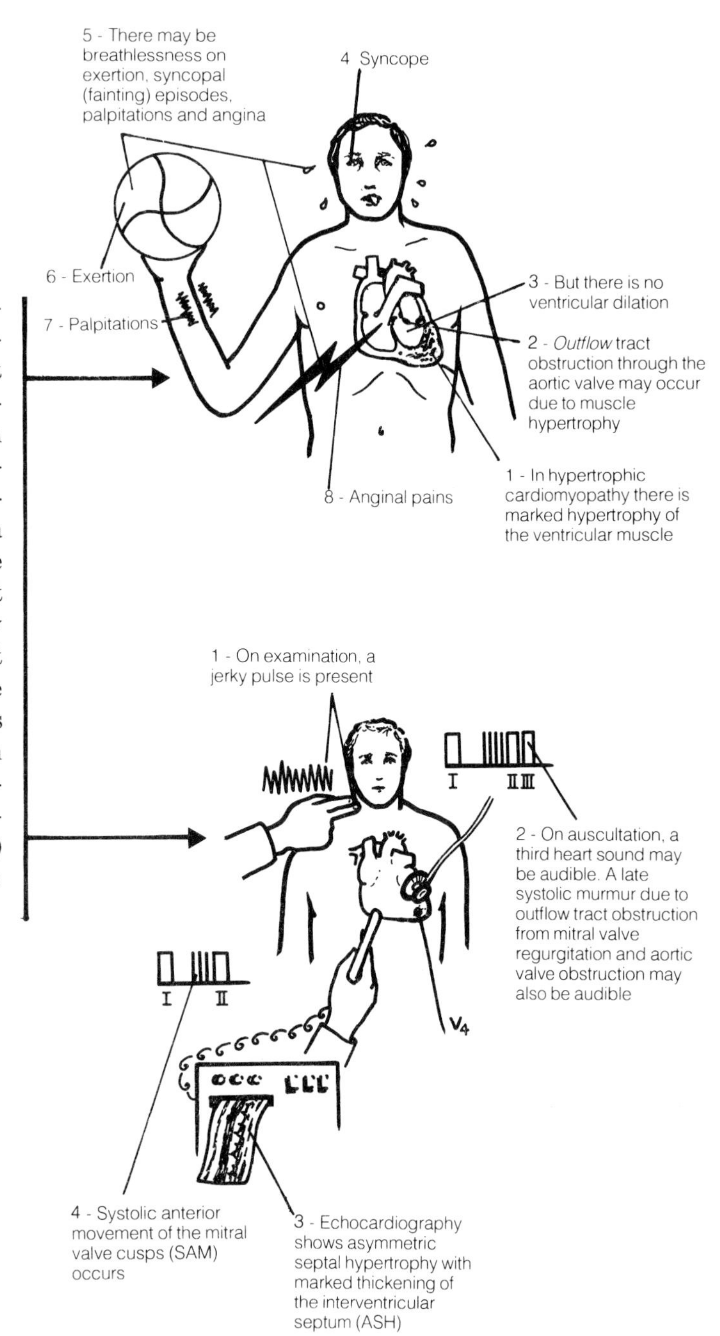

5 - There may be breathlessness on exertion, syncopal (fainting) episodes, palpitations and angina
4 Syncope
6 - Exertion
7 - Palpitations
3 - But there is no ventricular dilation
2 - *Outflow* tract obstruction through the aortic valve may occur due to muscle hypertrophy
8 - Anginal pains
1 - In hypertrophic cardiomyopathy there is marked hypertrophy of the ventricular muscle
1 - On examination, a jerky pulse is present
I
II III
2 - On auscultation, a third heart sound may be audible. A late systolic murmur due to outflow tract obstruction from mitral valve regurgitation and aortic valve obstruction may also be audible
I
II
V4
4 - Systolic anterior movement of the mitral valve cusps (SAM) occurs
3 - Echocardiography shows asymmetric septal hypertrophy with marked thickening of the interventricular septum (ASH)

Treatment

Betablockers are useful in improving left ventricular filling and reducing outflow tract gradients, as well as improving symptoms. Calcium antagonists such as Verapamil also produce symptomatic relief. Surgical septal resection can be performed in patients who fail to respond to medical treatment.

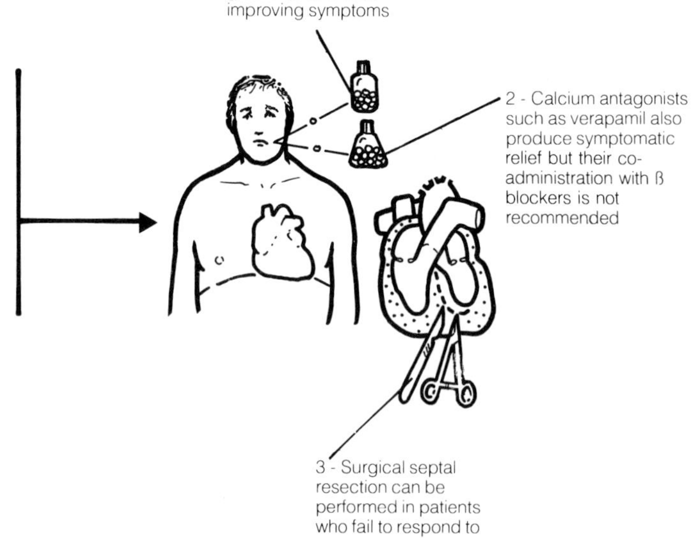
1 - Betablockers are useful in improving left ventricular filling and reducing outflow tract gradients as well as improving symptoms
2 - Calcium antagonists such as verapamil also produce symptomatic relief but their co-administration with ß blockers is not recommended
3 - Surgical septal resection can be performed in patients who fail to respond to medical treatment

PERICARDITIS

Inflammation of the pericardium has a variety of causes and may be classified pathologically into three different groups:
1. Acute fibrinous (dry) pericarditis.
2. Pericarditis with effusion.
3. Constrictive pericarditis.

1. Acute fibrinous pericarditis

Chest pain is a frequent presenting symptom. The pain is usually sharp, constant, and positional in nature. On examination, a **pericardial rub** may be present. This is a superficial scratchy sound which is usually transient and often heard best at the left sternal edge. The **ECG** characteristically shows raised ST segments which are concave upwards. T wave flattening or inversion may occur, but pathological Q waves are not present.

1 - In acute fibrinous pericarditis chest pain is the presenting symptom. It is usually sharp, constant and positional in nature
2 - On examination, a pericardial rub may be present. This is a superficial scratchy sound which is usually transient and often heard best at the left sternal edge
3 - The ECG characteristcally shows raised ST segments which are concave upwards. T wave flattening or inversion may occur, but pathological Q waves are not present
4 - Concave 'ST' elevation in acute pericarditis
R
T

2. Pericarditis with effusion

A small pericardial effusion may present with the symptoms described above. Larger effusions can compress the heart with impairment of ventricular filling and venous return (**cardiac tamponade**)

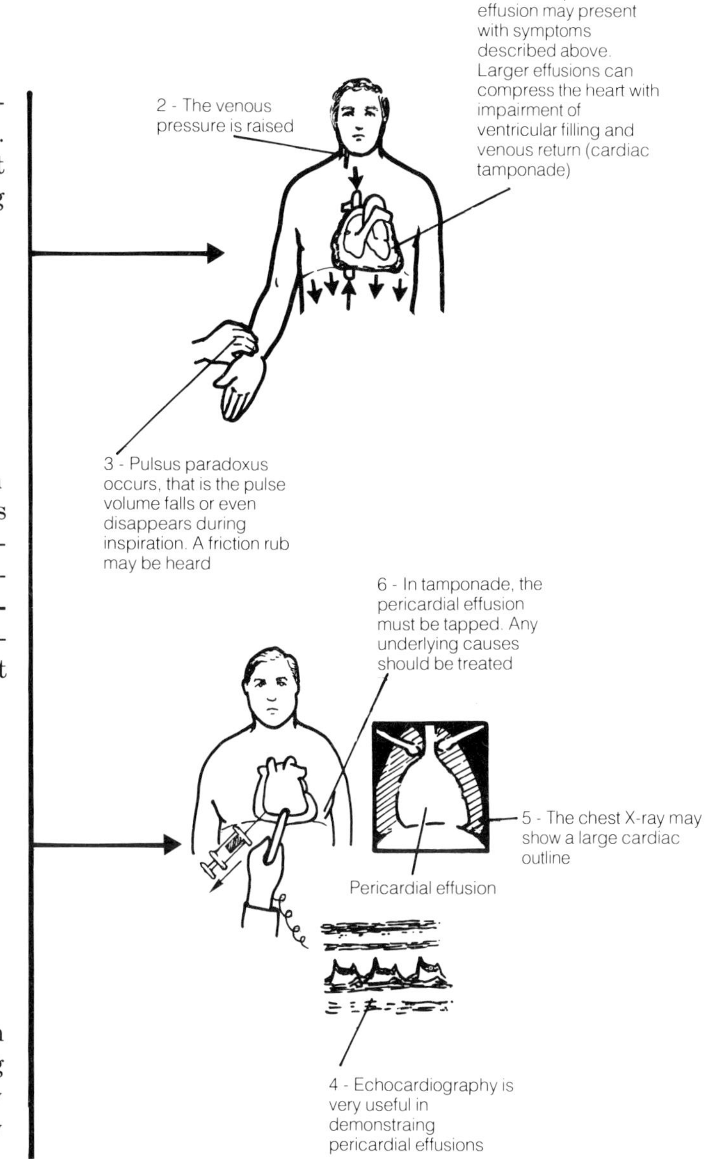
1 - A small pericardial effusion may present with symptoms described above. Larger effusions can compress the heart with impairment of ventricular filling and venous return (cardiac tamponade)
2 - The venous pressure is raised
3 - Pulsus paradoxus occurs, that is the pulse volume falls or even disappears during inspiration. A friction rub may be heard
6 - In tamponade, the pericardial effusion must be tapped. Any underlying causes should be treated
5 - The chest X-ray may show a large cardiac outline
Pericardial effusion
4 - Echocardiography is very useful in demonstraing pericardial effusions

Physical signs of a pericardial effusion

The venous pressure is raised. Pulsus paradoxus occurs, that is the pulse volume falls or even disappears during inspiration. A friction rub may be heard. **Echocardiography** is very useful in demonstrating pericardial effusions. The chest X-ray may show a large cardiac outline.

Treatment

In tamponade, the pericardial effusion must be urgently tapped. Any underlying causes should be treated. At open surgery a formal pericardial "window" is usually resected.

Cause of pericarditis
"Benign" pericarditis

This is presumed to be viral in origin, and there is often a preceding history of upper respiratory tract infection. Cocksackie B, mumps, and echo viruses have been implicated.

Myocardial infarction

A transient pericardial rub is heard in 20% of patients with acute myocardial infarction. In Dressler's syndrome, pericarditis occurs 10-21 days after infarction. A similar syndrome occurs in post-cardiotomy patients, 2-4 weeks after surgery. Treatment with steroids provides prompt relief of symptoms.

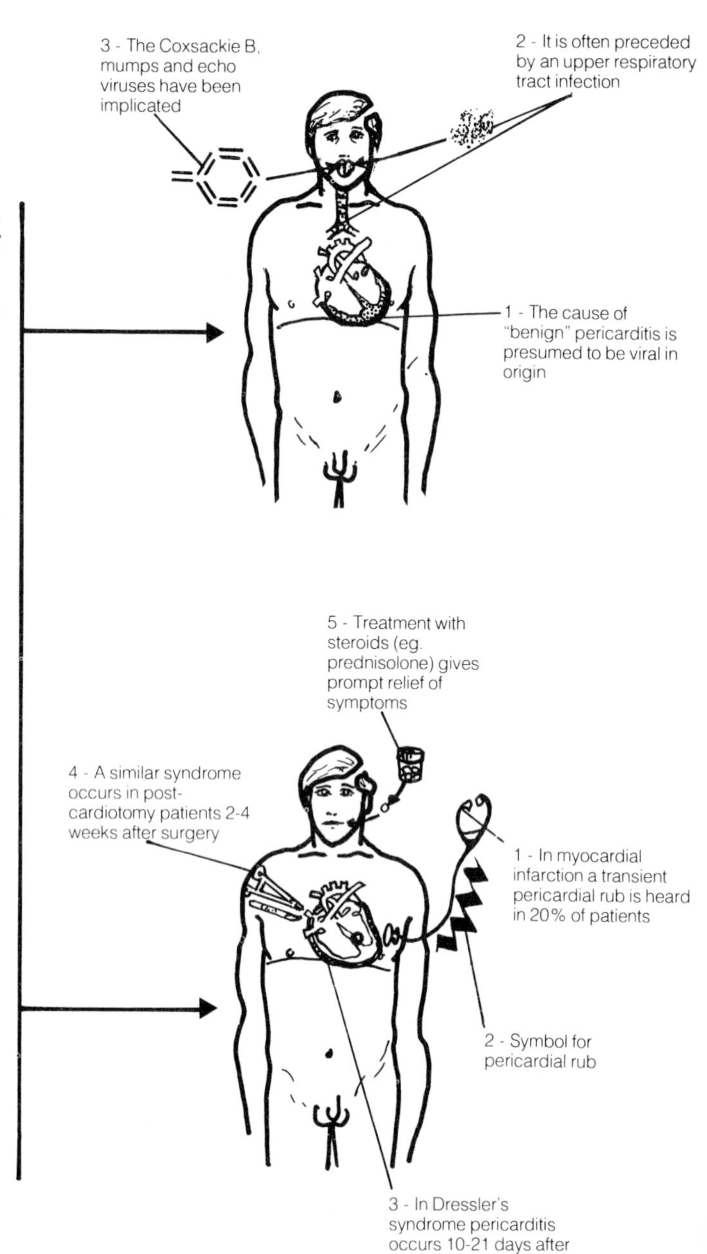

3 - The Coxsackie B, mumps and echo viruses have been implicated
2 - It is often preceded by an upper respiratory tract infection
1 - The cause of "benign" pericarditis is presumed to be viral in origin
5 - Treatment with steroids (eg. prednisolone) gives prompt relief of symptoms
4 - A similar syndrome occurs in post-cardiotomy patients 2-4 weeks after surgery
1 - In myocardial infarction a transient pericardial rub is heard in 20% of patients
2 - Symbol for pericardial rub
3 - In Dressler's syndrome pericarditis occurs 10-21 days after infarction

"Malignant" pericarditis

Can occur especially with carcinoma of the bronchus and breast, Hodgkin's disease and leukaemias. It is often a terminal event.

2 - It also occurs with Hodgkin's disease and leukaemias

1 - Malignant pericarditis occurs especially with carcinoma of the bronchus and breast

Tuberculous pericarditis

Is rare in the west, and more common in under-developed countries. A pericardial effusion is common. TB bacilli may be difficult to isolate from it, and a pericardial biopsy may be necessary to provide the diagnosis. Treatment is by aspiration and intensive anti-tuberculous therapy.

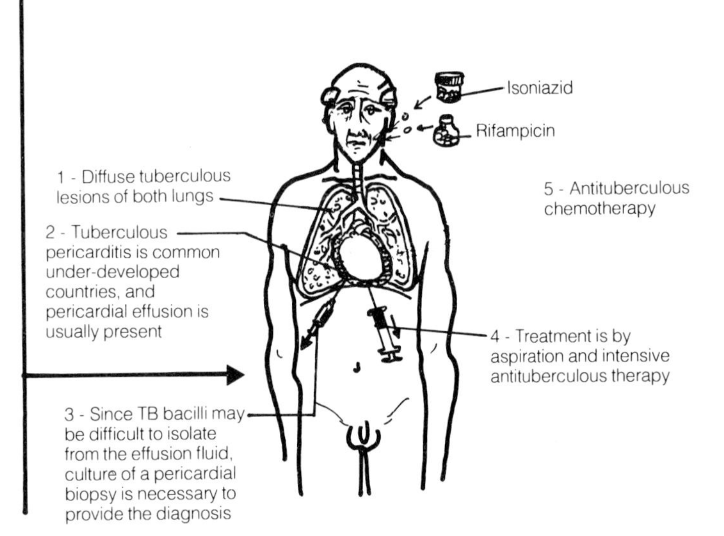
Isoniazid
Rifampicin
1 - Diffuse tuberculous lesions of both lungs
5 - Antituberculous chemotherapy
2 - Tuberculous pericarditis is common under-developed countries, and pericardial effusion is usually present
4 - Treatment is by aspiration and intensive antituberculous therapy
3 - Since TB bacilli may be difficult to isolate from the effusion fluid, culture of a pericardial biopsy is necessary to provide the diagnosis

Rheumatic fever

Pericarditis may occur as a complication of rheumatic fever, and can be associated with an effusion.

Bacterial pericarditis

Is rare, but can be caused by haemolytic streptococci, pneumococci or staphylococci.

Connective tissue disorders

Such as polyarteritis nodosa, giant cell arteritis, and especially systemic lupus erythematosus, may be associated with pericarditis.

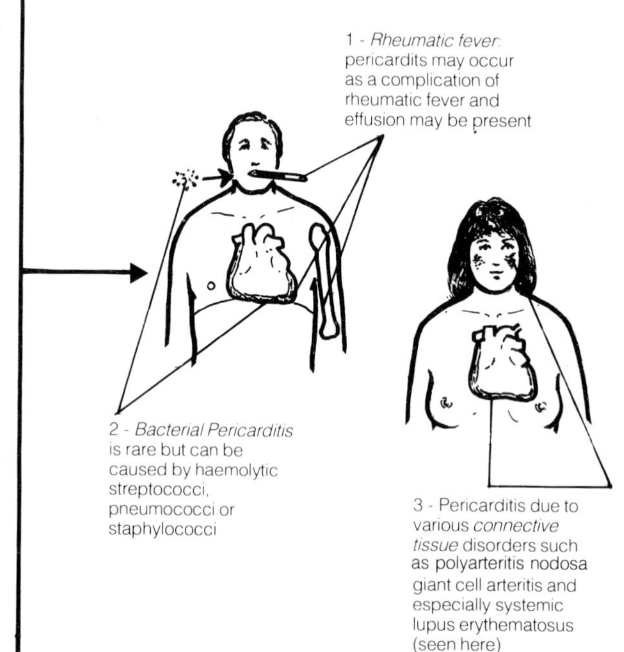
1 - *Rheumatic fever*: pericardits may occur as a complication of rheumatic fever and effusion may be present
2 - *Bacterial Pericarditis* is rare but can be caused by haemolytic streptococci, pneumococci or staphylococci
3 - Pericarditis due to various *connective tissue* disorders such as polyarteritis nodosa giant cell arteritis and especially systemic lupus erythematosus (seen here)

3. Constrictive pericarditis

This may result from previous tuberculous, bacterial or viral infection. The heart becomes encased in a thick layer of fibrous tissue, resulting in restricted ventricular filling.

Symptoms

The patient may complain of breathlessness on exertion, fatigue and abdominal swelling.

On examination

The venous pressure is raised with a rapid "y" descent on opening of the tricuspid valves. Atrial fibrillation may be present, but the spleen is not usually enlarged. The **chest X-ray** may show pericardial calcification especially on the lateral views. The heart size is usually normal. The **ECG** may show widespread T wave inversion. **Cardiac catheterization**; left and right atrial pressures are similar, and there is a rapid "y" descent in the venous wave form.

1 - The venous pressure is raised with a rapid "Y" descent of the venous pulse on opening of the tricuspid valve

"Y" descent

2 - There is often a loud third heart sound or "pericardial knock"

3 - Ascites and hepatomegaly are present, but the spleen is not usually enlarged

4 - Chest X-ray may show pericardial calcification especially on the lateral views. The heart size is usually normal

2 - Cardiac catheterisation shows that left and right atrial pressures similar and there is a rapid "Y" descent in the venous wave form as illustrated earlier

1 - The ECG may show widespread T wave inversion

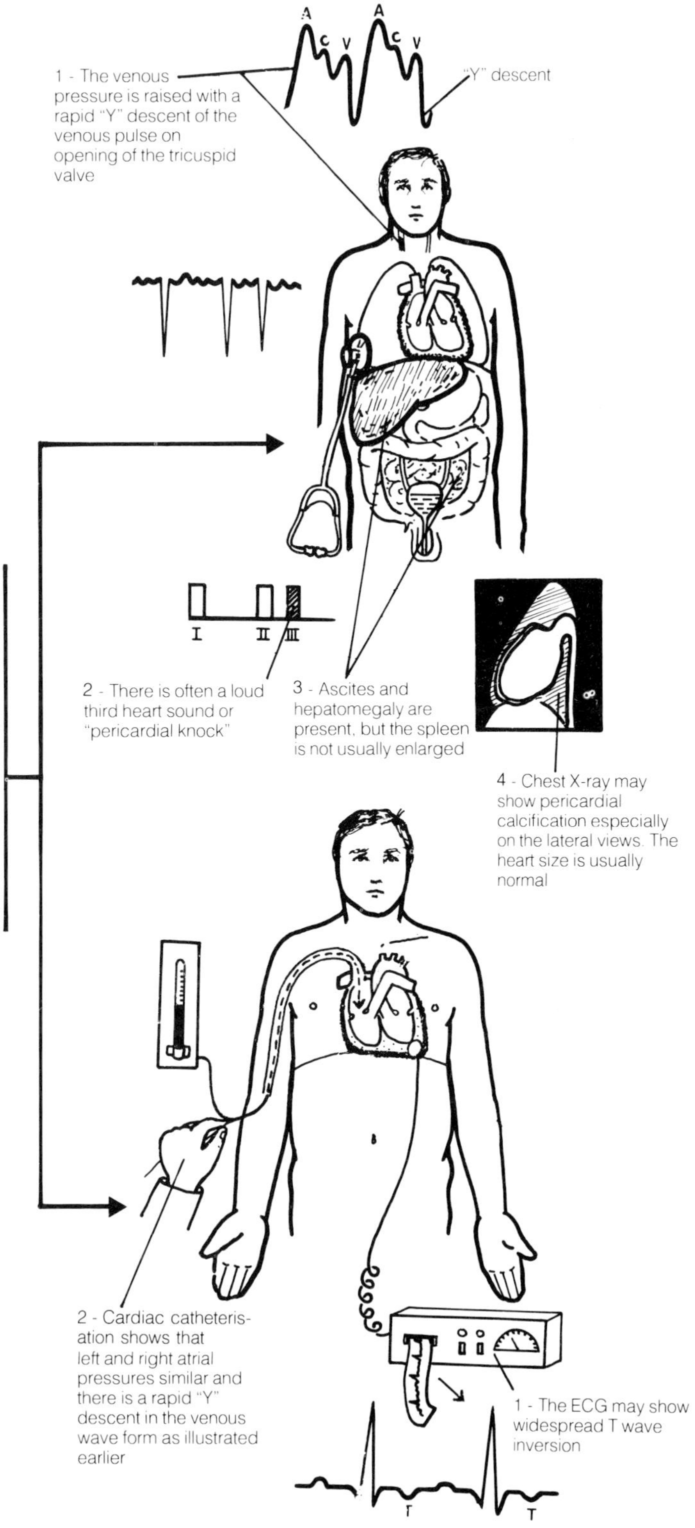

Treatment

Pericardiectomy, where the thickened pericardium is dissected off the heart, may provide a great clinical and symptomatic improvement.

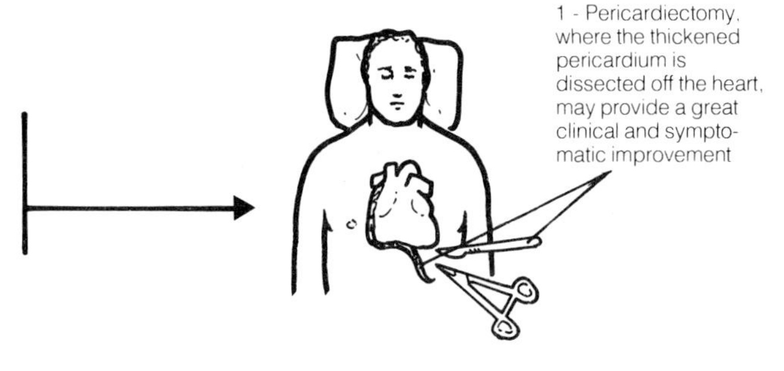
1 - Pericardiectomy, where the thickened pericardium is dissected off the heart, may provide a great clinical and sympto-matic improvement

VENOUS THROMBOSIS AND PULMONARY EMBOLISM

Immobility, dehydration and vessel wall injury are thought to contribute to deep venous thrombosis, but in many cases, no cause can be found. There is increased risk of deep vein thrombosis in patients who are immobile for example after surgery, in pregnancy, in the elderly, in patients on the contraceptive pill, and patients with malignancy.

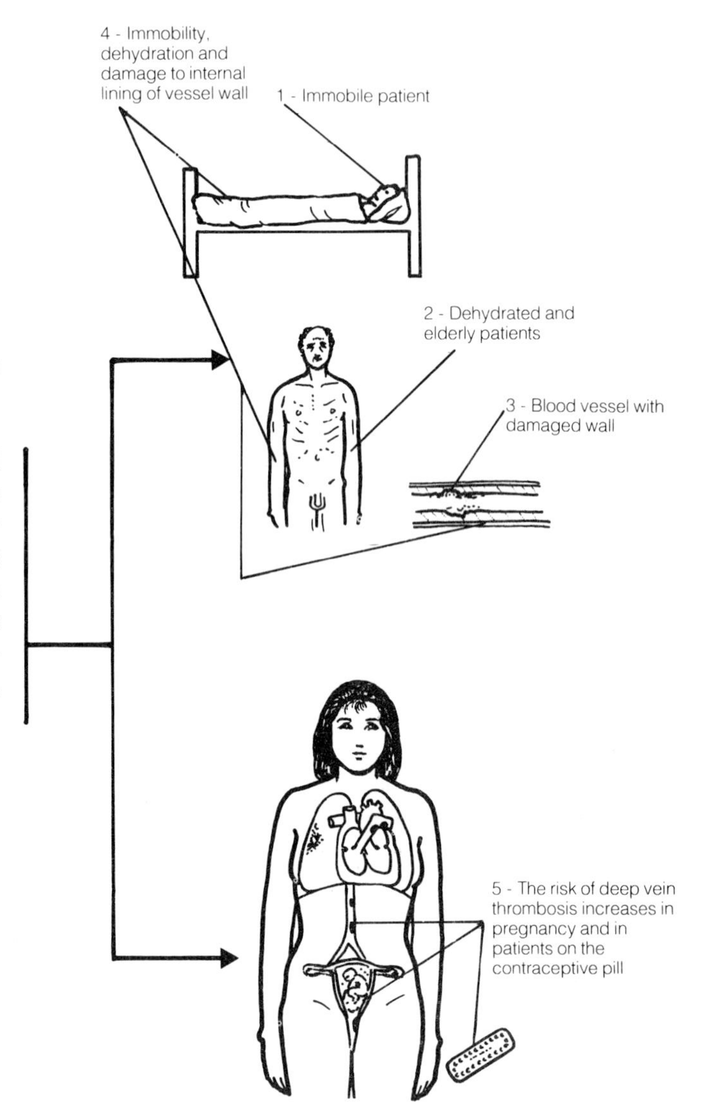
4 - Immobility, dehydration and damage to internal lining of vessel wall
1 - Immobile patient
2 - Dehydrated and elderly patients
3 - Blood vessel with damaged wall
5 - The risk of deep vein thrombosis increases in pregnancy and in patients on the contraceptive pill

Clinical features

There is often pain, swelling, redness and increased temperature in the affected limb. Homan's sign (pain in the calf on dorsiflexion of the ankle) may be positive. However, the clinical signs are often unreliable.

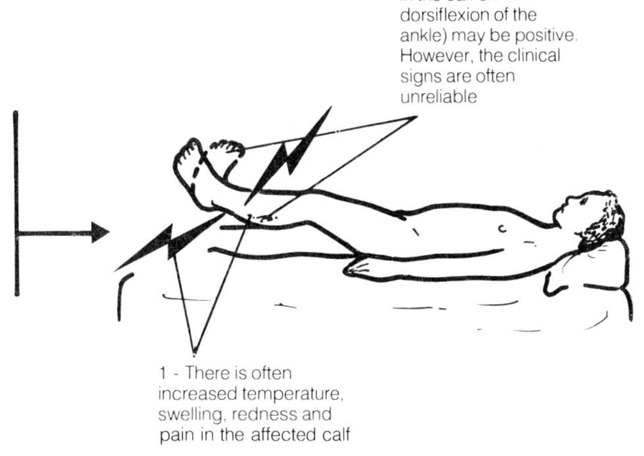
2 - Homan's sign (pain in the calf on dorsiflexion of the ankle) may be positive. However, the clinical signs are often unreliable
1 - There is often increased temperature, swelling, redness and pain in the affected calf

Investigations

As anticoagulation is not without risk, it is important to establish the diagnosis. **Venography:** Contrast medium is injected into a vein in the foot, and will accurately localise the clot. The method is painful and time consuming, but is the investigation of choice when there is doubt about the diagnosis.

Iodine 125 labelled fibrinogen can be injected intravenously, and is taken up by developing thrombus. There is therefore a delay before results are available, and the investigation is contraindicated in pregnancy.

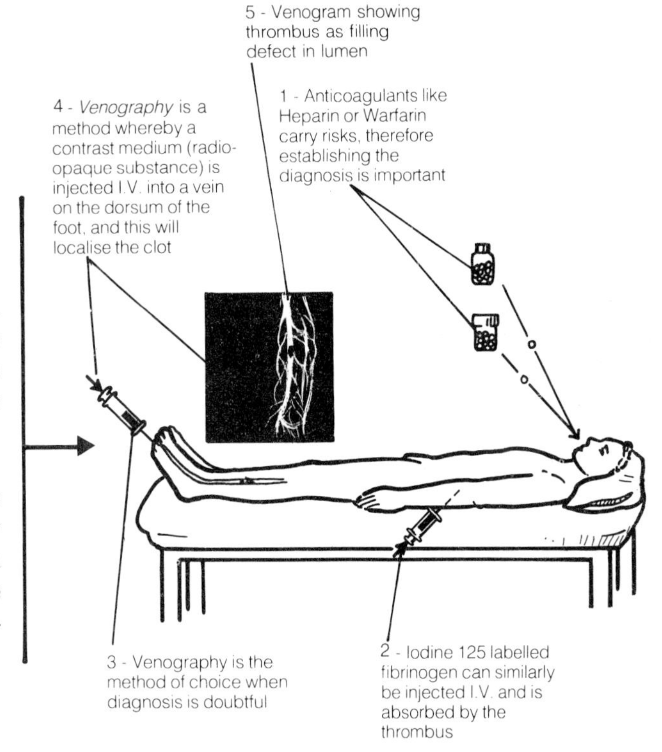
5 - Venogram showing thrombus as filling defect in lumen
4 - *Venography* is a method whereby a contrast medium (radio-opaque substance) is injected I.V. into a vein on the dorsum of the foot, and this will localise the clot
1 - Anticoagulants like Heparin or Warfarin carry risks, therefore establishing the diagnosis is important
3 - Venography is the method of choice when diagnosis is doubtful
2 - Iodine 125 labelled fibrinogen can similarly be injected I.V. and is absorbed by the thrombus

Doppler ultrasound

When a limb is compressed, blood flow through a patent vein can be detected by Doppler ultrasound. This technique is useful in detecting thromboses in the femoral, iliac or popliteal veins.

Impedance plethysmography

Conduction of electricity through blood is reduced when there is a deep vein thrombosis. In impedance plethysmography, conduction of electricity through the cuffed limb is measured. The technique is easy to perform, but there is a high incidence of false positive results.

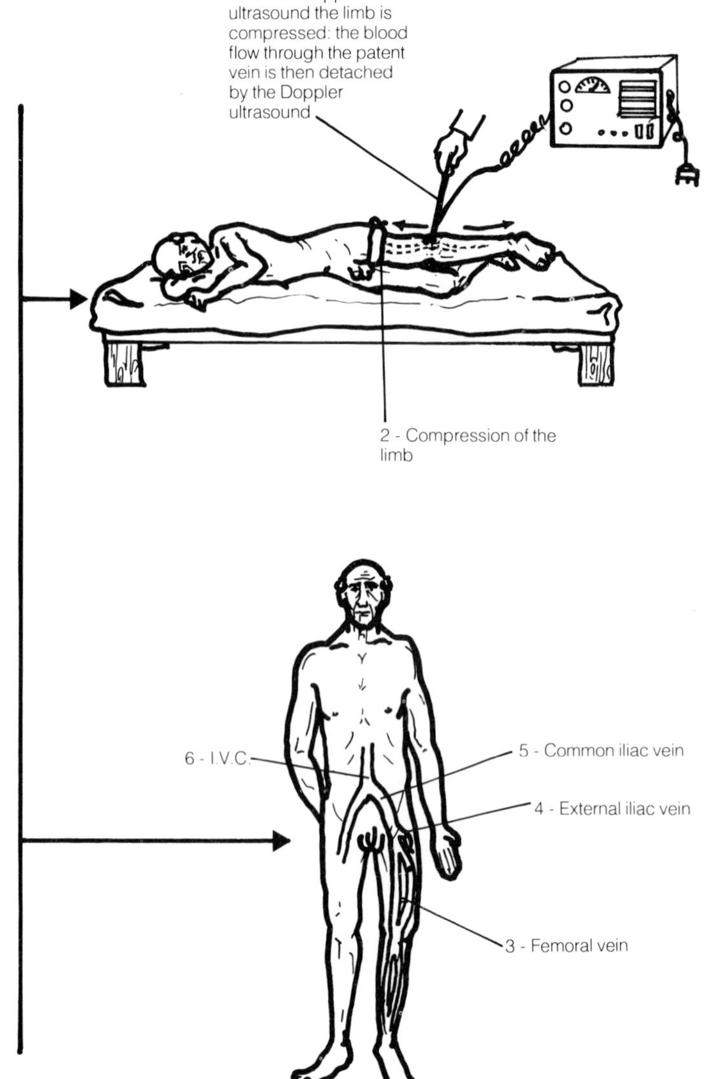
1 - In the Doppler ultrasound the limb is compressed: the blood flow through the patent vein is then detached by the Doppler ultrasound
2 - Compression of the limb
6 - I.V.C.
5 - Common iliac vein
4 - External iliac vein
3 - Femoral vein

Treatment

Patients with established deep venous thrombosis are at risk of pulmonary embolism and should be treated with anticoagulants. Heparin should be given as an initial bolus of 10,000 units, followed by an infusion of 10,000 units 6 hourly. The partial thromboplastin time (PTT) should be measured daily and the heparin dosage adjusted to keep the PTT at 2-3 times the control value. The patient is started on warfarin at the same time, as oral anticoagulants usually take about 3 days to act. Dosage regimens vary, but an adult could be given 10mg on three successive days, while a small female might be given 6mg on three successive days. After this, the heparin is stopped, and the warfarin dosage adjusted to give a prothrombin ratio of 2 to 3. If bleeding occurs while the patient is on warfarin, fresh frozen plasma should be given intravenously. Intravenous vitamin K is rarely used nowadays because it makes the patient resistant to anticoagulants for at least 2 weeks.

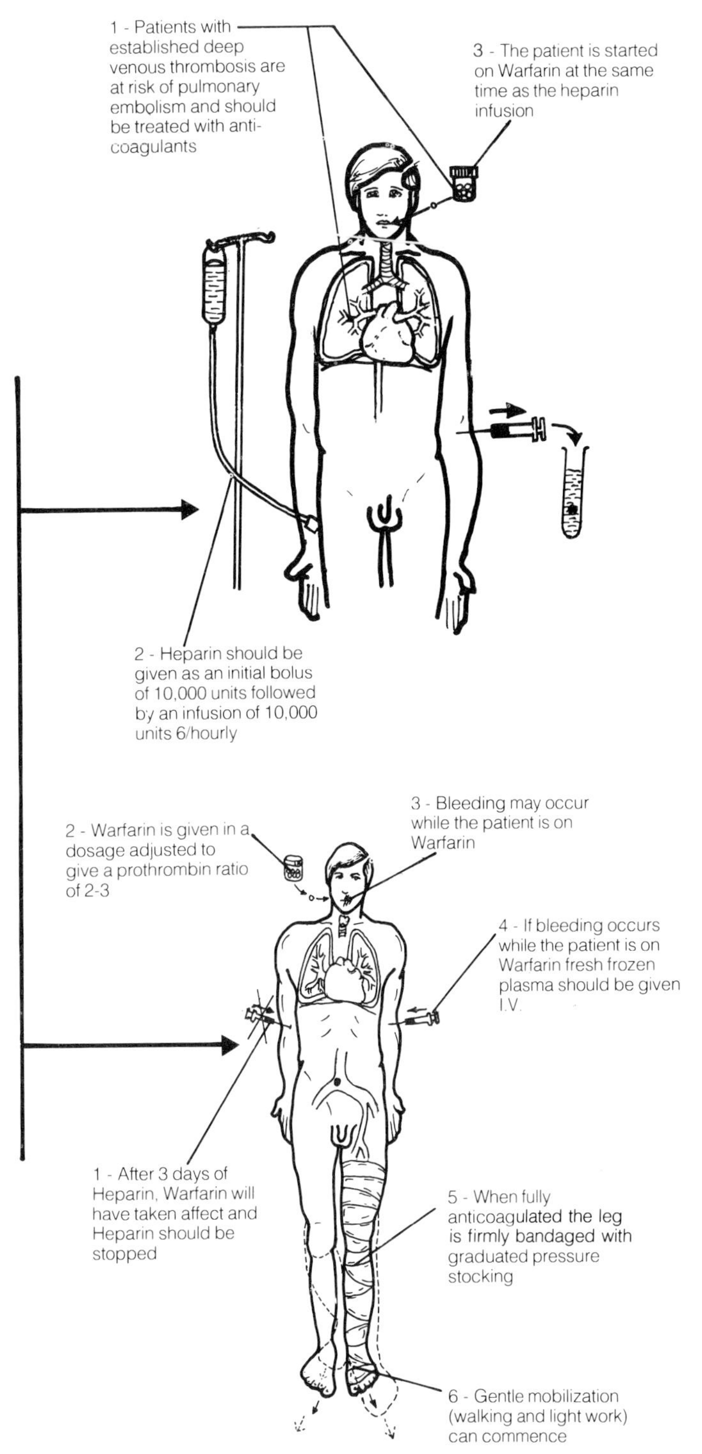
1 - Patients with established deep venous thrombosis are at risk of pulmonary embolism and should be treated with anti-coagulants
3 - The patient is started on Warfarin at the same time as the heparin infusion
2 - Heparin should be given as an initial bolus of 10,000 units followed by an infusion of 10,000 units 6/hourly
2 - Warfarin is given in a dosage adjusted to give a prothrombin ratio of 2-3
3 - Bleeding may occur while the patient is on Warfarin
4 - If bleeding occurs while the patient is on Warfarin fresh frozen plasma should be given I.V.
1 - After 3 days of Heparin, Warfarin will have taken affect and Heparin should be stopped
5 - When fully anticoagulated the leg is firmly bandaged with graduated pressure stocking
6 - Gentle mobilization (walking and light work) can commence

The patient should initially have bed rest. Once he is adequately anticoagulated, the whole leg is firmly bandaged with a graduated pressure stocking and gentle mobilization can commence. Calf vein thrombosis is usually treated with anti-coagulants for about 3 months, while iliofemoral thrombosis is treated for 6 months. The patient must be carefully followed up during this time, and the prothrombin ratio checked at least once a month.

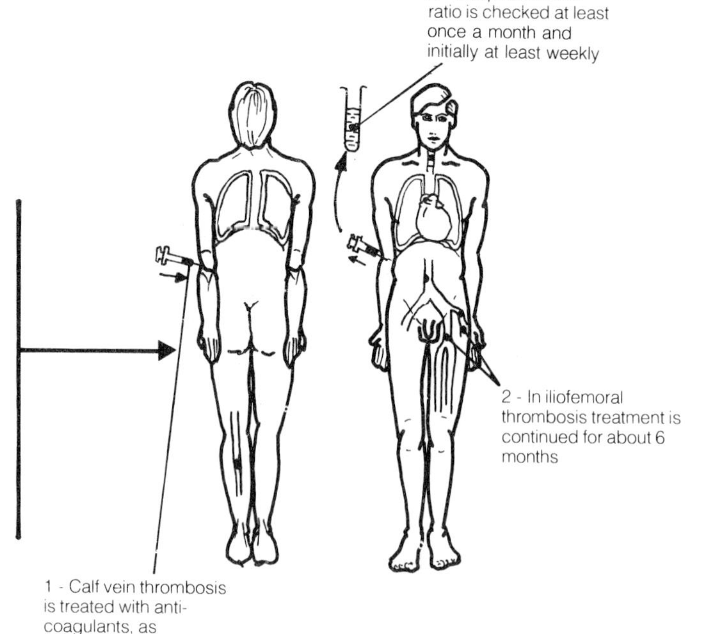
3 - The prothrombin ratio is checked at least once a month and initially at least weekly
2 - In iliofemoral thrombosis treatment is continued for about 6 months
1 - Calf vein thrombosis is treated with anti-coagulants, as explained already, for about 3 months

Pulmonary embolism

Pulmonary emboli can result from peripheral deep vein thrombosis in the popliteal or iliac veins, or from the heart such as in atrial fibrillation. Often, no source for the embolus can be found.

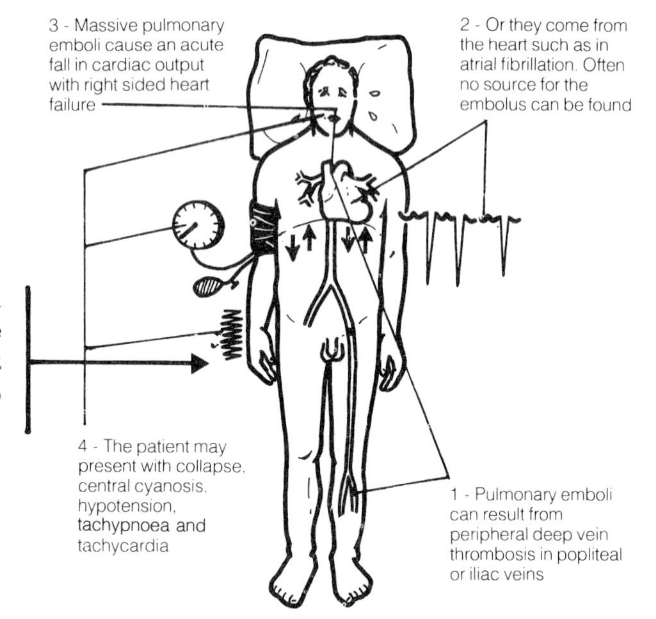
3 - Massive pulmonary emboli cause an acute fall in cardiac output with right sided heart failure
2 - Or they come from the heart such as in atrial fibrillation. Often no source for the embolus can be found
4 - The patient may present with collapse, central cyanosis, hypotension, tachypnoea and tachycardia
1 - Pulmonary emboli can result from peripheral deep vein thrombosis in popliteal or iliac veins

Massive pulmonary emboli cause an acute fall in cardiac output with right-sided heart failure. The patient may present with collapse, hypotension, tachypnoea, tachycardia and central cyanosis. The venous pressure will be raised, and a gallop rhythm may be audible at the left sternal edge. Cardiac arrest can occur. The chest X-ray may show areas of oligaemia or wedge shaped areas of pulmonary infarction. The **ECG** may show sinus tachycardia, a Q wave and inverted T wave in lead III, T wave inversion in leads VI and V2 and right bundle branch block and an S wave in lead I.

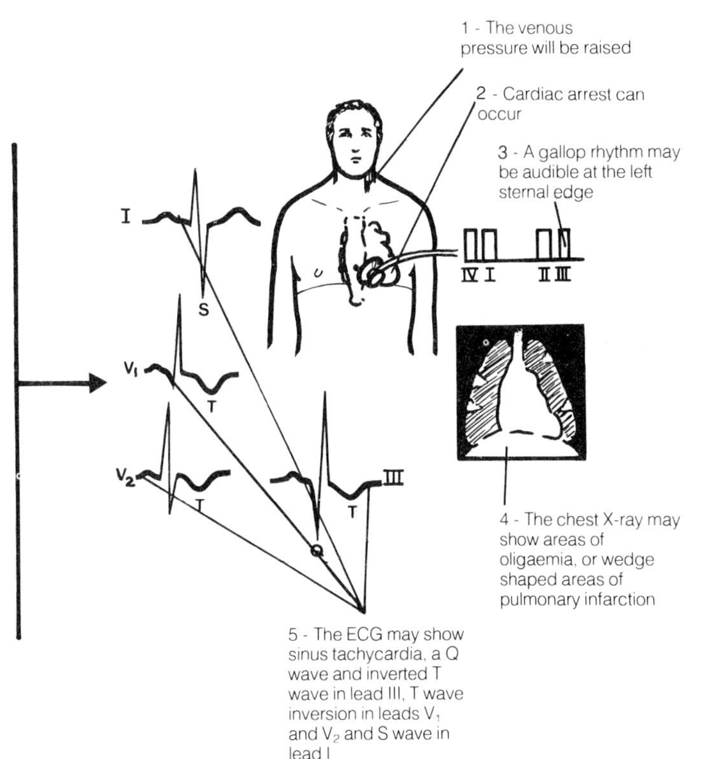
1 - The venous pressure will be raised
2 - Cardiac arrest can occur
3 - A gallop rhythm may be audible at the left sternal edge
IV I
II III
I
S
V_1
T
V_2
T
III
T
Q
4 - The chest X-ray may show areas of oligaemia, or wedge shaped areas of pulmonary infarction
5 - The ECG may show sinus tachycardia, a Q wave and inverted T wave in lead III, T wave inversion in leads V_1 and V_2 and S wave in lead I

Blood gases There is usually hypoxia, the PO2 being low. **The ventilation/perfusion lung scan** is useful in making the diagnosis; areas which are not perfused but are ventilating normally are characteristic of pulmonary emboli. **Pulmonary angiography** is useful in very

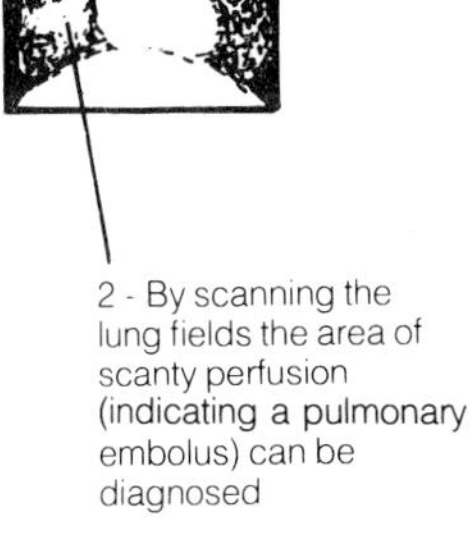

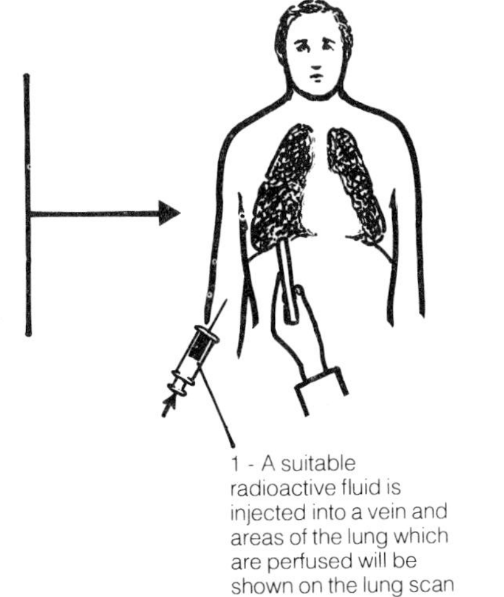
2 - By scanning the lung fields the area of scanty perfusion (indicating a pulmonary embolus) can be diagnosed
1 - A suitable radioactive fluid is injected into a vein and areas of the lung which are perfused will be shown on the lung scan

ill patients where the diagnosis is in question, especially if pulmonary embolectomy is being considered. A cannula is introduced into the pulmonary artery via the femoral or brachial veins; pulmonary emboli will produce characteristic filling defects after injecting contrast into the pulmonary arteries. This is a safe procedure in experienced hands.

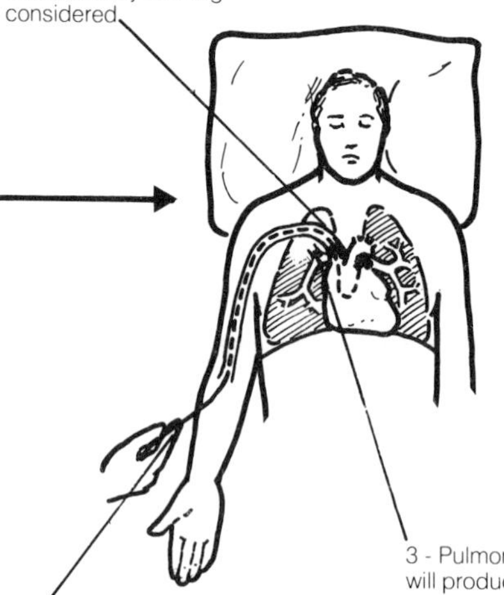

1 - Pulmonary angiography is useful in very ill patients where the diagnosis is in question, especially if pulmonary embolectomy is being considered

2 - A cannula is introduced into the pulmonary artery via the femoral or brachial veins

3 - Pulmonary emboli will produce characteristic filling defects after injecting contrast into the pulmonary arteries. This is a safe procedure in experienced hands

Minor pulmonary emboli: These often do not produce symptoms. However, the patient may present with haemoptysis, pleuritic chest pain and tachycardia. On examination a localised pleural rub may be audible. The chest X-ray may show wedge shaped areas of infarction or a small pleural effusion. It is often normal however. The ECG may show some of the features described above, but again can often be normal. Blood gases are usually normal, but the ventilation/perfusion scan will show mismatched defects.

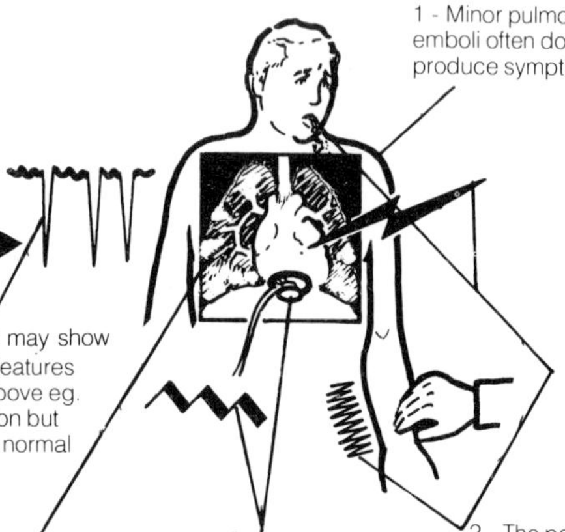

1 - Minor pulmonary emboli often do not produce symptoms

5 - The ECG may show some of the features described above eg. atrial fibrillation but often can be normal

4 - The chest X-ray may show wedge shaped areas of infarction, or a small pleural effusion. It is often normal however

3 - On examination a localised pleural rub may be audible

2 - The patient may present with haemoptysis, pleuritic chest pain and tachycardia

4 - The patient is given oxygen to correct hypoxia and should be rested in bed

3 - Cardiac massage is given to aid resuscitation

O_2

1 - Massive pulmonary emboli may cause collapse, cardiac arrest and shock

2 - Thrombolytic agents such as streptokinase can be life saving in massive pulmonary embolism

5 - If all measures fail, urgent pulmonary embolectomy should be attempted

Management

Massive pulmonary emboli may cause collapse and cardiac arrest, and resuscitation with external cardiac massage and correction of hypoxia and acidosis is necessary. If the patient remains shocked, the thrombolytic agents such as streptokinase can be used. If these fail, the patient may require urgent pulmonary embolectomy. In less severe cases, anticoagulants are required; initially heparin and then warfarin (see section on deep vein thrombosis). The patient should be rested in bed and given oxygen as necessary. Following pulmonary emboli anticoagulation is usually required for at least 6 months.

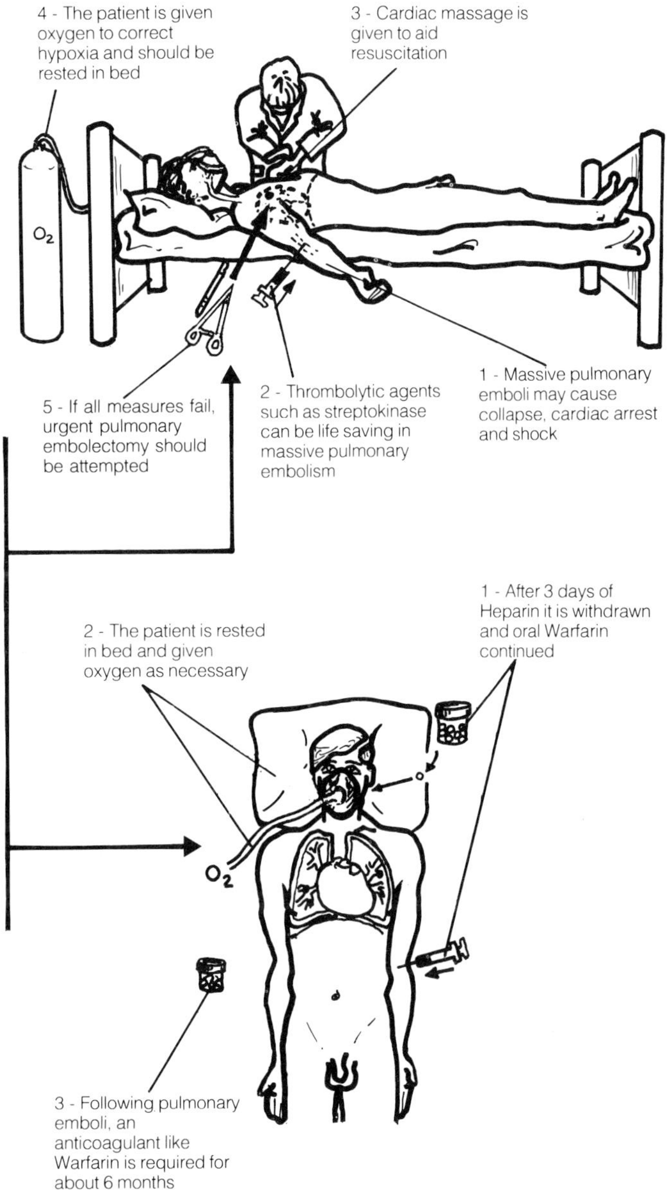

1 - After 3 days of Heparin it is withdrawn and oral Warfarin continued

2 - The patient is rested in bed and given oxygen as necessary

O_2

3 - Following pulmonary emboli, an anticoagulant like Warfarin is required for about 6 months

Prevention of pulmonary emboli

The risk of venous thrombosis and pulmonary emboli can be reduced by giving surgical patients small subcutaneous doses of heparin (usually 5,000 units twice daily) pre and post operatively. Females who have had venous thrombosis or pulmonary emboli should not take the oral contraceptive pill.

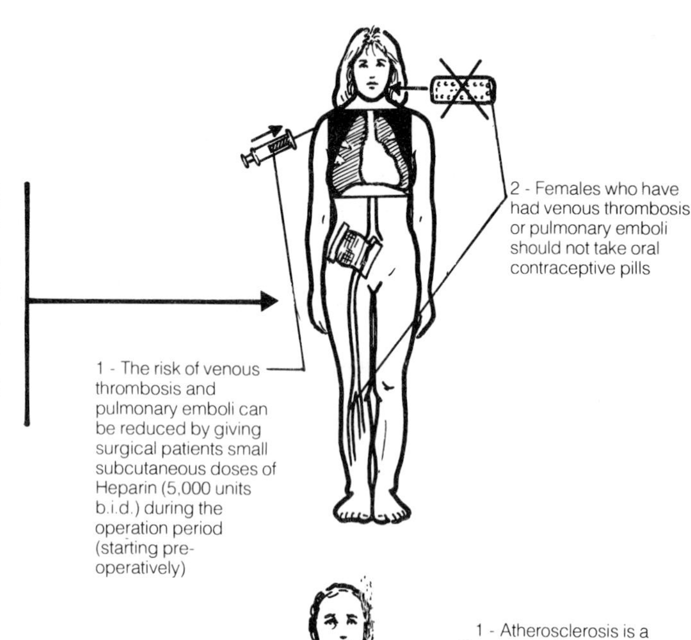
2 - Females who have had venous thrombosis or pulmonary emboli should not take oral contraceptive pills
1 - The risk of venous thrombosis and pulmonary emboli can be reduced by giving surgical patients small subcutaneous doses of Heparin (5,000 units b.i.d.) during the operation period (starting pre-operatively)

Peripheral vascular disease

Atherosclerosis is a common disease leading to progressive narrowing of the arteries. Risk factors include diabetes mellitus, hyperlipidaemias, hypertension and cigarette smoking. When the superficial femoral artery is affected, the patient develops **intermittent claudication**; a pain in the back of the calf which occurs on exertion and is relieved by rest.

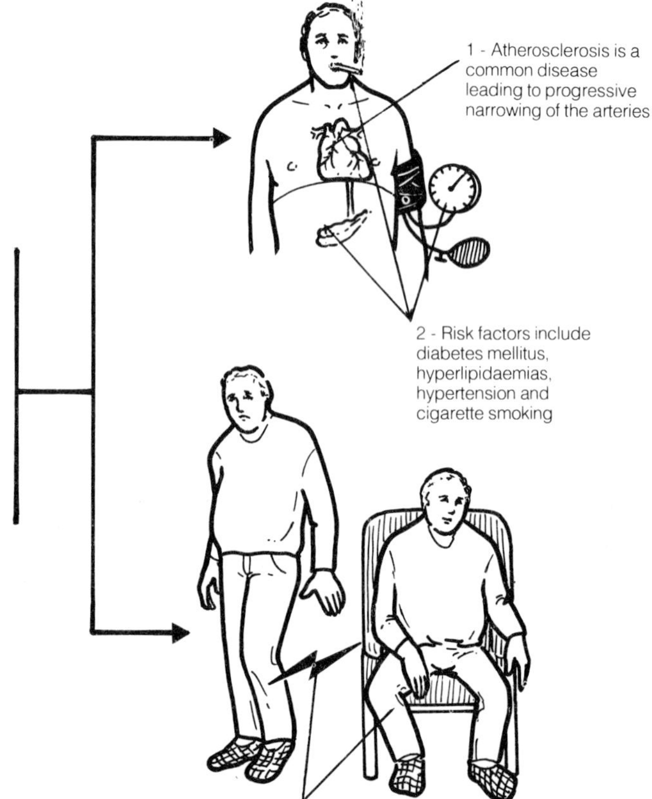
1 - Atherosclerosis is a common disease leading to progressive narrowing of the arteries
2 - Risk factors include diabetes mellitus, hyperlipidaemias, hypertension and cigarette smoking
3 - When the superficial femoral artery is affected, the patient develops intermittent claudication; a pain in the back of the calf which occurs on exertion and is relieved by rest

On examination, peripheral pulses in the leg are absent or reduced and a bruit may be heard over the femoral artery. Capillary return is impaired in the limb, and in severe cases gangrene may result. If the lower aorta and iliac arteries are affected, the patient may complain of pain in the buttocks on exertion and impotence (Leriche's syndrome).

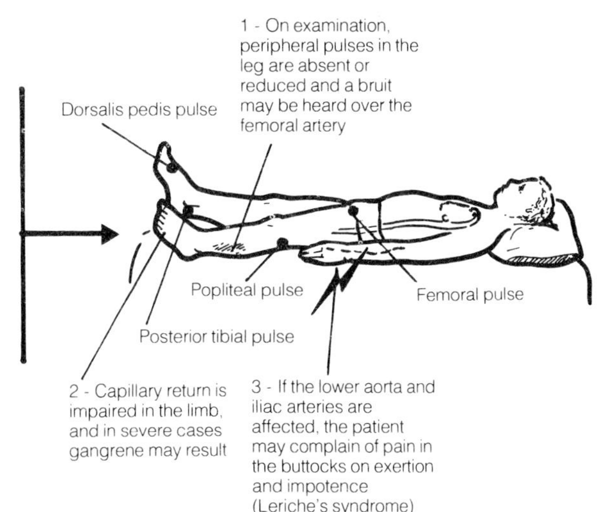
1 - On examination, peripheral pulses in the leg are absent or reduced and a bruit may be heard over the femoral artery
Dorsalis pedis pulse
Popliteal pulse
Femoral pulse
Posterior tibial pulse
2 - Capillary return is impaired in the limb, and in severe cases gangrene may result
3 - If the lower aorta and iliac arteries are affected, the patient may complain of pain in the buttocks on exertion and impotence (Leriche's syndrome)

Investigations

To exclude risk factors, a full blood count, blood glucose, serum lipids, chest X-ray and ECG should be performed. If surgery is being considered, angiography is necessary.

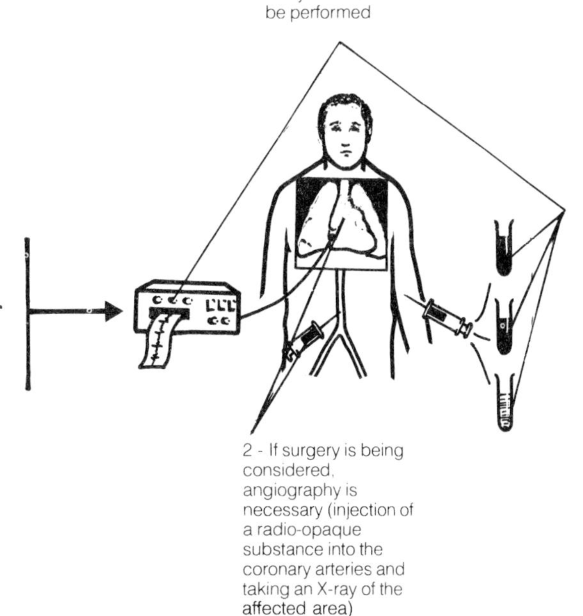
1 - To exclude risk factors a full blood count, blood glucose and serum lipids, chest X-ray and ECG should be performed
2 - If surgery is being considered, angiography is necessary (injection of a radio-opaque substance into the coronary arteries and taking an X-ray of the affected area)

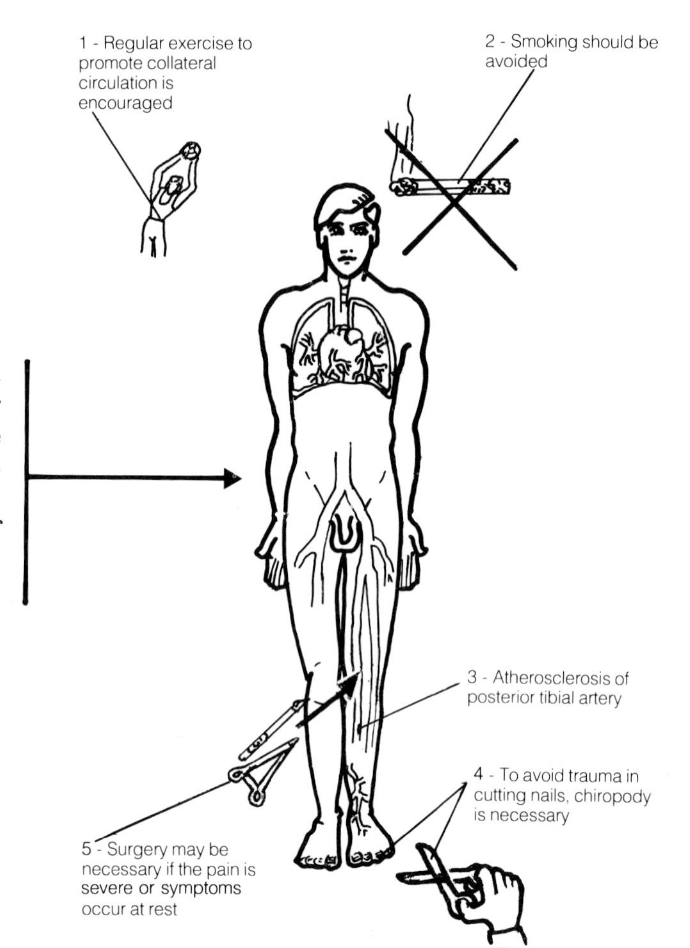
1 - Regular exercise to promote collateral circulation is encouraged
2 - Smoking should be avoided
3 - Atherosclerosis of posterior tibial artery
4 - To avoid trauma in cutting nails, chiropody is necessary
5 - Surgery may be necessary if the pain is severe or symptoms occur at rest

Management

Patients with stable symptoms should be advised to stop smoking, and regular exercise should be encouraged to promote collateral circulation. Chiropody is necessary, and trauma to the feet should be avoided. Surgery may be necessary if symptoms are severe if there is pain in the leg at rest.

Index

Adrenaline 75
Afterload 98
Airway obstruction, severe, pulsus paradoxus 13
Aldosterone 96
Amiloride 104
Amiodarone 72
Angina pectoris 51-60
  investigations 52-4
    coronary arteriography 54
    exercise ECG 53
    isotope scanning 54
  prophylaxis 55
  recurrent 78
  treatment 54-8
    beta blockers 56-7
    calcium antagonists 55
    coronary artery bypass grafting 58
    nitrates 55
  unstable 58
Angiotensin I/II 96
Aortic dissecting aneurysm 67, 135
Aortic regurgitation 9, 131-5
  acute 135
  Austin Flint murmur 23, 134
  causes 131
  clinical features 132-3
  immediate diastolic murmur 22, 133
  investigations 134
  management 135
Aortic stenosis 8
  acquired 128
  apex beat 15
  clinical features 128-9
  congenital 127, 143-4
  ejection clicks 19, 129
  ejection systolic murmur 20
  investigations 129-30
  management 130-1
  systolic thrill 16, 129
Aortic valve
  bicuspid 127
  mixed disease 9
Arteriovenous fistula 9
Aspartate aminotransferase 66
Atherosclerosis 50, 172-4
  investigations 173
  management 174
Atrial fibrillation 8, 36-7, 72
  jugular venous pulse 11
Atrial flutter 36
Atrial myxoma 118
  echocardiography 47, 118
Atrial septal defect 138-9
  second heart sound 18, 138
Atrial tachycardia (supraventricular tachycardia) 36
Atrio-ventricular node 5

Bendrofluazide 91, 104
Beta blockers 56-7, 82, 91-2
Blood pressure 14
Bundle of His 5
  block 32-3

Calcium antagonists 57, 93
Calcium gluconate 75
Captopril 95
Carcinoid syndrome 125
Cardiac arrest 74-5
Cardiac catheterization 49
  aortic regurgitation 134
  aortic stenosis 129-30
  constrictive pericarditis 163
  Fallot's tetralogy 146
  mitral stenosis 117
Cardiac massage 74, 75
Cardiac tamponade 159
Cardiogenic shock 77
Cardiomyopathy
  alcoholic 156
  congestive 156
  hypertrophic 21, 157-8
Cardioversion 39, 71, 72, 73
Carotid sinus massage 71
Chorea 108
Cigarette smoking 50, 172
Clonidine 94
Coarctation of aorta 141-3
  bicuspid aortic valve associated 127
  complications 143
  hypertension due to 84, 142, 143
  late systolic murmur 21
  radial/femoral pulses 13, 87
  signs 142
  treatment 143
Congenital heart disease 136-47
  acyanotic 136
    without shunt 141-2
  cyanotic 145-7
  see also individual diseases
Connective tissue disorders 162
Conn's syndrome (primary hyperaldosteronism) 84, 86, 90
  aldosterone 96
  renin 96
Coronary arteries 3
Coronary arteriography 54
Coronary artery bypass grafting 58
Coronary artery disease 50-60
  complete heart block 42
  exercise ECG 47, 53
  first degree heart block 40
  risk factors 50-1
Coronary spasm (Prinzmetal's variant angina) 31, 60
Creatine kinase 65
Cushing's syndrome 84
Cyanosis 6

Defibrillation 75
Diamorphine 69, 76, 102
Digoxin 105, 117
  first degree heart block due to 40
Disopyramide 72
Diuretics 104
  thiazide 76, 91, 92, 104
Dobutamine 77, 105
Dopamine 77, 105
Doppler ultrasound 166
Dressler's syndrome 79, 160

Echocardiography 46-7
  aortic regurgitation 47, 134
  aortic stenosis 129
  atrial myxoma 118
  congestive cardiomyopathy 156
  Fallot's tetralogy 146
  hypertrophic cardiomyopathy 157
  infective endocarditis 47, 151
  mitral regurgitation 121
  mitral stenosis 117
  pericardial effusion 159

Ectopic beats (extrasystoles) 8, 35
Effort syncope 128
Eisenmenger's syndrome 147
Electrocardiogram (ECG) 24-43
  acute fibrinous pericarditis 158
  angina pectoris 52
    exercise 53
  aortic regurgitation 134
  aortic stenosis 129
  atrial fibrillation 36-7
  atrial tachycardia (supraventricular tachycardia) 36
  bundle branch block 32-3
    left anterior hemiblock 33
    left bundle branch block 33
    left posterior hemiblock 33
    right bundle branch block 32
  cardiac axis 27
  coarctation of aorta 142
  constrictive endocarditis 163
  electrode leads 25-6
  exercise 47, 53
  extrasystoles (ectopic beats) 35
  heart block (AV block) 39-40
  hypertension 89
  hypertrophic cardiomyopathy 157
  left ventricular strain pattern 28
  mitral regurgitation 121
  mitral stenosis 116
  myocardial infarction 30, 63-5
  myocardial ischaemia 31
  PR interval 24
  Prinzmetal's variant angina 31
  pulmonary embolism 169, 170
  QRS complex 24
  right ventricular strain pattern 28, 29
  R on T phenomenon 35, 73
  sinus arrhythmia 34
  sinus bradycardia 34
  sinus tachycardia 34
  subendocardial infarction 30
  ventricular aneurysm 80
  ventricular fibrillation 39
  ventricular hypertrophy 28
  ventricular tachycardia 38, 72
  Wolff-Parkinson syndrome 37
Erythema marginatum 107
Ewart's sign 120
Examination of cardiovascular system 6-23
  auscultation 17
  inspection 6-7
  palpation 7-16
  percussion 16
Extrasystoles (ectopic beats) 8, 35

Fallot's tetralogy 145-6
  X-ray 45, 146
Fibrosis of conducting system, complete heart block 42
Finger clubbing 7
Frank-Starling effect 97
Frusemide 76, 102, 104

Giant cell arteritis 162
Glyceryl trinitrate 55

Haemosiderosis 116
Heart
  anatomy 3
  apex beat 15-16
    tapping 16
  auscultation 17
  blood supply 3
  conducting system 5
  cycle 5
  diastole 5
  gallop rhythm 19
  percussion 16
  physiology 4
  sounds *see* Heart sounds
  systole 5
  thrills 16
Heart block (atrio-ventricular block) 39-40
  complete 68
  first degree 40
  pulse rate 7
  second degree AV 40
  third degree AV (complete) 42
  type II AV 41
Heart failure 76, 97-100
  afterload 98
  clinical features 98-100
  gallop rhythm 19
  myocardial contractility 97
  preload 97
  see also Left ventricular failure; Right ventricular failure
Heart sounds 17-23
  ejection clicks 19
  first 17-18
  fourth 19
  murmurs 20-3
    Austin Flint 23, 134
    Carey-Coombs 108
    continuous (machinery) 23
    ejection systolic 20
    Graham Steele 115
    immediate diastolic 22
    innocent systolic 21
    late systolic 21
    mid-diastolic 22-3
    pansystolic 20
    rheumatic fever 107-8
    systolic 20
  opening snap 19
  second 18
  third 19
Heparin 69, 104, 167, 172
Homan's sign 165
Hydralazine 92, 105
Hypercholesterolaemia 51
Hyperlipidaemia 51
Hypertension 82-96
  adrenal 84
  causes 83-5
    coarctation of aorta 84
    drugs 85
    toxaemia of pregnancy 85
  clinical features 86-8
  coronary artery disease associated 50
  essential 83
  investigations 89-90
  malignant 91
    treatment 95
  renal 83
  secondary 83
  treatment 91-5
    beta blockers 91-2
    calcium antagonists 93
    captopril 95
    centrally acting drugs 93
    clonidine 94
    hydralazine 92, 95
    labetalol 95
    methyldopa 93
    minoxidil 94
    nitroprusside 95
    prazosin 92
    thiazide diuretics 91, 92
    vasodilators 92-3
Hypertensive retinopathy 87—8

Impedance plethysmography 166
Indomethacin 137
Infective endocarditis 148-54
  causes 149-50
  clinical features 150-1
    finger clubbing 7, 150
    Osler's nodes 7, 150
    Roth's spots 150
    splinter haemorrhages 7, 150
  echocardiography 47, 151
  incidence 148
  investigations 151
  predisposing factors 149
  prevention 154
  treatment 152-3
    culture negative 153
Intermittent claudication 172
Iodine 125 labelled fibrinogen 165
Ischaemic heart disease, atrial fibrillation 37
Isosorbide dinitrate 55, 105

Jugular venous pulse 10-12
  cannon waves 12
  in: atrial fibrillation 1

Jugular venous pulse *(cont.)*
  in: complete heart block 12
    constrictive pericarditis 12
    tricuspid regurgitation 12
    tricuspid stenosis 11

Labetalol 95
Lactate dehydrogenase 66
Left ventricular failure 100-2
  pulsus alternans 13
  treatment 102
Left ventricular hypertrophy, ECG 28
Leriche's syndrome 173
Lignocaine 72, 75
Low-density lipoprotein 51

Maladie de Roger 140
Marfan's syndrome 127, 131
Methyldopa 93
Minoxidil 94
Mitral regurgitation 119-22
  clinical features 120
  investigations 121
  pansystolic murmur 20, 120
  rheumatic 119
  systolic thrill 16
  treatment 121
  valve replacement 122
Mitral stenosis 113-18
  atrial fibrillation 37
  clinical features 114-15
  diastolic thrill 16
  differential diagnosis 118
  investigations 116-17
  management 117
  mid-diastolic murmur 22
  opening snap 19
  tapping apex beat 16
  surgery 118
Mitral valve disease 112, 112-23
  first heart sound 18
  mixed 123
  valve replacement 122
Mitral valve prolapse 122-3
  late systolic murmur 21, 122
Murmurs *see under* Heart sounds
Myocardial contractility 97
Myocardial infarction 50, 60-82
  clinical features 62-3
  complications 69-80
    acute mitral regurgitation 79
    acute ventricular septal defect 79
    atrial fibrillation 72
    cardiac arrest 74-5
    cardiogenic shock 77
    Dressler's syndrome 79, 160
    heart block 68, 70
    heart failure 76
    papillary muscle dysfunction/rupture 79
    pericarditis 78
    recurrent angina 78
    sinus bradycardia 69
    sinus tachycardia 68
    supraventricular tachycardias 71
    ventricular aneurysm 80
    ventricular fibrillation 73
    ventricular tachycardia 72
  investigations 63-7
    cardiac enzymes 65-6
    ECG 30, 63-5
    electrolytes 67
    X-ray of chest 67
  pathology 61
  prognosis 60
  rehabilitation 80-1
  secondary prevention 82
  treatment 69
    anticoagulant 82
Myocardial ischaemia
  ECG 31
    exercise 47
Myocarditis
  acute 155
  diphtheritic 155
  rheumatic 107-8
Myxoedema 7

Nifedipine 57, 93
Nitrates 55, 76, 102, 105

Oedema, peripheral 7
Orthopnoea 99, 100, 128
Osler's nodes 7, 150

Pacemaker 42
Papillary muscle
  dysfunction 79, 123
  rupture 79, 123
Paroxysmal nocturnal dyspnoea 99, 100, 128
Patent ductus arteriosus 9, 136-7
  continuous (machinery) murmur 23, 137
Pericardial effusion 159
  echocardiography 47
  X-ray 45
Pericardial rub 158
  transient 160
Pericardial tamponade 13, 159
Pericardiectomy 164
Pericarditis 78, 158-64
  acute fibrinous 158
  bacterial 162
  benign 160
  connective tissue disorders 162
  constrictive 162-4
    pulsus paradoxus 13
    treatment 164
  malignant 161
  myocardial infarction associated 160
  rheumatic 108, 162
  tuberculous 161
  with effusion *see* Pericardial effusion
Phaeochromocytoma 84, 86, 90
Pleural rub 170
Polyarteritis nodosa 162
Practolol 71, 72
Prazosin 92
Pregnancy, systolic murmurs 21
Preload 97
Primary hyperaldosteronism *see* Conn's syndrome
Prinzmetal's variant angina (coronary spasm) 31, 60
Prochlorperazine 69
Propranolol 56, 82, 91
Prostaglandin inhibitors 137
Pulmonary angiography 169-70
Pulmonary embolism 168-72
  acute right ventricular strain 29
  massive 169-70
  management 171
  minor 170
  prevention 172
Pulmonary oligaemia 46
Pulmonary plethora 45
Pulmonary regurgitation, immediate diastolic murmur 22
Pulmonary stenosis
  congenital 144-5
  ejection clicks 19, 144
  ejection systolic murmur 20
Pulse 7-13
  bisferiens 9
  character 8
  collapsing (water hammer) 9, 133
  pulsus alternans 13
  pulsus paradoxus 13
  radial/femoral compared 13
  rhythm 8
  rate 7
  slow rising 8
  venous *see* Jugular venous pulse

Radionuclide investigation 48
Renal artery stenosis 83, 87, 90
Renin-angiotensin-aldosterone system 96
Rheumatic fever 106-12
  aetiology 106
  clinical features 107-8
  complications 112
  Duckett Jones criteria 110
  first degree heart block 40
  investigations 109
  prevention 112
  treatment 111
Rheumatic heart disease 106-12
Right bundle branch block, second heart sound 18

Right ventricular failure 103
treatment 104-5
Right ventricular hypertrophy
apex beat 16
ECG 28
Roth's spots 150

Selective angiography 49
Sick sinus syndrome 38
Sino-atrial node 5
Sinus arrhythmia 34
Sinus bradycardia 34, 69
Sinus tachycardia 34, 68
Sodium nitroprusside 95, 102
Spironolactone 104
Splinter haemorrhages 7, 150
Starling's curve 4
Starling's law 4
Stokes-Adams attacks 43
Subendocardial infarction 30
Supraventricular tachycardia
(atrial tachycardia) 36, 71
Syphilitic aortitis 132
Systemic lupus
erythematosus 162

Tachycardia 7
Tachypnoea 7
Technetium 99M 48
Thallium-201 48
Thyrotoxicosis 7
atrial fibrillation 37
systolic murmurs 21
Timolol 82
Tocainide 72
Toxaemia of pregnancy 85
Tricuspid regurgitation 125-6
pansystolic murmur 20
treatment 126
Tricuspid stenosis 124
jugular venous pulse 11
mid-diastolic murmur 22-3
treatment 126
Turner's syndrome 127

Valves
mechanical 122
tissue (xenografts) 122
Vasodilators 92-3, 105
Venography 165
Venous thrombosis 164-8
calf vein 168
clinical features 165
iliofemoral 168
investigations 165-6
Doppler ultrasound 166
impedance
plethysmography 166
iodine 125 labelled
fibrinogen 165
treatment 167-8
Ventricular aneurysm 80
Ventricular fibrillation 39, 73
R on T phenomenon 35, 73
Ventricular septal defect 140
pansystolic murmur 20, 140
systolic thrill 16
Ventricular tachycardia 38, 72
Verapamil 57, 71, 72
Vitamin K 167

Warfarin 167
Wenkebach phenomenon 40
Wolff-Parkinson-White
syndrome 37

Xenografts 122
X-ray, chest 43-6
lung fields 45
pulmonary oligaemia 46
pulmonary plethora 45